ORGANIC
CHEMISTRY
VOLUME I

UNIVERSITY SCIENCE BOOKS COLLEGE OUTLINE SERIES

ORGANIC CHEMISTRY
VOLUME 1

Roger Macomber

Department of Chemistry
University of Cincinnati

University Science Books
Sausalito, California

DEDICATION

To my parents, my daughter Roxanne, and my new
family, Barbara, Dan, and Juliann. Nothing in life is
as important as the people we love.

University Science Books
55D Gate Five Road
Sausalito, CA 94965
Fax: (415) 332-5393

This book is printed on acid-free paper.

ISBN: 0-935702-90-3

Library of Congress Catalog Card Number: 96-60221

Printed in the United States of America

10   9   8   7   6   5   4   3   2   1

PREFACE

It is an unfortunate fact of life, for both students *and* professors, that the majority of college students enrolled in organic chemistry courses are *not* chemistry majors. Instead, most of them are pre-professionals (pre-meds, pre-dents, pre-vets) because such professional schools have dictated that their students should have completed a one-year sequence of organic chemistry *at the chemistry majors' level.* The result is that these non-chemistry majors are exposed to a vast amount of material, some of which (it can be argued) is of limited value to their ultimate career objectives. On the other side of the coin, it can also be argued that *everyone* should have some understanding of organic chemistry in order to be informed about some of the important challenges we all face in today's world.

When you take an organic chemistry course, there is both good news and bad news. The bad news is the sheer amount of material you have to learn. A typical organic chemistry course often seems overwhelming because the coverage of the subject in most modern textbooks is very thorough—so comprehensive, in fact, that rarely is more than two-thirds of the book covered in a normal year-long sequence. The good news is that, if properly presented, most undergraduate organic chemistry can be seen to develop from a limited set of understandable principles. If you can accept this fact, and are disciplined enough to learn *and understand* these principles, you will almost certainly reach your goals in the course. It is my purpose in writing this book to focus on these important fundamental principles, presenting them in as clear and logical a way as possible.

This book is primarily concerned with the *structure* of organic molecules. Beginning with a review of atomic structure, we discuss chemical bonding, factors that determine the shape of molecules, electron delocalization, nomenclature of functionalized organic molecules, properties of molecules, and stereochemistry. Then, in the last two chapters, we describe in a general way factors that control chemical reactions, how we investigate reaction mechanisms, and how we classify organic reactions. Volume II is primarily concerned with the chemical behavior of the common classes of organic compounds, and our abilities to manipulate these functional groups to synthesize organic compounds of all types.

There is more good and bad news. The good news is that organic chemistry at the undergraduate level is presented in a qualitative (as opposed to quantitative) way, with few mathematical equations or arithmetic calculations. The bad news is that this qualitative approach involves a pyramid of concepts and principles, each one of which serves as a foundation for what comes next. So it is most important that you fully understand each topic before proceeding on to the next. Furthermore (more bad news), there is a certain amount of unavoidable memorization (mainly terminology) that must be accomplished along the way, though probably less than you might have feared. In this series, I have tried to minimize the things that must be memorized, focusing mainly on concepts that are to be *understood.*

Perhaps the single most important piece of advice I can give you is this: *DO THE PROBLEM!* It is my conviction that more than half of the learning that you do in any chemistry course is a result of working the problems. As you begin reading this book, you will soon see that there are many examples and problems in each chapter, each with the answer appearing immediately after the question. You must develop enough discipline to try and work each problem *without looking at the answer!* Too many students read the question and the answer, then satisfy themselves that they could have arrived at the same answer. Do not adopt that approach. Work the problem yourself, and look at the answer only after you've given it your best shot. If your answer is not fully correct, you have missed something, and you should review the preceding material before going on.

But above all, the process of learning new things can be an exciting adventure. I have tried to write a book that will stimulate your curiosity and your thinking about organic chemistry. Now it's up to you to make the effort to expand your horizons. Good luck!

Universty of Cincinnati
Cincinnati, Ohio

Roger S. Macomber

ACKNOWLEDGMENTS

This book would never have been more than an idea had it not been for the encouragement, advice, and able assistance of several very special people. Emily Thompson, Elinor Williams, and Susan McColl have my deep gratitude for bringing an earlier version of this book to fruition. Professors Al Pinhas (University of Cincinnati) and Marinus Bardolph (emeritus, Southern Illinois University) labored mightily over each detail in the manuscript and proofs. Every reader will benefit greatly from their careful reviews and suggestions. And because this book approaches organic chemistry somewhat differently than other mainstream textbooks on the subject, it required a publisher who has courage and confidence in this novel approach. This person is Bruce Armbruster, president of University Science Books, along with his most able editor, Jane Ellis.

Finally, I would like to express my gratitude to all my organic chemistry students and faculty colleagues at the University of Cincinnati over the past 27 years, who've given me the opportunity to develop my own philosophy of teaching organic chemistry.

Cincinnati, Ohio   ROGER S. MACOMBER

CONTENTS

1 THE STRUCTURE OF ATOMS

THIS CHAPTER IS ABOUT

- ☑ **Atoms**
- ☑ **Electrons: The Classical Picture**
- ☑ **The Electron as a Wave; Atomic Orbitals**
- ☑ **Quantum Numbers and Orbital Shapes**
- ☑ **Electron Spin and Orbital Occupancy**
- ☑ **Electron Configurations and the Aufbau Principle**
- ☑ **Properties of Atoms**

1-1. Atoms

Before we begin to discuss organic molecules, it'll be useful to review the structure and properties of atoms, the building blocks of all molecules. An **atom** is the simplest particle of an element that retains the properties of that element. Although there are more than a hundred known elements, this doesn't mean that there are more than a hundred uniquely different types of chemical properties. In fact, chemists have found that all elements can be divided into groups according to similarities in their chemical properties—and these recurring similarities among elements have been found to correlate with patterns in atomic structure.

A. The periodic table

The patterns in atomic structure become most apparent when the elements are arranged in a **periodic table** (Appendix 1). Here, the elements are labeled by an **atomic number (Z)**, the integer beside each element's symbol in the table. Further, the elements are arranged in vertical columns called **groups**. Of greatest interest to us will be the **main-group elements**, those in groups IA to VIIIA [which excludes the **transition groups** (IIIB to VIIIB plus IB and IIB) of elements]. Finally, the horizontal rows of elements in the periodic table are called **periods**.

note: Before you go any further, please *memorize* (if you haven't already) the symbol, atomic number, and position (group and period) of the first eighteen elements (H to Ar).

EXAMPLE 1-1 Fill in the symbol and atomic number of each element in the abbreviated periodic table below *without looking at Appendix 1:*

Group

	IA							VIIIA
1		IIA	IIIA	IVA	VA	VIA	VIIA	
2								
3								

Period

Solution

Group

		IA								VIIIA
1		H 1	**IIA**	**IIIA**	**IVA**	**VA**	**VIA**	**VIIA**	He 2	
2		Li 3	Be 4	B 5	C 6	N 7	O 8	F 9	Ne 10	
3		Na 11	Mg 12	Al 13	Si 14	P 15	S 16	Cl 17	Ar 18	

Now be honest: Did you memorize them all?

> *caution:* The second-period elements (Li to Ne) are sometimes referred to as first-row elements, the third-period elements as second-row, and so on. We'll stick to the period numbers in our discussions.

B. The nucleus and atomic mass

Each and every atom has a nucleus, which usually has some number of electrons around it. The **nucleus** is at the center of the atom and contains virtually all of the atom's mass, even though it occupies only about one-trillionth (10^{-12}) of the atom's volume. The nucleus itself can be further subdivided into some number of positively charged particles called **protons** and some number (N) of equally massive but uncharged particles called **neutrons**. Now remember this:

- The number of protons in the nucleus of an atom is the same as its atomic number (Z), so this number uniquely identifies the element.

- The charge on a proton is +1, therefore the total charge on the nucleus of an atom always equals +Z.

EXAMPLE 1-2 How many protons are there in the atomic nucleus of each element below? (No fair looking at the periodic table; you're supposed to have memorized the atomic numbers already.)

H, Li, B, C, N, O, F, Si

Solution   H 1   Li 3   B 5   C 6   N 7   O 8   F 9   Si 14

EXAMPLE 1-3 What element has atoms with exactly 13 protons in their nuclei? How about 15 protons? 17 protons?

Solution Al has 13 protons, P has 15, and Cl has 17.

EXAMPLE 1-4 What is the charge on the nucleus of each atom below?

He, Be, Ne, Na, S

Solution   He +2   Be +4   Ne +10   Na +11   S +16

The **nominal mass** (A) of a nucleus is an integer equal to the sum of its protons (Z) and neutrons (N):

NOMINAL MASS OF A NUCLEUS
$$A = Z + N \tag{1-1}$$

But there's a complication. Although all nuclei of a given element have exactly the same number of protons (namely Z), these nuclei can differ in the number of neutrons. Nuclei that have the same Z but different N are called **isotopes** of the element. To distinguish one isotope from another, we often add the nominal mass A (the sum of the protons and neutrons) as a preceding superscript to the element's symbol; e.g., ^{13}C is the symbol for the isotope of carbon that has six protons and seven neutrons.

EXAMPLE 1-5 Two naturally occurring isotopes of chlorine are ^{35}Cl and ^{37}Cl. Describe the makeup of each nucleus.

Solution: We know that ^{35}Cl has a nominal mass $A = 35$ and atomic number (protons) $Z = 17$. Then, from Eq. (1-1),

$$N = A - Z$$
$$= 35 - 17$$
$$= 18$$

For ^{37}Cl, $A = 37$ and $Z = 17$, so

$$N = A - Z$$
$$= 37 - 17$$
$$= 20$$

Take another look at the periodic table (Appendix 1). The number below each element's symbol is its **average atomic mass** $(\overline{A})$, sometimes called **atomic weight**. The value $\overline{A}$ is a *weighted average* over all naturally occurring isotopes. In this averaging calculation the *exact* (rather than nominal) atomic mass of each isotope is used.

> *note:* Atomic mass (nominal or average) is measured in *atomic mass units* (amu); 1 amu = 1.66×10^{-24} g.

EXAMPLE 1-6 If natural chlorine is 76.0% ^{35}Cl (exact mass 34.98 amu) and 24.0% ^{37}Cl (exact mass 36.98 amu), what is the average atomic mass of natural chlorine?

Solution

$$\overline{A} = \frac{(\%\ ^{35}Cl)(\text{mass } ^{35}Cl) + (\%\ ^{37}Cl)(\text{mass } ^{37}Cl)}{100} = \frac{76.0(34.98) + 24.0(36.98)}{100} = 35.5 \text{ amu}$$

Now that we understand the makeup of the nucleus, remember this:

- The nucleus *never* changes during a chemical reaction.

Well, if the nucleus doesn't change during a chemical reaction, what *does* change? Obviously, we have to look outside the nucleus—to the electrons—to understand chemical reactions.

1-2. Electrons: The Classical Picture

A. In atoms and ions

All chemical reactions involve chemical bonds, which (as we'll see in Chapter 2) are made up of electrons. So, to understand chemical bonds and reactions, we have to find out a little about electrons.

In the early development of atomic theory, the electron was pictured as a negatively charged (−1) particle with almost no mass (about one two-thousandth the mass of a proton). Electrons were believed to follow circular paths around the nucleus, much like the orbits that planets follow when they revolve around the sun (though faster, of course!). This orbit-like motion was attributed to the strong electrostatic attraction between the electron and nucleus, a direct consequence of their opposite charges. And the kinetic energy of the electron was what prevented it from "falling" into the nucleus. This picture, though naive, is nonetheless a useful one, so we'll stick with it for a while to make some observations about electrons.

Because the charge (–1) on an electron is equal but opposite to that of the proton,

- An electrically *neutral* atom has as many electrons as it has protons, namely Z.

However, if one or more electrons are added to or removed from a neutral atom, the atom is converted to an **ion**, which has an *un*equal number of electrons and protons. An ion is therefore electrically charged. A positively charged ion (called a **cation**) has more protons than electrons, while a negatively charged ion (an **anion**) has more electrons than protons.

EXAMPLE 1-7 (a) How many electrons are there in H^+, Li^+, C, O^{2-}, Ne, and P^{3-}? (b) Which of them are cations? anions? Which are neutral?

Solution

(a) Start with the atomic number of each element; then subtract one electron for each positive charge or add one electron for each negative charge. Thus, since Z for H is 1, for H^+ the number of electrons is $1 - 1 = 0$. Similarly, for Li^+, $3 - 1 = 2$; for neutral C, just 6; for O^{2-}, $8 + 2 = 10$; for neutral Ne, just 10; and for P^{3-}, $15 + 3 = 18$.

(b) Cations: H^+ and Li^+. Anions: O^{2-} and P^{3-}. C and Ne are neutral atoms.

---

note: Although the *mass* of an atom is determined mainly by the mass of the nucleus, the *size* of an atom—that is, the volume it occupies—depends on the distance between the outermost electrons and the nucleus; see below.

B. Energy levels

We can learn a lot about electrons by studying their behavior in atoms. One way to do this is by examining **atomic emission spectra**. Such spectra show that:

- The electrons in an atom can occupy only certain discrete (fixed) states called **energy levels**.

This is true even though each atom has an infinite number of such levels. Because each energy level in a given atom has a certain (fixed) energy, transitions between any two specified levels will always involve *exactly* the same energy change, which in turn gives exactly the same spectroscopic emission. For example, the (only) electron in a hydrogen atom can occupy any one of an infinite number of energy levels. Each of these levels is labeled by what's called a **principal quantum number** (n), which can assume any integer value from 1 (the lowest energy level) to infinity. When the electron is in the $n = 1$ level, the atom is said to be in its electronic ground state. If the electron is in the $n = 2$ level, the atom is in its first excited state. But, most importantly, the energy *difference* between the $n = 1$ level and the $n = 2$ level for any hydrogen atom is always *exactly* the same.

The energy level of the electrons in an atom affects its size. Since size is usually expressed in terms of the atom's radius, which can be viewed as a measure of the orbital path of the outermost electron(s), it may not surprise you to learn that (all other things being equal) the larger the value of n, the farther the electron is from the nucleus and the larger the atom is. Thus, a hydrogen atom in its first excited state, with its electron in the $n = 2$ level, occupies a larger volume than a hydrogen atom in its ground state, with its electron in the $n = 1$ level. To summarize,

- The principal quantum number (n) is related both to the *energy* of an electron and to the *size* of the volume it occupies.

1-3. The Electron as a Wave; Atomic Orbitals

The trouble with electrons is, they have some properties not usually associated with particles. For instance, scientists have found that electrons can be diffracted just as electromagnetic waves can be. Moreover, at about the same time that this non-particulate behavior was first observed, **Heisenberg** enunciated his **uncertainty principle**, which says that we cannot *simultaneously* measure the exact position *and* momentum of a particle such as an electron. This principle suggests it may be more realistic to picture the electron as a spread-out, nebulous entity rather than as a discrete particle.

Finally, when the field of quantum mechanics (the study of matter and energy at the atomic scale) emerged, one of its basic tenets was that electrons are best viewed as *waves*. And, as with other wave phenomena, we can measure the energy of a wave, but we can't pinpoint its exact location. So, we speak of **electron density**, the *probability* of finding the electronic wave in a certain volume of space.

Schrödinger was able to write a differential wave equation that describes the various interactions between electrons and nuclei. But even for the simplest atom, hydrogen, the number of exact solutions to the **Schrödinger equation** is infinite. Each solution is a **wave function** (equation) that describes the energy of the electron. The square of the wave function is related to the probability of finding the electron in a certain volume of space. These wave equations have come to be known as *atomic orbitals*. Each atomic orbital has the nucleus at its center.

- An **atomic orbital** is a wave equation that describes the energy of, and the volume occupied by, an electron in an atom.

By adopting the wave picture of electrons, we can find out even more about their behavior.

1-4. Quantum Numbers and Orbital Shapes

A. Quantum numbers

Each hydrogen orbital—that is, each solution to the Schrödinger equation for the hydrogen atom—can be characterized by *three* quantum numbers. This set of quantum numbers can be regarded as the "address" of an electron in a given orbital. The first of these is our old friend n, the principal quantum number, which describes the energy and size of the orbital. The next one, called the **angular momentum** (or azimuthal) **quantum number**, l, relates to the *shape* of the orbital. And the third one, the **magnetic quantum number**, m, describes the *orientation in space* of the orbital. Let's see how these three quantum numbers are related.

We know from Section 1-2 that the principal quantum number n can assume integer values from 1 to infinity. It turns out that the angular momentum quantum number l can only adopt integer values from 0 to $n - 1$. Thus, when $n = 1$, l must equal zero. But when $n = 2$, l can be either 0 or 1. The values of the magnetic quantum number m depend, in turn, on the values of l; m can only assume the integer values from $-l$ to $+l$.

EXAMPLE 1-8 Write out all possible combinations of l and m values for n values of 1, 2, and 3.

Solution

n	l	m
1	$n - 1 = 0$	0
2 $\}$	0 $n - 1 = 1$	0 $-1, 0, +1$
3 $\}$	0 1 $n - 1 = 2$	0 $-1, 0, +1$ $-2, -1, 0, +1, +2$

Notice that for each value of n there are n different values of l, and for each value of l there are $(2l + 1)$ different values of m.

B. Orbital shapes

By graphing each wave function we can get an idea of the three-dimensional shape of each orbital. But before we can draw pictures of orbitals, we have to decide what probability limits we wish to depict. Each orbital extends to infinity in all directions, but the electron density decreases as we get farther from the nucleus. So, let's arbitrarily agree to draw orbitals in such a way as to denote a 95% relative probability of finding the electron (or electronic wave distribution, to be exact) inside. Of course, we could equally well pick a 90% limit (an orbital of the same shape, but smaller volume), or a 99% limit (same shape, but larger volume).

Now, what about the shapes?

1. s orbitals An orbital with $l = 0$ is *spherical* (Figure 1-1a) and is called an *s* **orbital**. Each principal quantum level (each value of n) has one *s* orbital associated with it, and these are labeled with the value of n (e.g., $1s$, $2s$, $3s$, etc.). Recall that if $l = 0$, then $m = 0$ — a reminder that there is no orientation or directionality to a spherical *s* orbital. Remember also that, *in a given atom*, the $1s$ orbital is lower in energy and smaller than the $2s$, which in turn is lower in energy and smaller than the $3s$, and so on.

2. p orbitals An orbital with $l = 1$ is *dumbbell-shaped*, having a cross section that resembles a figure eight (Figure 1-1b). Such an orbital is called a *p* **orbital**. Notice that, because $l = 1$, m can have values of -1, 0, or $+1$. This means that p orbitals always occur in sets of *three*. Each value of m can be assigned (arbitrarily) to one of the Cartesian coordinate directions, x, y, or z. This requires that the three p orbitals in a set be orthogonal, i.e., mutually perpendicular. There is one set of (three) p orbitals in each principal quantum level *except* the lowest ($n = 1$) level (why?). And remember that, although each p orbital has two "lobes," it is still just *one* orbital.

3. d orbitals When $l = 2$, the resulting orbital is called a *d* **orbital**, and it has *four lobes* (Figure 1-1c). With $l = 2$, m can assume values of -2, -1, 0, $+1$, or $+2$; so d orbitals always occur in sets of *five*. The orientations and corresponding labels of the five d orbitals in a set are shown in Figure 1-1c. (Again, the assignment of labels to individual m values is arbitrary.) There is a set of d orbitals in each principal quantum level, except when n equals 1 or 2.

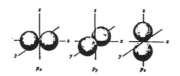

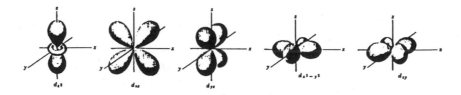

Figure 1-1. Representation of hydrogen atomic orbitals: **(a)** an *s* orbital, **(b)** a *p* orbital, **(c)** a *d* orbital. In each case the nucleus is at the origin of the coordinate system. *(Reprinted with the permission of R. G. Pearson.)*

EXAMPLE 1-9 Complete the hydrogen atomic orbital energy diagram below by drawing a small horizontal line for each orbital at the appropriate relative energy. Label each orbital and give its quantum numbers. To help you get started, the 1*s* orbital is already shown.

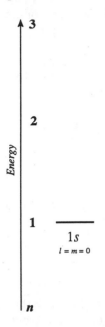

Solution

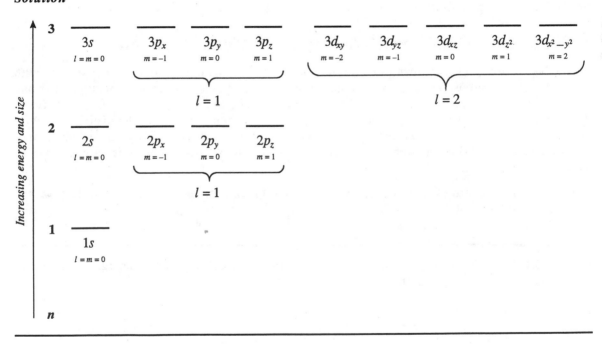

Notice in the hydrogen atom, Example 1-9, that all the orbitals in a given principal quantum level are **degenerate**, which means that each has *exactly* the same energy. For example, in the hydrogen atom the 3*s*, 3*p*, and 3*d* orbitals are degenerate.

1-5. Electron Spin and Orbital Occupancy

We know from Section 1-2 that the ground state of a hydrogen atom has the electron in the $n = 1$ principal quantum level. Further, from Section 1-4, we know this level has only one orbital, the 1*s*. Therefore, we can write the ground-state **electron configuration** (the arrangement of electrons) of hydrogen as $(1s)^1$.

Now, what is the ground-state electron configuration of helium? You know (from memory, right?) that helium has atomic number 2 ($Z = 2$), and therefore has two electrons in the neutral atom. Certainly, the first electron goes into the $1s$ orbital around the helium nucleus. But what about the second electron? Surprise! It, too, goes in the $1s$ orbital, giving a $(1s)^2$ configuration. Yet, the **Pauli exclusion principle** says that no two electrons *in the same atom* can have the same set of quantum numbers. Certainly, if the two electrons in a helium atom occupy the same orbital, they have identical values of n (1), l (0), and m (0). Where's the difference?

The difference has to do with one more property of electrons that we haven't yet addressed; that is, **magnetic spin**. Each electron is like a miniature magnet, which can adopt either of two spin orientations, up or down in a vertical magnetic field. Therefore, we need to define one more quantum number, a **spin quantum number** (s, not to be confused with s orbitals), which can be either $+1/2$ or $-1/2$. If two electrons in an atom are in the *same* orbital, they must have *different* spin quantum numbers and hence, *opposite* spins. We describe this situation by saying that the electrons are **paired**. Remember that, because s has only two possible values,

- No orbital can contain more than two electrons.

We show this situation pictorially by drawing the electrons as arrows (pointing up or down) in an orbital energy diagram. Thus, helium would look like this:

$$\frac{\uparrow\downarrow}{1s}$$

There is one other thing you should be aware of. A helium $1s$ orbital is *lower* in energy than a hydrogen $1s$ orbital. This is because the helium nucleus has *two* protons pulling on the electrons, while a hydrogen nucleus has only one proton. Therefore, the helium electrons are held more tightly by the nucleus. (More on this in Section 1-7.)

1-6. Electron Configurations and the Aufbau Principle

While the energy diagram shown in Example 1-9 is valid for hydrogen and any other species with just one electron, atoms and ions with more than one electron possess other complications. For one thing, because all electrons are negatively charged, they tend to repel each other, and this affects their energy. Also, the s orbital in any given principal quantum level is, on the average, slightly closer to the nucleus than are p orbitals of the same level. Because of these effects, the Schrödinger equation can be solved only approximately. The resulting solutions (orbitals) are similar to hydrogen orbitals, except that,

- In multi-electron atoms, the energy of orbitals in a given principal quantum level follows the order $s < p < d$.

EXAMPLE 1-10   Redraw the diagram in Example 1-9 to reflect the relative energies of s, p, and d orbitals in multi-electron atoms.

Solution

Now let's "build-up" (in German, *aufbau*) the ground-state electron configuration of the rest of the first 18 elements using the following **aufbau** principle:

- Always fill the lowest available orbital before populating a higher one.

Element number three (lithium, right?) will have its first two electrons filling the 1*s* orbital. The third electron will go into the lowest unfilled orbital, the 2*s*. So, its configuration can be written either as $(1s)^2(2s)^1$ or as $He(2s)^1$, where He stands for the electron configuration of helium. Similarly, beryllium ($Z = 4$) has configuration $He(2s)^2$. In terms of orbital energy diagrams, these two configurations can be drawn like this:

Li

$He(2s)^1$

Be

$He(2s)^2$

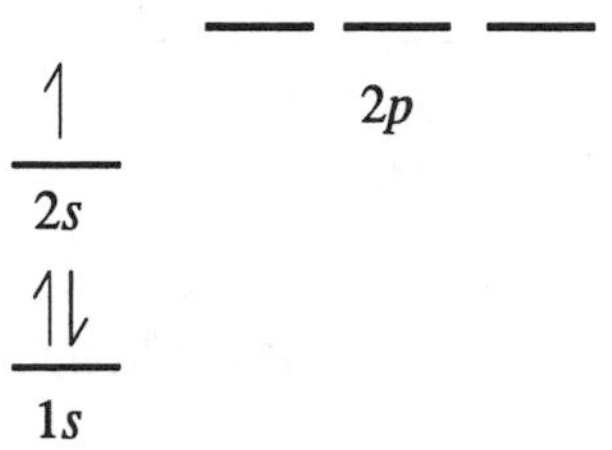

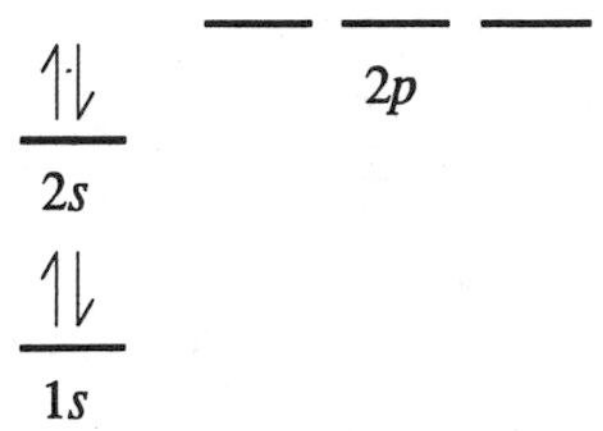

The next six elements of the second period (boron through neon) fill the 2*p* sublevel. But there is a correct way to do this. **Hund's rule** says that,

- When electrons are distributed among *degenerate* orbitals, they are spread among as many orbitals as possible, with spins *parallel* (unpaired).

Thus, boron, carbon, nitrogen, and oxygen look like this:

B

$He(2s)^2(2p)^1$

C

$He(2s)^2(2p)^2$

N

$He(2s)^2(2p)^3$

O

$He(2s)^2(2p)^4$

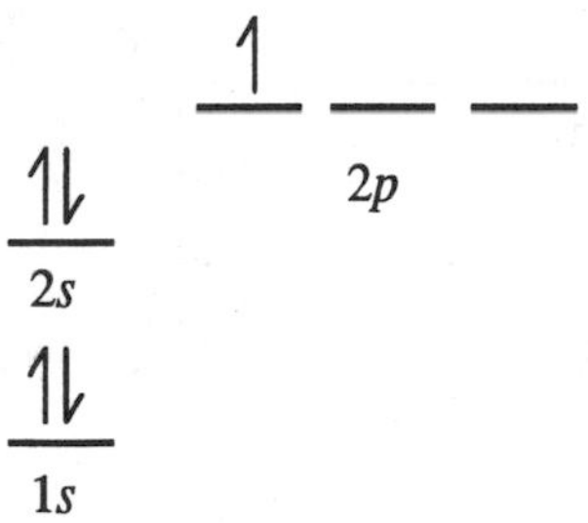

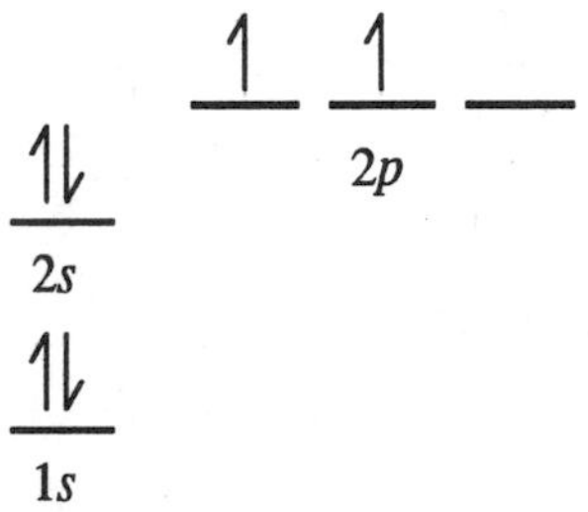

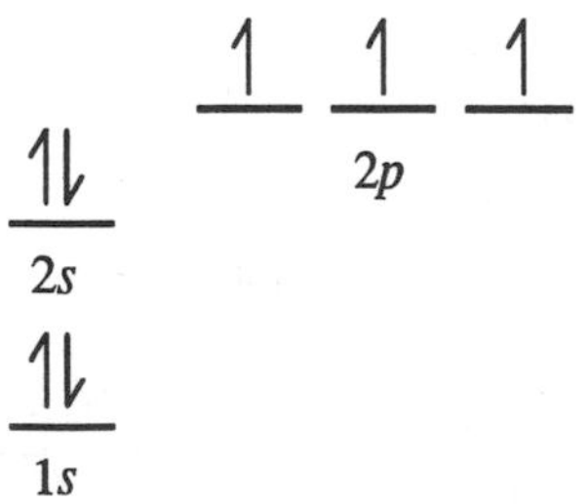

For reasons we'll discuss in Chapter 2, the electrons in the outermost (highest *n* value) principal quantum level are called **valence electrons.** The rest are called **subvalence electrons.** Thus, boron, carbon, and nitrogen each possess *two* subvalence electrons along with three, four, and five valence electrons, respectively.

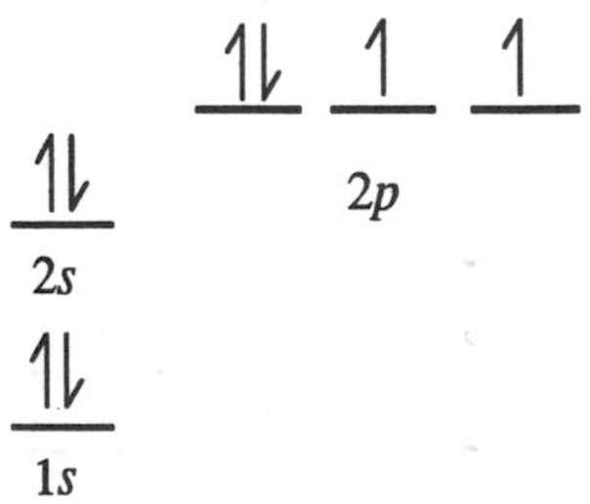

EXAMPLE 1-11 Using orbital energy diagrams like the ones above, draw the electron configuration of the next two second-period elements.

Solution

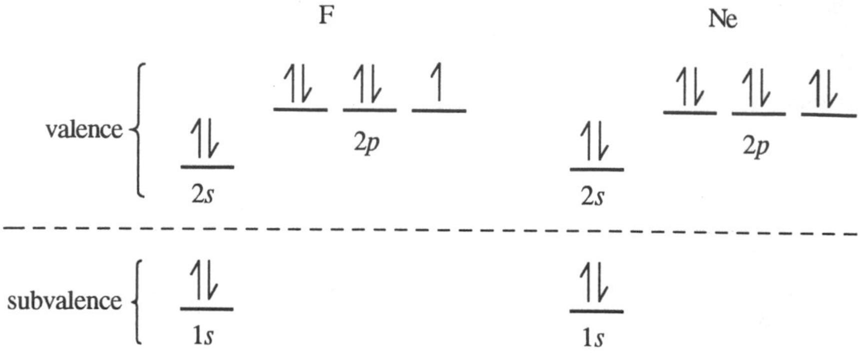

F   Ne

valence $\begin{cases}\end{cases}$ 2s 2p   valence 2s 2p

subvalence $\begin{cases}\end{cases}$ 1s   subvalence 1s

$He(2s)^2(2p)^5$ $He(2s)^2(2p)^6$

EXAMPLE 1-12 How many valence and subvalence electrons are there in each of the atoms in Example 1-11?

Solution Each has two subvalence electrons, while F has seven and Ne has eight valence electrons.

EXAMPLE 1-13 (a) Using orbital energy diagrams, draw the *valence* electron configurations of elements with $Z = 11$ through $Z = 18$. (b) How many subvalence electrons does each possess?

Solution

(a)   Na   Mg   Al

3s 3p   3s 3p   3s 3p

$Ne(3s)^1$ $Ne(3s)^2$ $Ne(3s)^2(3p)^1$

Si   P   S

3s 3p   3s 3p   3s 3p

$Ne(3s)^2(3p)^2$ $Ne(3s)^2(3p)^3$ $Ne(3s)^2(3p)^4$

Cl   Ar

3s 3p   3s 3p

$Ne(3s)^2(3p)^5$ $Ne(3s)^2(3p)^6$

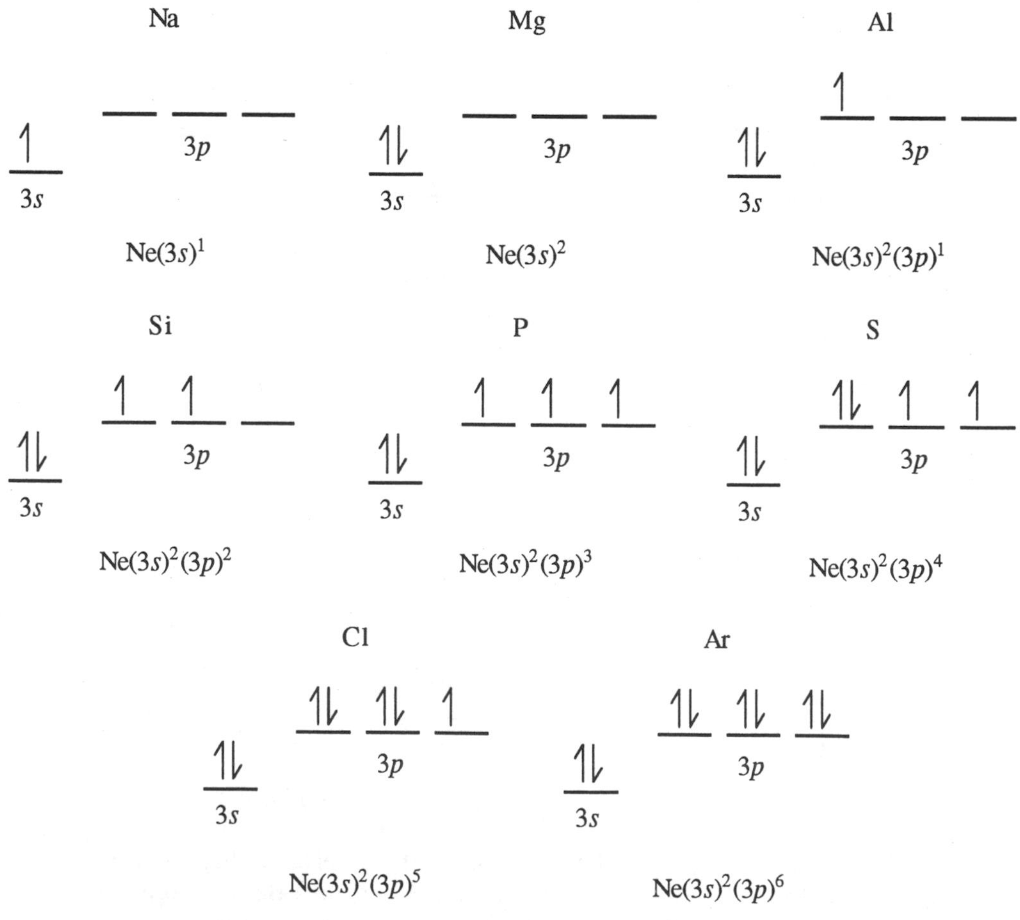

(b) Each has ten subvalence electrons, corresponding to the configuration of neon.

By now I hope you've noticed the parallel between electron configuration of the elements and their location in the periodic table. Notice how each group-IA element (e.g., H, Li, Na) has just one valence electron, in an s orbital $[(ns)^1]$, while each group-IIA element has two $[(ns)^2]$. Each group-IIIA element has three valence electrons in configuration $(ns)^2(np)^1$, and so on through group VIIIA $[(ns)^2(np)^6]$. [Helium's location is somewhat anomalous because its configuration is $(1s)^2$, as with group-IIA elements. Still, its chemical properties—and the fact that it has a completely filled valence level—make it more similar to the group-VIIIA elements.]

Notice that the first-*period* elements (H and He) have the $n = 1$ level as their valence level, second-period elements (Li through Ne) have their valence electrons in the $n = 2$ level, while third-period elements (Na though Ar) have a valence level with $n = 3$. Therefore,

- The number of valence electrons in a neutral main-group atom equals its group number, and the valence level of each such atom equals its period number.

There is one other thing to point out. Unlike the elements in the first two periods, each third-period element has a set of *empty d* orbitals in its valence shell. These empty d orbitals have a significant effect on the chemistry of these elements, as we'll see during our study of organic compounds.

Even though we've agreed to picture the electron as wavelike, it is still a common practice to depict *valence* electrons as dots when we draw them in structures. Thus, we can represent the second-period elements this way:

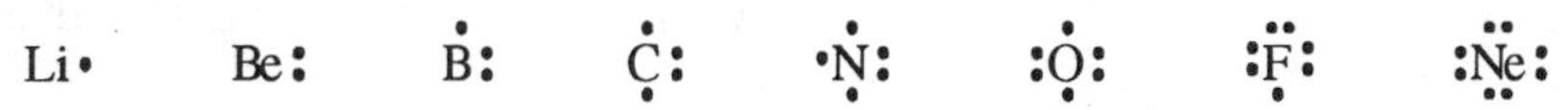

$$\text{Li}\cdot \qquad \text{Be}\colon \qquad \dot{\text{B}}\colon \qquad \dot{\text{C}}\colon \qquad \cdot\dot{\text{N}}\colon \qquad \colon\dot{\text{O}}\colon \qquad \colon\ddot{\text{F}}\colon \qquad \colon\ddot{\text{Ne}}\colon$$

Notice how two dots together (as in a colon, :) denote *paired* electrons in a single orbital.

EXAMPLE 1-14 Draw electron-dot structures for the third-period elements.

Solution

$$\text{Na}\cdot \qquad \text{Mg}\colon \qquad \dot{\text{Al}}\colon \qquad \dot{\text{Si}}\colon \qquad \cdot\dot{\text{P}}\colon \qquad \colon\dot{\text{S}}\colon \qquad \colon\dot{\text{Cl}}\colon \qquad \colon\ddot{\text{Ar}}\colon$$

1-7. Properties of Atoms

A. Kernel charge

To make useful generalizations about trends in the properties of atoms, it's helpful to define the **kernel** (or core) of an atom as the nucleus plus all *subvalence* electrons. The **kernel charge** of an atom equals the number of protons in the nucleus (Z) minus the number of subvalence electrons.

EXAMPLE 1-15 Calculate the kernel charge for each of the first 18 elements *without looking at the periodic table*. Also indicate the n value for the *valence* electrons in each atom.

Solution The number of *subvalence* electrons in an atom always equals the number of electrons in the group-VIIIA element of the preceding period. Thus for H and He the number of subvalence electrons is zero, and the kernel charges are +1 and +2, respectively. Each second-period element has two subvalence electrons (those in the configuration of helium), and therefore the kernel charges are Li, $3 - 2 = +1$; Be, $4 - 2 = +2$, etc. The kernel charge of each of the first 18 elements is shown below:

$n = 1$	H (1)							He (2)
$n = 2$	Li (1)	Be (2)	B (3)	C (4)	N (5)	O (6)	F (7)	Ne (8)
$n = 3$	Na (1)	Mg (2)	Al (3)	Si (4)	P (5)	S (6)	Cl (7)	Ar (8)

- The kernel charge of a main-group element is numerically equal to its number of valence electrons, which in turn equals its group number.

It is the kernel charge (that is, the nuclear charge as modified by the subvalence electrons) that is actually responsible for "holding on" to the valence electrons.

B. Atomic size

Now, let's examine some trends in the properties of atoms. For example, size: Which atom is larger, lithium or neon? The answer is...lithium! This is because the kernel charge of the neon nucleus is greater (+8 vs. +1), allowing neon to hold onto its valence electrons more tightly, leaving them at a lower energy and contracting the orbital. Thus,

- In general, for elements in the same period (i.e., whose valence electrons are in the same principal quantum level), the atomic radius *decreases* as the group number (and hence, kernel charge) *increases*. (Helium is an exception.)

EXAMPLE 1-16   What is the predicted order of size (atomic radius) of the third-period elements?

Solution   Na > Mg > Al > Si > P > S > Cl > Ar. (Actually, P, S, and Cl all have about the same radius.)

How about lithium vs. sodium: Which is larger? If you answered sodium, you're right! Because the valence electron of sodium is in the $3s$ orbital, while lithium's is in the $2s$ (which is closer to the nucleus), the kernel (+1 charge in both cases) holds the latter electron more tightly. So,

- When comparing elements in the same *group* (i.e., same kernel charge), the atomic radius *increases* as the period number (principal quantum number of the valence electrons) *increases*.

C. Ionization potential

The next atomic property of interest is the **ionization potential,** the energy required to remove an electron from an atom: $A \rightarrow A^+ + e^-$. The greater the ionization potential, the more difficult it is to remove the electron. If the atom has more than one electron, the outermost (highest energy) electron is the most easily removed, requiring an energy called the **first ionization potential.** Removing the second electron defines the **second ionization potential,** and so on. In general, the first ionization potential is less than the second, which is less than the third, etc. Because the ionization potential is a direct measure of how tightly an electron is held by the nucleus, it shouldn't surprise you to learn that the magnitude of the ionization potential, like atomic radius, is a function of kernel charge and period number:

- In general, the first ionization potential *increases* as group number (and kernel charge) *increases*, but *decreases* as period number (valence principal quantum level) *increases*.

EXAMPLE 1-17   Predict the order of first ionization potentials for all second-period elements. Do the same for third-period elements.

Solution   Li < Be < B < C < N < O < F < Ne; Na < Mg < Al < Si < P < S < Cl < Ar. (Actually, B, Al, and S have slightly anomalous values.)

By the way, what type of species results when an electron is removed from an atom? The term *ionization* should give you a clue: The answer is, a cation (Section 1-2). For example, the first ionization of Na gives Na^+, with electron configuration $He(2s)^2(2p)^6$. Notice that Na^+ is *isoelectronic* (that is, has the same electron configuration) with Ne.

EXAMPLE 1-18 With which neutral atom is each of the following ions isoelectronic?

$$Li^+, \quad Be^{2+}, \quad Cl^-$$

Solution Li^+ and Be^{2+} are both isoelectronic with He, and Cl^- is isoelectronic with Ar.

While we're on the subject, which do you suppose is greater, the *second* ionization potential of sodium ($Na^+ \rightarrow Na^{2+} + e^-$) or the *first* ionization potential of neon ($Ne \rightarrow Ne^+ + e^-$)? The answer is the second ionization potential of Na. Although Na^+ and Ne are isoelectronic (as are Na^{2+} and Ne^+), the sodium nucleus has one more proton than a neon nucleus, so the electrons in Na^+ are more tightly held. For the same reason, Mg^{2+} (kernel charge +2) is smaller than Na^+ (kernel charge +1), even though they are isoelectronic.

D. Electron affinity

The next property that's useful to discuss is **electron affinity,** the energy *released* when an atom *accepts* an electron: $A + e^- \rightarrow A^-$. The larger the electron affinity, the more strongly an electron is attracted to atom A, and the more stable is ion A^-. The product of this process is an anion, and because it has one more electron than the parent atom, A^- has a substantially larger radius than A. Unfortunately, electron affinities are not as uniformly correlated with group and period numbers as are atomic radii and ionization potentials. Nonetheless, it is a valid generalization that, neglecting the group-VIIIA elements,

- Electron affinities tend to *increase* as group number (and kernel charge) *increases*.

E. Electronegativity

The last property of atoms we'll discuss at this point is actually a property of atoms *in molecules*. It is called **electronegativity** and is a quantitative measure of how much "pull" an atom *in a molecule* exerts on neighboring *bonding* electrons. The electronegativity value for each of the group-IA through -VIIA elements is shown below (group VIIIA elements He, Ne, and Ar are not shown because they don't form molecules):

	IA	IIA	IIIA	IVA	VA	VIA	VIIA
1	H 2.1						
2	Li 1.0	Be 1.5	B 2.0	C 2.5	N 3.0	O 3.5	F 4.0
3	Na 0.9	Mg 1.2	Al 1.5	Si 1.8	P 2.1	S 2.5	Cl 3.0

Remember these two facts:

- The greater an atom's electronegativity, the more "pull" it exerts on neighboring bonding electrons.

- The most electronegative element is fluorine.

It will come as no surprise that, because it is a function of kernel charge and valence principal quantum level,

- Electronegativity (like ionization potential) *increases* as group number *increases* but *decreases* as period number *increases*.

Armed with these facts about atoms and their properties, we're ready to find out how size, ionization potential, electron affinity, and electronegativity of atoms determine the nature of the chemical bonds they form.

SUMMARY

1. The periodic table organizes all known elements on the basis of recurring patterns in electron configuration, properties, and reactivity.

2. The vertical columns in the periodic table are called groups. The main groups are those labeled IA through VIIIA.

3. The horizontal rows in the periodic table are called periods. The period number equals the principal quantum level of the valence electrons.

4. Each element has a unique atomic number (Z), which is equal to the number of protons in its nucleus; the number of electrons in the neutral atom is also equal to Z.

5. Isotopes of an element have the same atomic number (Z) but differ in the number of neutrons (N). The nominal atomic mass (A) of an atom is equal to the sum $Z + N$.

6. Ions are charged species in which the number of protons is not equal to the number of electrons. A cation has an excess of protons (and a positive charge), while an anion has an excess of electrons (and a negative charge).

7. In quantum mechanics, the electron is pictured as wavelike. The mathematical wave equations that describe the energy of an electron and its distribution around the nucleus (i.e., the probability of finding it in a certain region of space) are called atomic orbitals.

8. Atomic orbitals are characterized by three quantum numbers:
   (a) the principal quantum number (n), which can have integer values from one to infinity, describes the *energy* and *size* of the orbital;
   (b) the angular momentum quantum number (l), which can assume integer values from zero to $n - 1$, describes the *shape* of the orbital;
   (c) the magnetic quantum number (m), with integer values from $-l$ to $+l$, describes the *orientation* of the orbital.

9. No orbital can contain more than two electrons; and if there *are* two electrons in a single orbital, their spins must be *paired* (opposite), corresponding to different spin quantum numbers ($s = +1/2$ or $= -1/2$). This fact is called the Pauli exclusion principle.

10. When writing the electron configuration of an atom, we put each electron into the lowest energy orbital that is not yet filled (the aufbau principle). When filling sets of degenerate orbitals, we spread the electrons among as many orbitals as possible, with spins *parallel* (Hund's rule).

11. In neutral main-group atoms (except for helium), the number of valence electrons (those in the outermost principal quantum level) is equal to the group number.

12. The kernel charge of a main-group atom is the charge on the nucleus ($+Z$) minus the number of subvalence electrons and is equal to its group number (except for helium).

13. Among main-group elements (but excluding helium), the following generalizations can be made:
   (a) Atomic radii decrease as group number (and kernel charge) increase; atomic radii increase as period number increases.
   (b) First ionization potentials increase as group number increases, and decrease as period number increases.
   (c) Electronegativity increases as group number increases and decreases as period number increases (except for group-VIIIA elements).

RAISE YOUR GRADES

Can you define...?

- ☑ atomic number (Z)
- ☑ group number
- ☑ nominal atomic mass (A)
- ☑ ion; cation, anion
- ☑ quantum numbers n, l, m, s
- ☑ excited state
- ☑ electron density
- ☑ electron configuration
- ☑ valence and subvalence electrons
- ☑ electronegativity

- ☑ period number
- ☑ isotope
- ☑ average atomic mass ($\bar{A}$)
- ☑ electron energy level
- ☑ ground (electronic) state
- ☑ atomic orbital
- ☑ s, p, and d orbitals
- ☑ paired electrons
- ☑ degenerate orbitals
- ☑ kernel and kernel charge

☑ ionization potential ☑ isoelectronic
☑ atomic radius ☑ electron affinity

Can you explain...?

☑ how the periodic table is arranged
☑ the makeup of the nucleus and the electron configuration of any main-group atom in the first three periods
☑ how average atomic masses are computed
☑ what information is available from the values of n, l, m, and s
☑ the meaning of the aufbau principle, Pauli exclusion principle, and Hund's rule
☑ how periodic trends in atomic radius, ionization potential, electron affinity, and electronegativity depend on group and period numbers of an element

SOLVED PROBLEMS

PROBLEM 1-1 Draw a periodic table showing the first 18 elements. Include group and period numbers, and specify the atomic number and valence-electron configuration for each element. Then, using arrows along the bottom and side of the table, indicate the direction of increasing atomic radius, first ionization potential (IP), and electronegativity (EN).

Solution

Group

Period	IA	IIA	IIIA	IVA	VA	VIA	VIIA	VIIIA
1	H 1 $(1s)^1$							He 2 $(1s)^2$
2	Li 3 $(2s)^1$ $(2p)^0$	Be 4 $(2s)^2$ $(2p)^0$	B 5 $(2s)^2$ $(2p)^1$	C 6 $(2s)^2$ $(2p)^2$	N 7 $(2s)^2$ $(2p)^3$	O 8 $(2s)^2$ $(2p)^4$	F 9 $(2s)^2$ $(2p)^5$	Ne 10 $(2s)^2$ $(2p)^6$
3	Na 11 $(3s)^1$ $(3p)^0$	Mg 12 $(3s)^2$ $(3p)^0$	Al 13 $(3s)^2$ $(3p)^1$	Si 14 $(3s)^2$ $(3p)^2$	P 15 $(3s)^2$ $(3p)^3$	S 16 $(3s)^2$ $(3p)^4$	Cl 17 $(3s)^2$ $(3p)^5$	Ar 18 $(3s)^2$ $(3p)^6$

Atomic radius ↑ (up the side)
First IP and EN ↓ (down the side)

← Atomic radius
First IP and EN →

PROBLEM 1-2 Complete each statement with the appropriate term(s):

(a) The atomic number of an atom equals the number of _____ in its nucleus, and the number of _____ in the neutral atom.

(b) The group number of an atom equals its number of _____ electrons as well as its _____ charge.

(c) The period number of an atom equals the _____ of its valence electrons.

Solution **(a)** protons, electrons; **(b)** valence, kernel; **(c)** principal quantum level.

PROBLEM 1-3 **(a)** For each ion below, show its electron configuration by means of an orbital-energy diagram. In the diagram indicate which are valence electrons. **(b)** Draw an electron-dot structure for each species. **(c)** What is the order of size of these ions?

$$Mg^{2+}, \quad Al^{3+}, \quad S^{2-}, \quad Cl^-$$

Solution

(a)

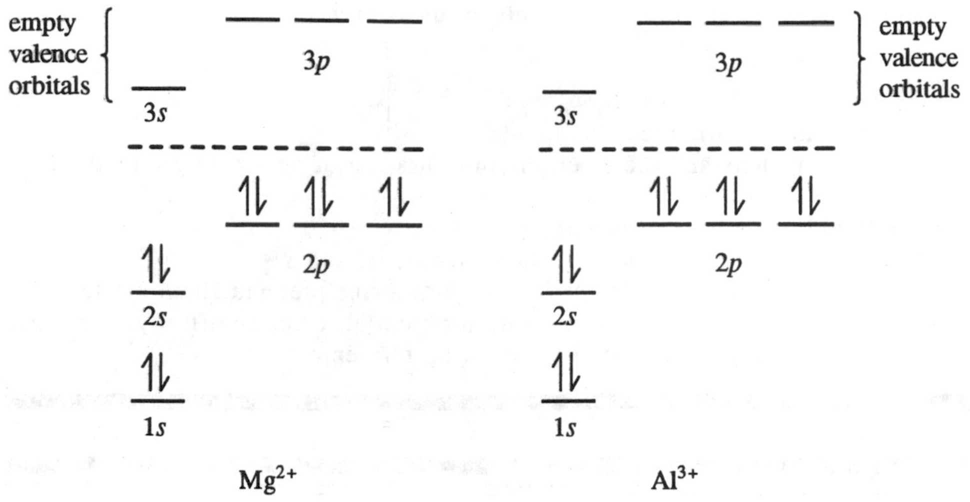

Note that these two cations, which are isoelectronic, have no valence electrons left after ionization; that is, each has an empty *valence shell*.

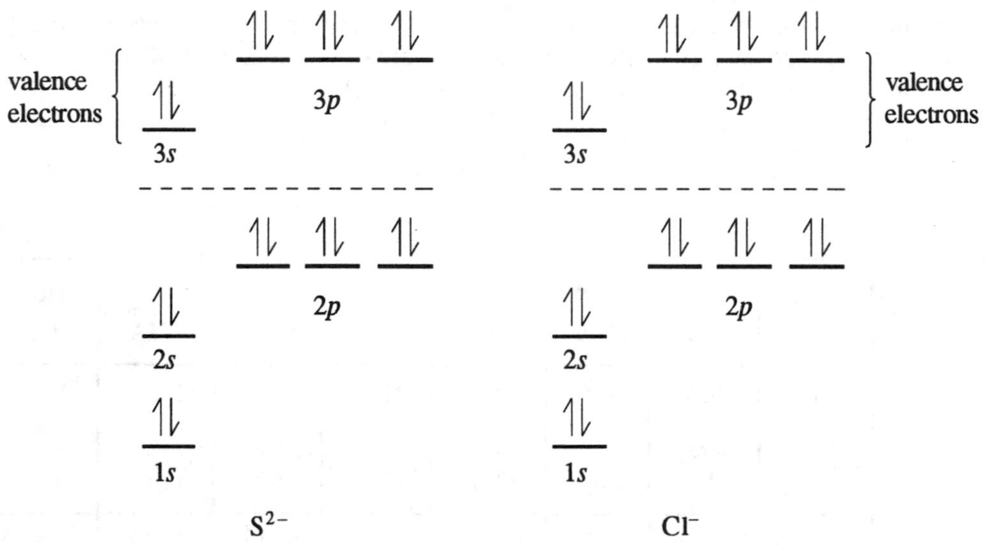

These two isoelectronic anions have filled valence shells.

(b) Since Mg^{2+} and Al^{3+} have no valence electrons, the electron-dot structures for these species are

$$Mg^{2+} \qquad Al^{3+}$$

And the electron-dot structures for S^{2-} and Cl^-, which both have 8 valence electrons, are

$$:\!\ddot{S}\!:^{2-} \qquad :\!\ddot{C}\!l\!:^-$$

(c) $Al^{3+} < Mg^{2+} \ll Cl^- < S^{2-}$

PROBLEM 1-4 **(a)** Give the number of valence electrons and the kernel charge for each of the first seven second-period elements. **(b)** Arrange these elements on the basis of increasing (1) atomic radius, (2) first ionization potential, and (3) electronegativity.

Solution

(a)

	Li	Be	B	C	N	O	F
valence electrons	1	2	3	4	5	6	7
kernel charge	+1	+2	+3	+4	+5	+6	+7

(b) (1) Increasing atomic radius: F < O < N < C < B < Be < Li
(2) Increasing first ionization potential: Li < Be < B < C < N < O < F
(3) Increasing electronegativity: Li < Be < B < C < N < O < F

PROBLEM 1-5 A sample of natural lithium is composed of two isotopes, ^{6}Li (exact mass 6.015 amu) and ^{7}Li (7.016 amu). **(a)** Describe the makeup of the nuclei of each of these isotopes. **(b)** From the average atomic mass of lithium (6.941 amu) calculate the percent of each isotope in natural lithium.

Solution
(a) ^{6}Li has 3 protons, $Z = 3$; so the number of neutrons N must be $6 - 3 = 3$. ^{7}Li also has $Z = 3$, so $N = 7 - 3 = 4$.
(b) Using $(\%)_6$ to represent the percent of ^{6}Li and $(\%)_7$ to represent the percent of ^{7}Li, we know that

$$\frac{(\%)_6(6.015) + (\%)_7(7.016)}{100} = 6.941$$

and

$$(\%)_6 + (\%)_7 = 100\%$$

So

$$(\%)_6(6.105) + [100 - (\%)_7](7.016) = (6.941)(100)$$

Therefore

$$(\%)_6 = 7.5; \qquad (\%)_7 = 92.5$$

PROBLEM 1-6 **(a)** Draw a typical s orbital, p orbital, and d orbital. **(b)** What is the value of l for each of these orbitals?

Solution
(a) See Figure 1-1 (p. 6).
(b) s orbital: $l = 0$
  p orbital: $l = 1$
  d orbital: $l = 2$

2 CHEMICAL BONDING

THIS CHAPTER IS ABOUT

☑ **Why Do Atoms Bond?**
☑ **Ionic Bonds and Ionic Compounds**
☑ **Molecular Compounds, Molecular Formulas, and Structural Formulas**
☑ **Covalent Bonds and Lewis Structures**
☑ **Multiple Bonds**
☑ **Polar Covalent Bonds**
☑ **The Molecular Orbital Picture of Single Bonds**

2-1. Why Do Atoms Bond?

A. Stability and reactivity

It's a fact of nature that atoms (unlike people) usually behave predictably. One manifestation of this fact is that a collection of atoms will always "prefer" to adopt the most stable arrangement. As it turns out, the most stable arrangement is also the one with lowest energy.

- Whenever a collection of atoms *gives off* energy, the collection itself is left at a lower-energy—and hence more stable—condition.

We've already seen an example of this principle in terms of electron configurations: The ground (most stable) state of an atom has the electrons distributed among the lowest-energy orbitals. Moreover, when enough energy is added to an atom to promote one of its electrons to a higher level, the resulting excited state is *less* stable than the ground state. One of the most fundamental rules of chemical behavior is this:

- Stability and reactivity are inversely related: The less reactive a species is, the more stable it is.

B. The noble-gas configuration and valence

It has long been recognized that the group-VIIIA elements (all of which are gases) are extremely unreactive. In fact, to this day there are no known compounds containing helium, neon, or argon. For this reason, the group-VIIIA elements are usually referred to as the **noble** (or inert) **gases.** In terms of atomic structure there is one thing that all atoms in this group share: Each has a *filled* set of *s* and *p* orbitals in its valence level. (Helium, of course, has no *p* orbitals in its valence level.) Filling one *s* and three *p* orbitals requires eight electrons (see Section 1-6), so we speak of a filled **octet** of electrons as a **noble-gas configuration.** The single most important generalization we can make about bonding between *second-period* atoms is this:

- Atoms tend to gain, lose, or share electrons in order to achieve a noble-gas electron configuration.

By making use of this relatively simple concept (which is often called the *octet rule*), we can rationalize and predict the nature of the chemical bonds in literally millions of compounds!

EXAMPLE 2-1 How many electrons must be gained or lost by each second-period element to attain a noble-gas configuration?

Solution

	Li	Be	B	C	N	O	F	Ne
to attain a helium configuration, *lose*	1	2	3	4	5	6	7	8
to attain a neon configuration, *gain*	7	6	5	4	3	2	1	0

1. The noble-gas configuration and electronegativity

Now, let's ask a question: Do you think fluorine—the most electronegative element—would prefer to give up seven electrons, or simply gain one? Clearly, it would prefer to gain one. By contrast, lithium is the least electronegative (or most *electropositive*) element in the period. Obviously, it will prefer to lose one electron rather than gain seven. We can generalize by saying:

- To attain a noble-gas configuration, electronegative elements tend to gain electrons, while electropositive elements tend to lose electrons.

Because the relatively electropositive elements in the second period (Li, Be, and B) tend to lose electrons, while the electronegative ones (N, O, and F) will gain them, we can condense the solution to Example 2-1 as shown below.

	Li	Be	B	C	N	O	F	Ne
change needed to attain noble-gas configuration	lose 1	lose 2	lose 3	±4	gain 3	gain 2	gain 1	— 0

2. Valence

We can now define an extremely important characteristic of an atom, its *valence:*

- The **valence** of an atom equals the number of electrons it gains, loses, or shares in order to attain a noble-gas configuration.

Let's compare the valence of each second-period element with its number of valence electrons.

	Li	Be	B	C	N	O	F	Ne
valence electrons	1	2	3	4	5	6	7	8
valence	1	2	3	4	3	2	1	0

Notice that the valence of an atom is equal to the number of valence electrons for atoms that lose electrons. But for atoms that gain electrons, the valence equals eight (the octet) minus the number of valence electrons. Valence is sometimes defined as the "combining capacity" of an atom, that is, the number of bonds it normally forms.

3. The special place of carbon

Finally, notice the special place that carbon occupies. A carbon atom has position and electronegativity midway between the extremes. It also has the highest valence (four) of any atom in the second period. Moreover, it forms strong bonds to a wide variety of elements, including itself. For these reasons, carbon forms the widest variety of different compounds of any element. And this is why it is given the special honor of having an entire discipline dedicated to its chemistry.

- Organic chemistry is the study of the compounds of carbon.

EXAMPLE 2-2 What is the predicted valence of each third-period element?

Solution

	Na	Mg	Al	Si	P	S	Cl	Ar
valence electrons	1	2	3	4	5	6	7	8
valence	1	2	3	4	3	2	1	0

note: We'll see later that because each of these elements has a set of empty *d* orbitals in its valence level (Section 1-6), they can "expand" their valence beyond eight electrons in some circumstances.

2-2. Ionic Bonds and Ionic Compounds

A. Ionic bonds

The simplest type of chemical bond is one involving the electrostatic attraction of two oppositely charged ions. These ions are formed by the transfer of one or more electrons between atoms. For example, what would happen if a sodium atom (Na·) should encounter a fluorine atom (:Ḟ:)? Recall that both of these atoms would prefer to have a noble-gas configuration, with either a completely filled or completely empty valence level. The answer is obvious: The sodium atom can transfer its one valence electron to the fluorine atom, simultaneously satisfying the electron configuration requirements of both. We show that transfer of electrons with a curved "fishhook" arrow:

$$Na \cdot \ + \ :\ddot{F}: \ \longrightarrow \ Na^+ \ + \ :\ddot{F}:^-$$

The two resulting ions (Na^+ and F^-) are both isoelectronic with neon. They are held together by their mutual electrostatic attraction, and this attraction is called an **ionic bond.** Of course, an electron transfer of this sort will occur only when the resulting ions are more stable than the original atoms. For this to be true, one atom must have a high electron affinity, the other must have a low ionization potential, and the electrostatic attraction between the ions must be strong. It is a useful generalization (though not without exceptions) that two atoms must differ by at least two "electronegativity units" (Section 1-7) for them to engage in ionic bonding. Thus, most ionic compounds involve a cation from groups IA or IIA, and an anion from groups VIA or VIIA.

EXAMPLE 2-3 Which of the bonds below are likely to be ionic?

$$Li-Cl, \quad H-O, \quad Na-H$$

Solution Look up the electronegativity values for each pair of elements in Section 1-7; then calculate the difference between the two values.

	Li	Cl	H	O	Na	H
electronegativity	1.0	3.0	2.1	3.5	0.9	2.1
difference		2.0		1.4		1.2
ionic?		yes		no		no

B. Ionic compounds

It is important to realize that ionic compounds such as NaF and LiCl are composed of *ion aggregates*, not molecules. In an ionic crystal the cations and anions adopt a highly ordered structural pattern, with each cation surrounded by a fixed number of anions, and vice versa. The result is that we can't identify one particular cation as belonging to one particular anion. All we can be sure of is that the ratio of cations to anions will assure an electrically neutral crystal.

EXAMPLE 2-4 **(a)** Describe the type of bonding in magnesium fluoride, magnesium oxide, and aluminum oxide. **(b)** For each compound write a chemical formula that shows the ratio of ions, assuming that each compound is ionic.

Solution

(a) Calculate the differences in electronegativity to determine whether or not the bonds can be ionic.

	Mg	F	Mg	O	Al	O
electronegativity	1.2	4.0	1.2	3.5	1.5	3.5
difference		2.8		2.3		2.0
ionic?		yes		yes		yes

(b) In writing a formula for each compound, we must decide on the basis of valence how many electrons have to be transferred. Magnesium has a valence of 2, while fluorine has a valence of 1. Therefore, one Mg atom can transfer two electrons, one to each of *two* F atoms:

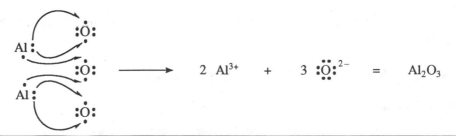

$$\text{Mg} \cdot^+ \longrightarrow \text{Mg}^{2+} + 2\, {:}\overset{\cdot\cdot}{\underset{\cdot\cdot}{\text{F}}}{:}^{-} = \text{MgF}_2$$

In like manner, one Mg atom (valence 2) will transfer both of its electrons to *one* O atom (also valence 2) to form MgO:

$$\text{Mg}{:} \; + \; {:}\overset{\cdot}{\underset{\cdot}{\text{O}}}{:} \longrightarrow \text{Mg}^{2+} + {:}\overset{\cdot\cdot}{\underset{\cdot\cdot}{\text{O}}}{:}^{2-} = \text{MgO}$$

And for aluminum oxide, a little thought will convince you that two Al atoms (valence 3) will react with three O atoms (valence 2):

$$\longrightarrow 2\,\text{Al}^{3+} + 3\, {:}\overset{\cdot\cdot}{\underset{\cdot\cdot}{\text{O}}}{:}^{2-} = \text{Al}_2\text{O}_3$$

Ionic compounds tend to be high-melting crystalline solids, which are usually soluble in water. They also tend to be strong electrolytes; that is, they dissociate into separate ions when dissolved in water.

- The vast majority of organic compounds are *not* ionic in nature.

2-3. Molecular Compounds, Molecular Formulas, and Structural Formulas

A. Molecular compounds and molecular formulas

By and large, organic chemistry deals with **molecular compounds.** Such compounds can be either solids, liquids, or gases at room temperature. Although some are water-soluble, the majority are not. And very few are electrolytes.

In a molecular compound, the simplest particle that still retains the properties of that compound is a **molecule.** (Remember that if a compound is ionic, its fundamental unit is an ion aggregate.) A molecule, in turn, is constructed from two or more atoms bonded together in a well-defined fashion. A shorthand way to describe the makeup of a molecule is its **molecular formula,** which lists the number and kinds of atoms present in the molecule. For example, the molecular formula C_2H_6O tells us that there are two carbon atoms, six hydrogen atoms, and one oxygen atom in the molecule. Notice that the molecular formula bears a strong resemblance to the chemical formulas of ionic compounds (see Example 2-4). But in fact, the formula alone does not tell us whether a compound is ionic or molecular until we know something about its structure and look at differences in electronegativity. Molecular formulas of organic compounds are usually given with the carbons listed first, hydrogens second, and the remaining elements listed in alphabetical order. The **molecular weight** (MW) of a compound is the sum of the (average) atomic masses of its constituent atoms. So, the molecular weight of C_2H_6O is $2(12.0) + 6(1.0) + 1(16.0) = 46.0$ amu.

EXAMPLE 2-5 A certain molecule is composed of ten carbons, twenty hydrogens, and two chlorines.
(a) What is its molecular formula?
(b) What is its molecular weight?

Solution
(a) $C_{10}H_{20}Cl_2$
(b) MW $= 10(12.0) + 20(1.0) + 2(35.5) = 211.0$ amu

B. Structural formulas

When there are only two atoms in a molecule (as in H_2, O_2, and F_2), there's no question of which atom is bonded to which. If there are three or more atoms in the molecule, there can be a question. For example, we all know that water is H_2O. But what is its *structure*? That is, what is the sequence of atoms and bonds in the structure of water? Is it H–H–O or H–O–H? Notice that the molecular formula does not give such structural information. Of course, most people know that in water both hydrogens are bonded to the oxygen rather than to each other, and we show this fact by writing H–O–H. Because this formula H–O–H gives us structural information about the sequence of atoms and bonds, we call it the *structural formula* of water. A **structural formula** shows which atoms are connected to which other atoms.

But how do we *know* that H–O–H is the structure of water, rather than H–H–O? That's where valence again comes to the rescue.

2-4. Covalent Bonds and Lewis Structures

A. Covalent bonds

The simplest molecule is H_2. What type of bonding holds the two hydrogens together? It can't be an ionic bond, because the difference in electronegativity between the two atoms is zero. Still, both atoms want to attain a filled valence level (the configuration of helium). How can this be accomplished *without* electron transfer? The answer is that the *pair* of electrons (one from each hydrogen atom) is *shared* between the two hydrogen nuclei:

$$H \cdot \qquad \cdot H \longrightarrow H : H$$

Such a bond, which results from sharing a *pair* of electrons, is called a **covalent bond.** Because each hydrogen nucleus "sees" two electrons around it, both atoms "feel" as though they have attained a filled valence shell. The reason two hydrogen atoms prefer to bond to each other rather than to exist as separate atoms is that they are more stable when bonded; that is because they have each attained a filled valence level by sharing electrons. The electron pair is the "glue" that holds the two nuclei together. Because the electron pair has a high density (probability) *between* the two nuclei, both nuclei are attracted *to* the electron pair, and are shielded from each other *by* the electron pair.

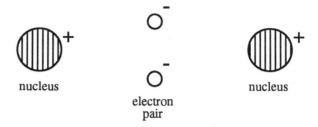

nucleus

electron pair

nucleus

B. Lewis and other structures

Covalent bonds can be drawn as pairs of electrons (two dots) in structural formulas called **Lewis structures.** (G. N. Lewis was the first person to draw covalent bonds this way; hence the name "Lewis" structures.) To draw the **Lewis electron-dot structure** of water, begin by drawing the three atoms with their valence electrons as dots. (Remember that subvalence electrons are not shown.) Then "pair up" (share) electrons in such a way as to satisfy the valence of each atom:

Notice how each hydrogen "sees" two electrons, while the oxygen "sees" eight. Only two of the pairs (four electrons) around oxygen are involved in bonding. The other two pairs are not. We call the latter pairs of electrons **nonbonding pairs** (or **unshared pairs**).

Another way to denote bonds is with lines instead of dots, as in H–O–H. Such structures are called **Kekulé structures.** In this book we'll use a hybrid method, drawing bonding pairs (bonds) as lines, and showing nonbonding pairs as dots. Thus, water would be drawn H–Ö–H. But beware! As we become more accustomed to writing structures this way, we'll sometimes get lazy and leave out the nonbonding pairs. But they're still present, and they have a profound effect on chemical behavior. Therefore, you might want to extend our table of valences, first seen in Section 2-1, to include information about nonbonding pairs.

	Li	Be	B	C	N	O	F	Ne
	Na	Mg	Al	Si	P	S	Cl	Ar
valence electrons	1	2	3	4	5	6	7	8
normal valence	1	2	3	4	3	2	1	0
usual number of nonbonding pairs	0	0	0	0	1	2	3	4

note: Some covalent compounds of boron and beryllium adopt bonding patterns somewhat more complex than the octet rule alone would predict.

EXAMPLE 2-6 For each molecular formula below draw both a Lewis structure and line/dot structure that satisfies the valence of each atom. Be sure to show all nonbonding pairs.

(a) F_2 (b) CH_4 (c) C_2H_6O

Hint: In (c) all hydrogens are bonded to carbons.

Solution

(a)

(b)

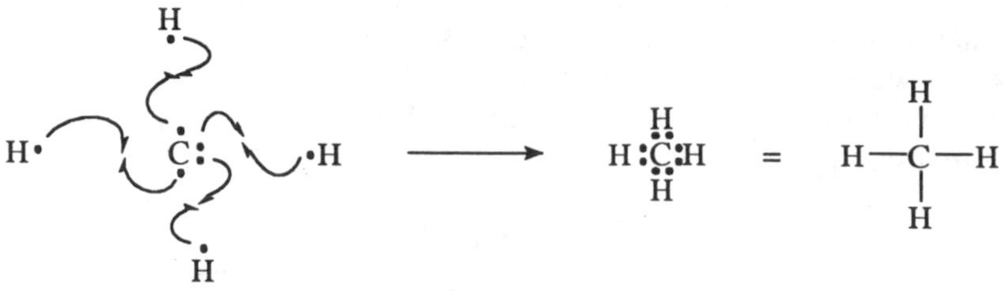

(c)

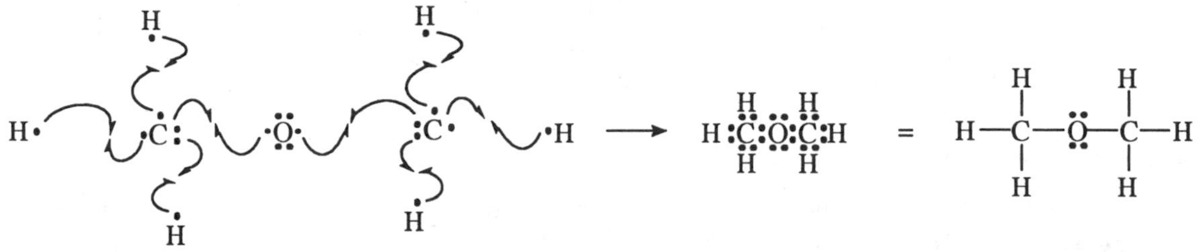

C. Structural isomers

Suppose there hadn't been the hint for part (c) in Example 2-6? Could you have written any other structures for C_2H_6O that satisfy the valence of each atom? How about this one?

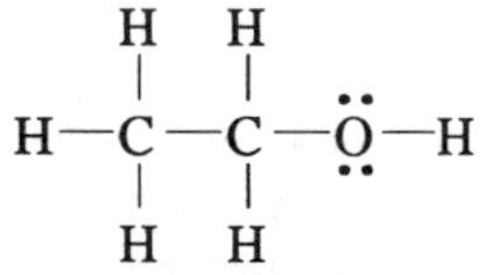

Note how each atom's valence is still satisfied: Each hydrogen has one bond (two electrons) around it, each carbon has four bonds (eight electrons), and the oxygen has two bonds and two nonbonding pairs (eight electrons). So, each atom has its noble-gas configuration. Yet, this structural formula is *different* from the one shown in Example 2-6c. That is, it has a different sequence of atoms and bonds. How can this be? The answer is that there can often be two or more *different structures* with the *same* molecular formula. Two molecules that have the same molecular formula, but a different sequence of atoms and bonds, are called **structural** (or **constitutional**) **isomers.** This is why you must remember this:

• Molecular formulas alone do not convey structural information.

EXAMPLE 2-7 Draw line/dot formulas for two structural isomers of $C_2H_4Cl_2$. Be sure the valence of each atom is satisfied.

Solution

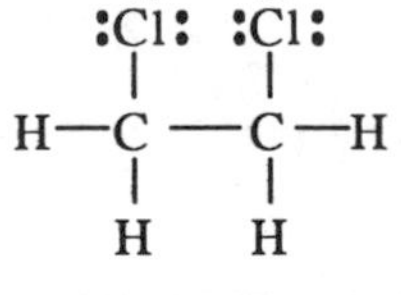

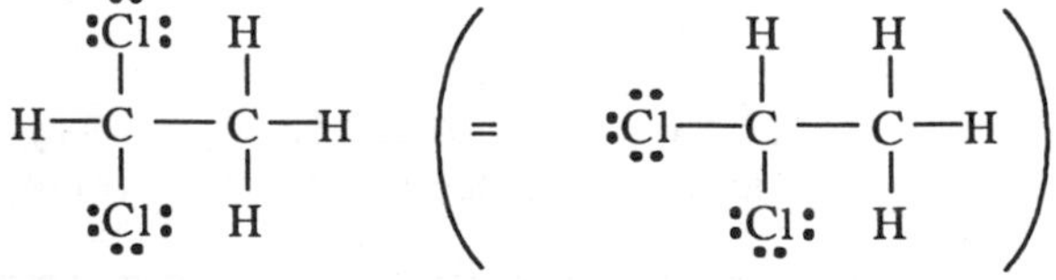

2-5. Multiple Bonds

Suppose you were asked to draw the structural formula of C_2H_4. You could start, as we have done before, by pairing up the electrons:

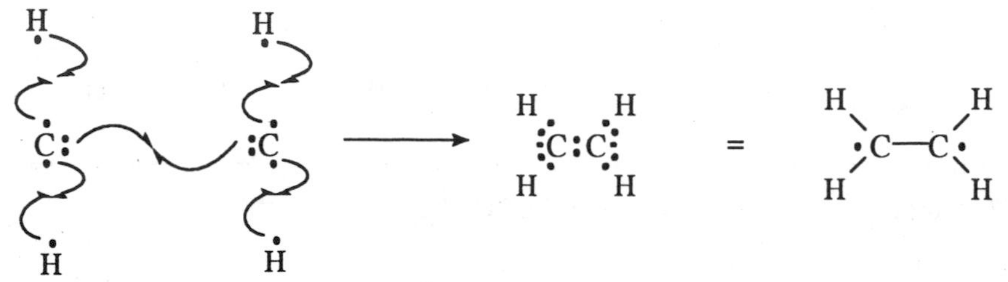

Notice, however, that neither of the carbons has satisfied its valence yet. Each has only seven electrons around itself. How can these electrons be further rearranged to provide a complete octet for both carbons? We simply have to pair up the odd (unpaired) electron on each carbon to make a second bond between the two carbons:

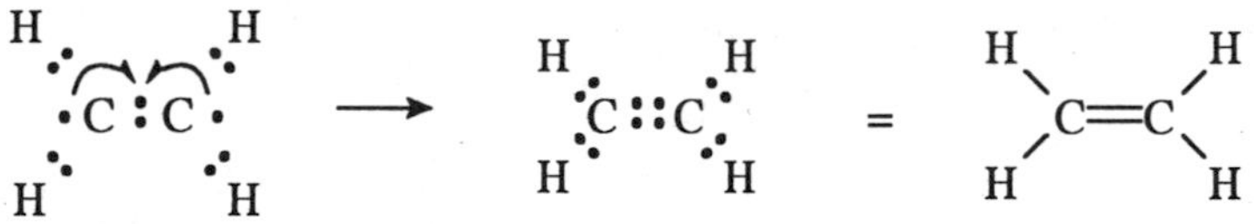

Such a bond, which involves the sharing of *two* electron pairs, is called a **double bond.** (All of the other bonds we've seen so far in this chapter involve just one pair of electrons and are described as **single bonds**). Note how the formation of the double bond has satisfied the valence of both carbons, since each now has eight electrons (four bonds) around it.

There can also be **triple bonds**, resulting from the sharing of *three* electron pairs. A good example is C_2H_2:

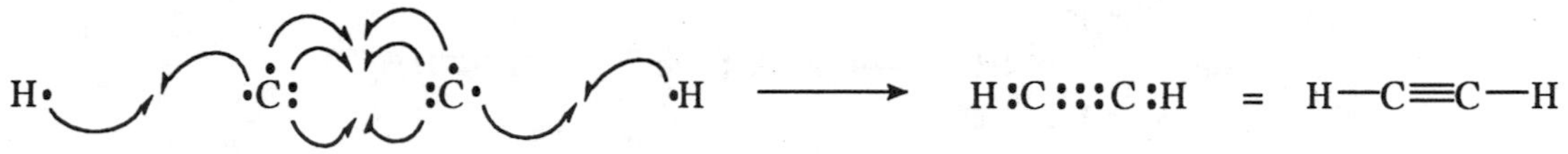

In this structure each carbon is again surrounded by eight electrons (four bonds), so all valences are satisfied.

note: There are no "quadruple" bonds between main-group elements.

EXAMPLE 2-8 Draw a structural formula for each of the molecular formulas shown below. Be sure to satisfy the valence of each atom, and show all nonbonding pairs.

(a) CH_2O (b) CH_3N (c) HCN

Solution

(a)

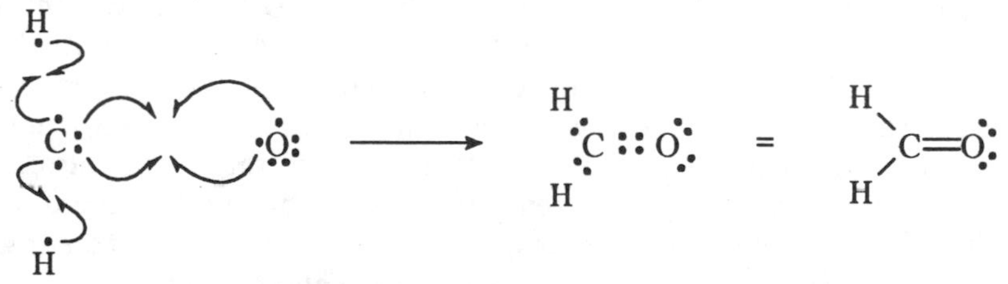

(b)

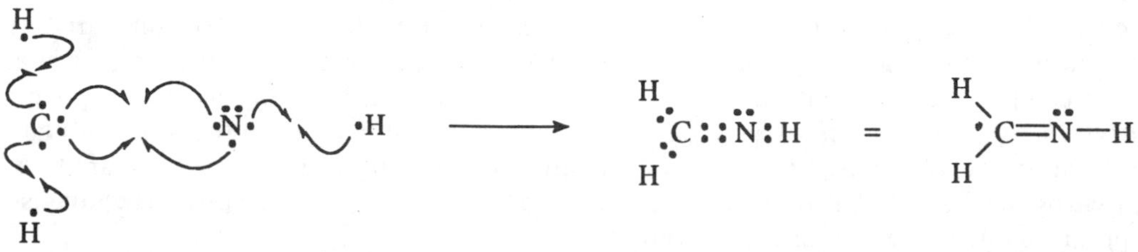

(c)   H·   ·C̈·   ·N̈:   ⟶   H:C:::N: =   H—C≡N:

EXAMPLE 2-9 Draw two structural isomers of C_3H_4.

Solution

$$H{\backslash}C{=}C{=}C{/}H \quad \text{and} \quad H—C—C≡C—H$$

Or you might draw

H H
 \ /
  C
 / \
H—C=C—H

(If you drew this cyclic structure, you must have had a little organic chemistry before!)

Now, suppose you were asked to explain why two helium atoms *don't* form He_2. To answer this question you might try to write a Lewis structure for He_2, which would look like this:

He: :He   —?→   He::He   =   He=He

But how many electrons does each helium "see"? Four! Yet, each helium can accommodate only two electrons to fill its valence level. So, since there is no benefit to sharing these electrons, the molecule doesn't exist. (We'll discuss this hypothetical molecule again in Section 2-7.)

It shouldn't surprise you that

● A triple bond is stronger than a double bond, which in turn is stronger than a single bond.

But it's also a fact that a double bond is only about 60% stronger than a single bond, while a triple bond is only a little more than twice as strong as a single bond.

2-6. Polar Covalent Bonds

So far, we've encountered two types of chemical bonds: **ionic bonds,** which are formed by complete electron transfer between atoms of very different electronegativity, and **covalent bonds,** which are formed by sharing electron pairs between atoms of comparable electronegativity. But the distinction between these two types is not always cut and dried. Actually, a *purely* covalent bond, one in which the electron pair is shared *equally* between the two nuclei, can occur only when the bonded atoms are equally electronegative (as in H_2, F_2, etc.). If there *is* a difference in electronegativity between the atoms, though not a large enough difference to make the bond ionic, there will be an *unequal* sharing of electron density between the atoms; that is, the electrons will be attracted more to one atom than to the other. Such a bond is called a **polar covalent bond.**

Let's look again at the two O–H bonds in water. The difference in electronegativity between oxygen and hydrogen is 1.4 (Example 2-3), with oxygen being the more electronegative. From the definition of electronegativity (Section 1-7), we know the oxygen will get a greater share of the bonding electron density than the hydrogens. For this reason, the oxygen will acquire a *partial negative charge,* since it possesses greater electron density than it did in the uncombined (neutral) state. Similarly, each hydrogen acquires a *partial positive charge;* but since there are two of them, each hydrogen gets only half the charge the oxygen does. We show these **partial charges** acquired in a polar covalent bond with the symbol δ.

$$1/2\,\delta+ \quad \overset{..}{\underset{..}{\text{O}}} \quad 1/2\,\delta+$$
$$\text{H}\!-\!\text{O}\!-\!\text{H}$$
$$\delta-$$

Another way to denote a polar covalent bond is with a crossed arrow (+———►), which shows the direction of charge by pointing toward the more electronegative atom. This arrow represents the **charge dipole** of the bond. The longer the arrow, the larger the electronegativity difference, and the larger the dipole. So we can write the structure like this:

$$\text{H}\!-\!\!-\!\overset{..}{\underset{..}{\text{O}}}\!-\!\!-\!\text{H}$$

EXAMPLE 2-10 Using crossed arrows, show the direction of the charge dipole of each bond in the structures shown in Example 2-6.

Solution

(a) $:\!\overset{..}{\text{F}}\!-\!\overset{..}{\text{F}}\!:$ has no dipole because there is no electronegativity difference between F and F.

(b) Carbon is *slightly* more electronegative than hydrogen (2.5 vs. 2.1), so you can draw the structure of CH_4 with crossed arrows pointing from each H to C:

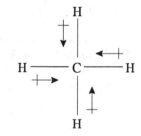

(c) Carbon is slightly more electronegative than hydrogen, but oxygen is more electronegative than carbon (3.5 vs. 2.5), so you can write the structure of C_2H_6O as

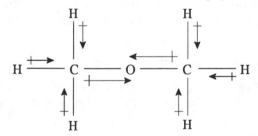

By now you may have noticed that an ionic bond is simply the extreme of a polar covalent bond, where the "sharing" of the electron pair is so unequal that there is *no* sharing! Thus ionic and covalent bonds represent the extremes of the bonding continuum. Most bonds are at least somewhat polar, and the degree of polarity is determined by their differences in electronegativity. The direction and magnitude of bond polarity are very important in determining the chemical properties of a bond. Furthermore, it is a useful generalization that,

● All other things being equal, the more polar a covalent bond is, the stronger it is.

2-7. The Molecular Orbital Picture of Single Bonds

Do you remember the definition of "atomic orbital"? It is a mathematical description (a solution to the Schrödinger equation) of how the electrons are distributed in an atom (see Section 1-4). We can perform a similar (though more complex) calculation to find out how the electrons *in a molecule* are distributed. But the solutions to this equation are called **molecular orbitals,** and they are spread out over two or more nuclei. In this book, we'll represent these molecular orbitals (MOs) pictorially, as we did for atomic orbitals (AOs).

Molecular orbitals can also be viewed as arising from the overlap (combination) of AOs. The set of AOs used to construct a set of MOs is called the **basis set.** Always remember this:

● A molecular orbital, like an atomic orbital, can contain no more than two electrons, and then only if the electron spins are paired.

A. The case of hydrogen

Let's return to the hydrogen molecule, H_2. Instead of drawing it in a Lewis or line/dot structure, try to draw the two hydrogen $1s$ AOs (the basis set) overlapping to form the MO of H_2. Here would be a logical guess:

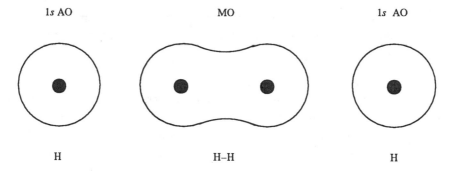

The resulting MO is a "peanut-shaped" volume surrounding *both* hydrogen nuclei. The two electrons (one from each hydrogen atom) share this orbital with their spins paired and spend a significant portion of their time *between* the two nuclei. Remember that the MO must be more stable (lower energy) than the AOs; otherwise, the bond wouldn't form!

B. The case of helium

Now, let's reconsider the hypothetical molecule He_2 from Section 2-5. The two $1s$ AOs again overlap to form the same type of MO we just described for hydrogen. But there is a problem: This MO will accommodate only *two* electrons. Where will the other two electrons go? The answer is that they go to another MO we haven't talked about yet. Suffice it to say:

● If there are n AOs in the basis set, their overlap will result in the formation of n MOs.

Thus the combination of two $1s$ AOs will result in the formation of *two* MOs. To describe the second one of these, we have to introduce the concept of **orbital phase.**

EXAMPLE 2-11 How many MOs can be constructed from a basis set of six AOs?

Solution Six!

C. Orbital phase

You have, no doubt, had some experience with sine waves, such as the one shown in Figure 2-1a. This wave function has a **phase** (the sign of the function) that varies from plus to minus as the wave goes from peak to trough. The point where the wave crosses the axis (and has zero amplitude) is called a **node.** When *two* sine waves of the same frequency overlap, the result depends on their relative phases. If the two waves are **in-phase** (that is, the two waves reach peaks and troughs at the same points on the horizontal axis), the combination will be a bigger sine wave, as shown in Figure 2-1b. This phenomenon is called **constructive interference.** However, if the two equally intense sine waves are **out-of-phase** (that is, one wave reaches a peak when the other reaches a trough), the resultant "wave" is a flat line of zero amplitude (see Figure 2-1c), because one wave cancels the other. This situation is called **destructive interference.**

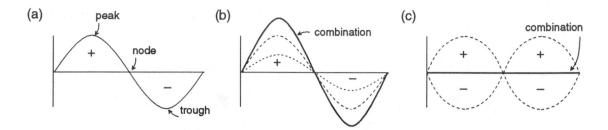

FIGURE 2-1. (a) A single sine wave; **(b)** constructive interference between two in-phase sine waves; **(c)** destructive interference between two equally intense out-of-phase sine waves.

It turns out that atomic orbital waves, like sine waves, have phases (signs). Whether a given AO is all of one phase, or has multiple phases, depends on whether it has any nodes. And this, in turn, is a function of its quantum numbers. For our present purposes, here is all you need to remember:

- A 1*s* orbital has *no* nodes and is therefore all one phase (plus or minus).

- A 2*p* orbital has *one* node (sometimes called a **nodal plane**) and two phases.

We show this by putting different signs (+ and −) in the two lobes of the orbital.

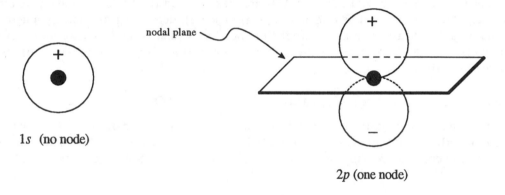

1*s* (no node)

2*p* (one node)

EXAMPLE 2-12 A 3*d* orbital has *two* nodal planes. Show the phase of each lobe.

Solution

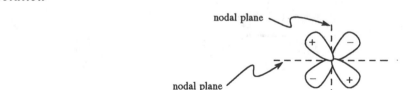

---

note: It is most important not to confuse these phases (signs) of an orbital with ionic charges caused by excesses of electrons or protons. In this book, any sign within the lobe of an orbital represents the phase of that lobe, and has nothing to do with charge.

Because 1*s* orbitals have phase (but not nodes), their overlap can result in either constructive or destructive interference. The peanut-shaped MO we discussed above is the result of *constructive* interference between two AOs of the *same* phase:

But if the two AOs have opposite phase, their overlap will result in *destructive* interference, and the MO will have a node *between* the nuclei:

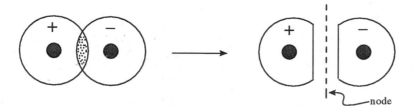

Note that this MO, though it has two separate lobes, is still a *single* MO, and can accommodate no more than two electrons. And because this MO has a node between the nuclei, there is no electron density there to hold the nuclei together. Moreover, the two nuclei are no longer shielded from each other and are repelled by their like positive charges. Therefore, this orbital is less stable (higher in energy) than its constituent 1s orbitals.

D. Bonding and antibonding MOs

To be able to discuss these two MOs more readily, we're going to give them labels. The original peanut-shaped one is called a **bonding MO** because it has no node between the nuclei and its energy is *lower* than that of the basis set AOs. The latter MO, which has a node between the nuclei, is called an **antibonding MO** and is designated with an asterisk (*). Its energy is *higher* than that of the basis set AOs. Just as AOs are classified on the basis of their symmetry (*s, p, d,* etc.), MOs are labeled as either σ (sigma) or π (pi), depending on their symmetry. An MO that is cylindrically symmetric around the bond axis (that is, one whose cross section is circular) is called a σ MO. Both the bonding and antibonding combinations of two 1s AOs are of this type.

E. Molecular orbital energy diagrams: σ MOs vs. 1s AOs

We can summarize everything we've said so far about the combination of two 1s AOs by looking at the **molecular orbital energy diagram** (MOED) shown in Figure 2-2. Notice how the σ MO is lower in energy (by Δ) than the AOs, while the σ* MO is higher in energy (by Δ).

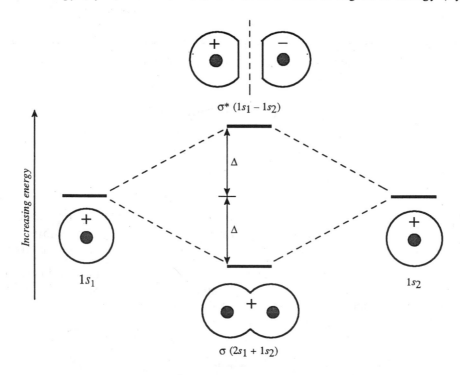

FIGURE 2-2. The molecular energy diagram (MOED) for overlap of two 1s atomic orbitals.

Now, look at Figure 2-3, where MOEDs for H_2 and He_2 are compared. In the case of H_2 (Figure2-3a), each hydrogen donates one electron. The resulting pair of electrons (whose spins are paired) goes into the σ orbital. The H_2 molecule is thus more stable than the separate atoms by 2Δ. The bond in this molecule is a σ bond. But look what happens in the case of He_2 (Figure 2-3b): There are now *four* electrons (two pairs) to be distributed into the two MOs. One pair goes into the σ MO, liberating 2Δ in energy. But the second pair must go into the σ^* MO, *costing* 2Δ in energy. Thus, the cost of antibonding exactly cancels the benefit of bonding. So, there is no net benefit to be gained from making the bond—and that's why He_2 does not form.

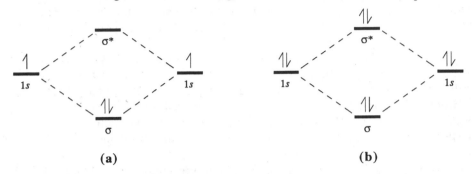

FIGURE 2-3. Electron occupancy in the molecular orbitals of **(a)** H_2 and **(b)** He_2.

F. Bond order

The net **bond order** between two bonded atoms equals the number of electron pairs in bonding MOs *minus* the number of electron pairs in antibonding MOs. (A single electron counts as one-half of a pair.)

EXAMPLE 2-13 What is the bond order in **(a)** H_2, **(b)** He_2, **(c)** He_2^{2+}, and **(d)** He_2^+?

Solution
(a) In the H_2 molecule there is one bonding pair and no antibonding pairs, so the bond order is $1 - 0 = 1$, a single bond.
(b) In He_2 there is one bonding pair and one antibonding pair, so the bond order is $1 - 1 = 0$, and no bond forms.
(c) He_2^{2+} results when the two highest energy electrons are removed from He_2. This ion is isoelectronic with H_2 and has a bond order of 1.
(d) He_2^+ has a total of three electrons, one pair in the bonding MO, and one electron (one half of a pair) in the antibonding MO. The bond order is therefore $1 - 1/2 = 1/2$.

You might be wondering why we need to discuss MOs at all. Why not just be satisfied with Lewis structures? The answer is that patterns in molecular structure and reactivity are often best understood in terms of MOs. Furthermore, in certain molecules the Lewis (valence-bond) structure gives an incomplete picture of the structure, while the MO approach gives a more satisfactory representation.

SUMMARY

1. Two atoms will bond together when the bonded atoms are more stable (lower in energy) than the separate atoms.
2. When forming bonds, second-period atoms tend to acquire a noble-gas electron configuration, i.e., a completely filled octet or completely empty valence shell by gaining, losing, or sharing electrons.
3. The normal valence of second-period atoms is equal to the number of valence electrons for groups IA to IVA and equal to eight minus the number of valence electrons for groups VA to VIIIA.
4. The discipline of organic chemistry is the study of carbon compounds. Carbon earns this unique place because it has a valence of four and because it forms strong bonds with a wide variety of elements, including itself.

5. Ionic bonds involve electrostatic attraction between anions and cations (with noble-gas configurations) formed as a result of complete electron transfer between atoms of highly different electronegativity.

6. The molecular formula of a compound shows only the types and ratios of atoms in the molecule. The structure of the molecule—the sequence of atoms and bonds—is shown in the structural formula.

7. Covalent bonds involve the sharing of electron pairs between atoms of comparable electronegativity. Such bonds are shown as lines in structural formulas. Each shared electron pair (as well as any nonbonding pairs) contributes to satisfying the valence requirements of the bonded atom.

8. Covalent bonds may consist of one, two, or three shared electron pairs, giving single, double, and triple, bonds, respectively.

9. Polar covalent bonds involve unequal sharing of the electron pair(s), with the more electronegative atom acquiring a slight excess of electron density (and a partial negative charge, $\delta-$) while the less electronegative atom acquires a partial positive charge ($\delta+$). The resulting charge dipole of the bond can be denoted with a crossed arrow ($\longmapsto$).

10. Molecular orbitals (MOs) are mathematical descriptions (shown pictorially) of how electrons are distributed in a molecule. An MO can accommodate no more than two electrons (with spins paired).

11. Molecular orbitals can be viewed as arising from the overlap of atomic orbitals (AOs). A bonding MO results from constructive interference between the AOs (like phases) and has no nodes between the nuclei. An antibonding MO results from destructive interference (unlike phases) between the AOs and has a node between the nuclei.

12. Sigma (σ) bonds involve MOs that are cylindrically symmetric around the bond axis.

13. The net bond order between two atoms equals the number of electron pairs in bonding MOs minus the number of pairs in antibonding MOs.

RAISE YOUR GRADES

Can you define...?

☑ noble-gas configuration
☑ valence
☑ organic chemistry
☑ ionic bond
☑ molecular compound
☑ molecular formula
☑ molecular weight
☑ structural formula
☑ Lewis structure
☑ nonbonding pairs
☑ structural isomers
☑ single/double/triple bonds
☑ polar covalent bond

☑ bond dipole
☑ molecular orbital (MO)
☑ basis set
☑ bonding MO
☑ antibonding MO
☑ σ MO
☑ σ* MO
☑ phase (of an orbital)
☑ constructive interference
☑ node
☑ MOED
☑ bond order

Can you explain...?

☑ the relationship between energy, stability, and reactivity
☑ how atoms attain a noble-gas configuration
☑ how the valence of an atom is determined
☑ how ionic and covalent bonds are formed
☑ how the properties of ionic compounds differ from those of molecular compounds
☑ the nature of polar covalent bonds
☑ how σ and σ* MOs are formed from the overlap of s AOs
☑ how electrons occupy MOs
☑ how to draw MOEDs
☑ how to calculate bond orders

SOLVED PROBLEMS

PROBLEM 2-1 Give the number of valence electrons, normal valence, and normal number of nonbonding pairs for the first 18 elements (excluding helium) in the bonded state.

Solution

	H							
	Li	Be	B	C	N	O	F	Ne
	Na	Mg	Al	Si	P	S	Cl	Ar
valence electrons	1	2	3	4	5	6	7	8
valence	1	2	3	4	3	2	1	0
nonbonding pairs	0	0	0	0	1	2	3	4

PROBLEM 2-2 For each formula below indicate whether the substance is ionic or covalent and, using line/dot structures, show how the electrons are either transferred or shared.

(a) $MgCl_2$ **(b)** BCl_3 **(c)** CH_4S **(d)** CH_2O_2

note: The electronegativity values you'll need for this and some of the other problems in this section are given below:

	IA						
1	H 2.1	**IIA**	**IIIA**	**IVA**	**VA**	**VIA**	**VIIA**
2	Li 1.0	Be 1.5	B 2.0	C 2.5	N 3.0	O 3.4	F 4.0
3	Na 0.9	Mg 1.2	Al 1.5	Si 1.8	P 2.1	S 2.5	Cl 3.0

Solution

(a) The difference in electronegativity between Mg and Cl is $3.0 - 1.2 = 2.8$, so $MgCl_2$ is ionic.

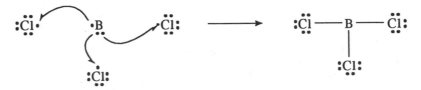

(b) The difference in electronegativity between B and Cl is $3.0 - 2.0 = 1.0$, so BCl_3 is polar covalent.

(c) The differences in electronegativity are as follows:

H–C: $2.5 - 2.1 = 0.4$
C–S: $2.5 - 2.5 = 0$
S–H: $2.5 - 2.1 = 0.4$

so each of these bonds is covalent.

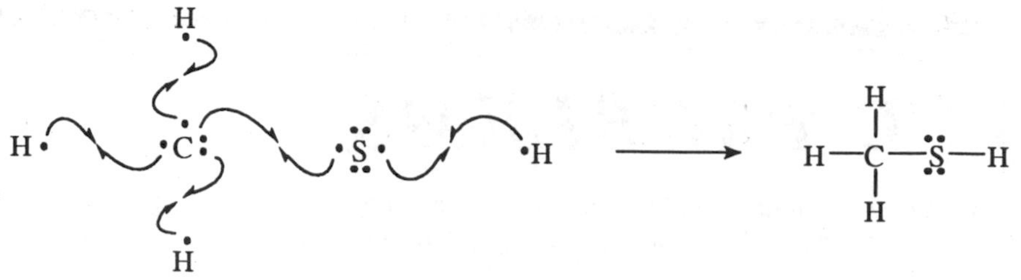

(d) The differences in electronegativity are as follows:

H–C: 2.5 – 2.1 = 0.4, so this bond is covalent.
C–O: 3.5 – 2.5 = 1.0, so this bond is polar covalent.
O–H: 3.5 – 2.1 = 1.4, so this bond is polar covalent.

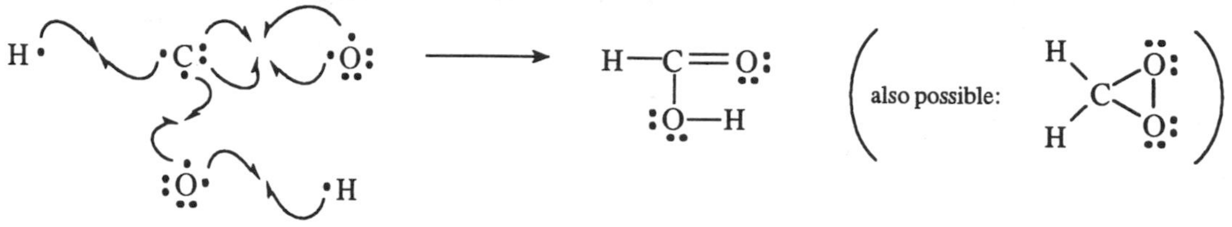

PROBLEM 2-3   Write line/dot structural formulas for three isomers of C_2H_2O. Be sure to satisfy the valence of each atom.

Solution

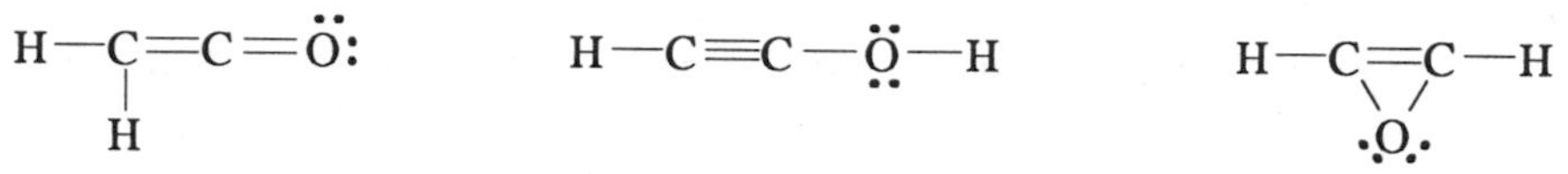

PROBLEM 2-4   Identify the most polar bond in each structure below; then indicate your answer with a crossed arrow in the proper orientation.

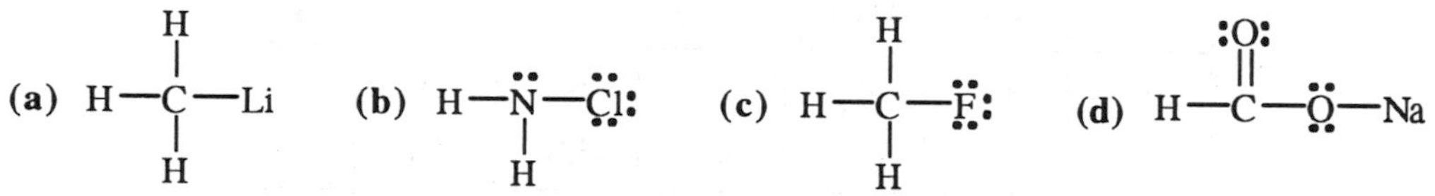

Solution   First, find the electronegativity differences:
(a) C–H: 2.5 – 2.1 = 0.4
   C–Li: 2.5 – 1.0 = 1.5
(b) N–H: 3.0 – 2.1 = 0.9
   N–Cl: 3.0 – 3.0 = 0
(c) C–H: 0.4
   C–F: 4.0 – 2.5 = 1.5
(d) C–H: 0.4
   C–O: 3.5 – 2.5 = 1.0
   O–Na: 3.5 – 0.9 = 2.6 (ionic!)

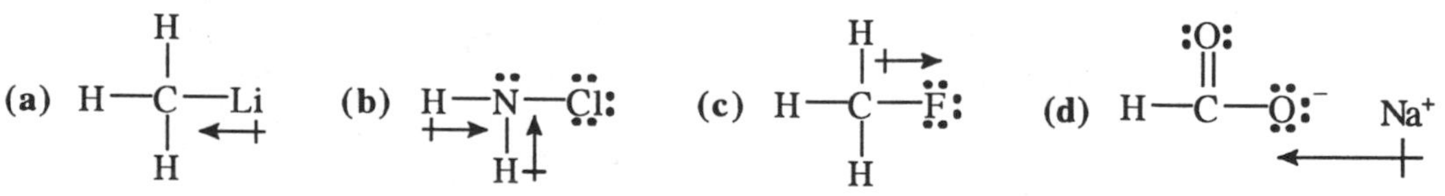

PROBLEM 2-5   Fill in all nonbonding pairs in the structures below:

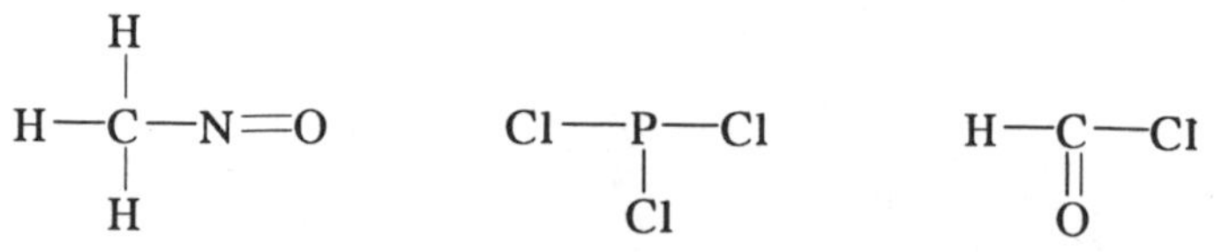

Solution

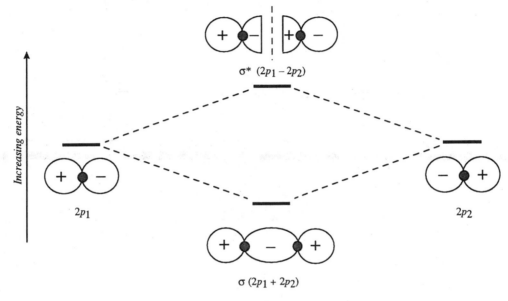

PROBLEM 2-6 When two $2p$ orbitals overlap end-to-end, they can undergo either constructive or destructive interference. Using an MOED, draw each resulting MO, showing both its shape and relative energy. Label each orbital appropriately.

Solution

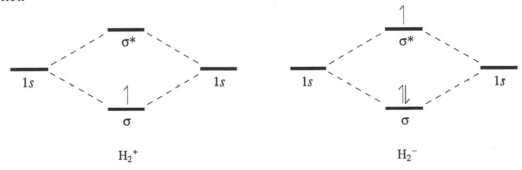

Note: These MOs are σ-type because they are cylindrically symmetric.

PROBLEM 2-7 Calculate the molecular weight of **(a)** $C_5H_8O_2$ and **(b)** C_2H_5N.

Solution
(a) MW $C_5H_8O_2$ = 5(12.0) + 8(1.0) + 2(16.0) = 100 amu
(b) MW C_2H_5N = 2(12.0) + 5(1.0) + 14.0 = 43.0 amu

PROBLEM 2-8 Using an MOED, show the orbital occupancy for H_2^+ and H_2^-. Calculate the bond order for each species.

Solution

Since H_2^+ has only one electron in its bonding MO and no electrons in the antibonding MO, its bond order is 1/2. H_2^-, however, has a pair of electrons in the bonding MO and a single electron in the antibonding MO, so its bond order is 1 − 1/2 = 1/2.

PROBLEM 2-9 Write a line/dot structural formula for each molecular formula below:

(a) CO_2 **(b)** CH_2O_3 (two structural isomers)   **(c)** C_2H_4O (three structural isomers)

Solution

(a)

$$:\ddot{O}=C=\ddot{O}:$$

(b)

H—Ö—C—Ö—H   and   H—C—Ö—Ö—H
            ‖   ‖
           :O: :O:

$$\left(\text{also} \quad \begin{array}{c} :\ddot{O}—\ddot{O}: \\ | \quad \quad | \\ H—C—\ddot{O}: \\ | \\ H \end{array} \quad \text{or} \quad \begin{array}{c} H \\ | \\ C \\ H—\ddot{O} \end{array} \right)$$

(c)

H   ·Ö·
 \ ‖
H—C—C—H
 |
 H

H   H
 \ /
H—C=C—Ö—H
        ··

  ·Ö·
  / \
H—C—C—H
 | |
 H   H

3 THE SHAPE OF MOLECULES

3-1. What Constitutes Molecular Shape?

A. Dipoles and bond angles

In Chapter 2 we developed structural formulas to show the sequence of atoms and bonds in a molecule. We decided to represent the structure of water as shown below, where the crossed arrows indicate the direction of the charge dipoles in the O–H bonds:

$$H \longrightarrow \overset{..}{\underset{..}{O}} \longleftarrow H$$

Now we need to talk a little about the "dipole" concept. Molecules have a property called **dipole moment** which is the sum (or *vector sum*, to be exact) of all the individual bond dipoles. The structure of water shown above *suggests* that the two bond dipoles, being equal in magnitude but opposite in direction, should cancel each other, leaving water with a substantial moment of zero. Yet, water has a substantial dipole moment! How can this be?

The problem is that our structural formula for water is misleading in one very important respect: In fact, the water molecule is *not* linear, even though the above structural formula suggests that it is. The water molecule is actually bent, with a **bond angle** (the **internuclear H–O–H angle**) of 104.5°. Thus it is more accurate to draw the structure of water this way:

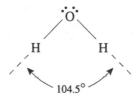

Now it becomes clear that the two O–H bond dipoles do *not* cancel; rather, they add (vectorially) to produce a large net dipole moment. The dipole moment of water is also augmented by the two unshared electron pairs on oxygen.

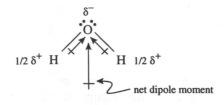

EXAMPLE 3-1 If the structure of water *were* linear (which it is *not*), what would its bond angle be?

Solution 180°.

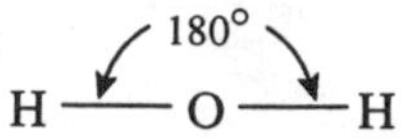

$$H \xrightarrow{\quad} O \xrightarrow{\quad} H$$

$$180°$$

It may be obvious to you that a molecule must have at least three atoms for its shape to be anything but linear. (A diatomic molecule *must* be linear.) In the molecule A–B–C there is only one bond angle. End atoms like A and C are called **terminal atoms**, while inside atoms like B are called **internal atoms**.

EXAMPLE 3-2 Circle all the internal atoms in the structure below:

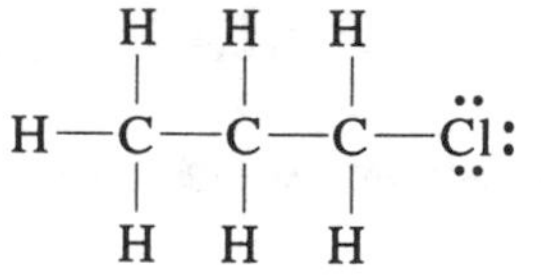

$$
\begin{array}{ccccccc}
 & H & & H & & H & \\
 & | & & | & & | & \\
H\!-\!\!&C&\!\!-\!\!&C&\!\!-\!\!&C&\!\!-\!\ddot{C}\ddot{l}: \\
 & | & & | & & | & \\
 & H & & H & & H & \\
\end{array}
$$

Solution

$$
\begin{array}{ccccccc}
 & H & & H & & H & \\
 & | & & | & & | & \\
H\!-\!\!&\boxed{C}&\!\!-\!\!&\boxed{C}&\!\!-\!\!&\boxed{C}&\!\!-\!Cl \\
 & | & & | & & | & \\
 & H & & H & & H & \\
\end{array}
$$

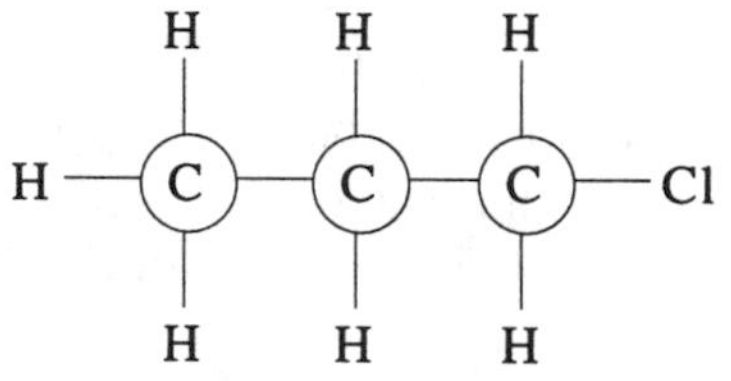

All uncircled atoms are terminal atoms.

B. Three-dimensional character of molecules

Another molecule we encountered in Chapter 2 was **methane**, CH_4. We drew its structural formula in Example 2-6b this way:

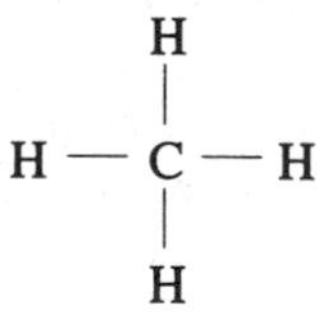

$$
\begin{array}{c}
H \\
| \\
H-C-H \\
| \\
H \\
\end{array}
$$

This drawing suggests that methane is flat, a so-called **square planar** molecule, with H–C–H bond angles of 90° and 180°. But methane is *not* flat! Instead, it is **tetrahedral**, with each hydrogen located at the corner of the imaginary three-dimensional tetrahedron and carbon at the center. Each of the H–C–H bond angles is 109.5, and all C–H bond lengths are equal. This tetrahedral shape can be shown like this:

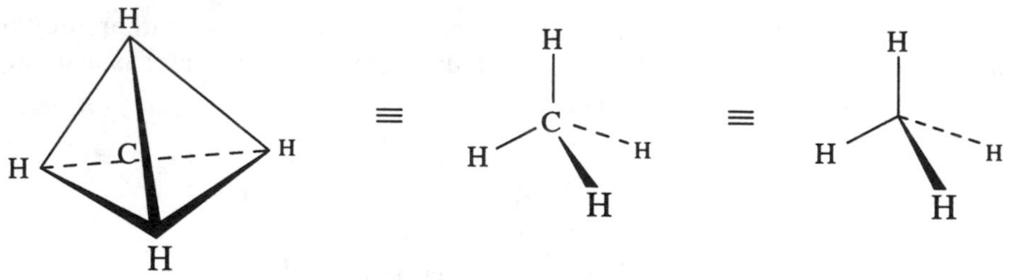

With this structure we are introducing a very important convention used to portray three-dimensional structures on a two-dimensional page. Bonds that lie *in* the plane of the paper are drawn as simple lines. Bonds that are to be viewed as coming out from the page toward the viewer are drawn as solid wedges (with the wide end of the wedge toward you). And bonds that recede *behind* the plane of the page are shown as dashed lines (or dashed wedges in some books). This perspective is emphasized in the above structures with H's of different sizes to aid you in visualizing the perspectives. But normally we'll use letters that are all the same size, and perspective will be shown only by the way the bonds are drawn.

Here are three other structural drawings that show different (but equivalent) ways to depict the tetrahedral shape of methane:

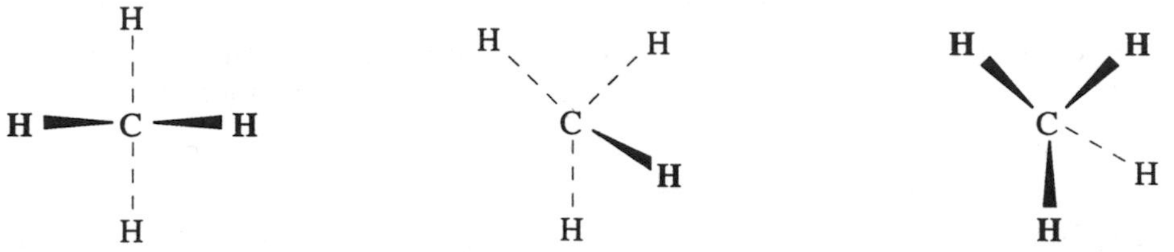

Which view is the most meaningful to you?

note: At this point, let me give you a very important bit of advice. If you have not already done so, go out and buy yourself a set of molecular models. It doesn't have to be an expensive set, just one that allows you to build models of the molecules we discuss. The importance of being able to see these structures in three dimensions will make the discussions that follow throughout the rest of this book *much* easier to comprehend.

C. Determining the shapes of molecules

The **shape** (geometry) of a molecule is fully defined only when we know the relative three-dimensional position of every atom in it. Chemists can obtain this information for a molecule *in the crystalline state* by using a technique known as **X-ray crystallography,** which also provides the values of all bond angles and **bond lengths** (the distance between bonded nuclei). It is true that all of the bonds in a molecule are in constant vibration, even when the molecule is in the crystalline state. And in the liquid and gaseous states molecules are constantly tumbling around in both rotational and translational motion. Nonetheless, when we speak of the most stable shape of a molecule, we are usually referring to the shape as determined crystallographically. While it is difficult for us to predict *exact* values of bond angles in molecules, there are some very useful generalizations we'll develop in the next section to help us predict the approximate shapes of most organic molecules. For now, the most important thing to remember is that

- Structural formulas alone do not usually depict the shape of molecules accurately.

EXAMPLE 3-3 We saw the structural formula of BCl_3 in Problem 2-2b. You may assume for the moment that the molecule is planar (flat), with all Cl–B–Cl bond angles equal. (a) What is the value of the Cl–B–Cl bond angle? (b) Draw the shape of the molecule as viewed both from *above* the molecular plane and *along* the molecular plane.

Solution

(a) If all three bond angles are equal *and* the molecule is planar, the sum of the three Cl–B–Cl angles must equal 360°, so each angle must be 360°/3 = 120°.

(b)

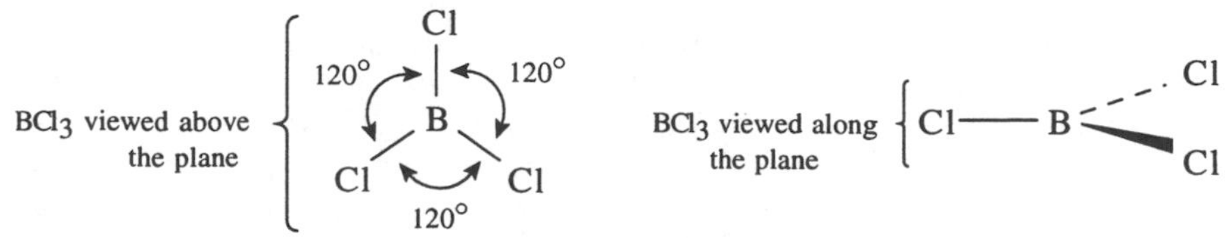

This type of structure is called **trigonal planar.**

3-2. The Nominal Shapes of Molecules: VSEPR

Predicting the approximate shapes of molecules whose internal atoms are from the second period is as easy as counting to four! The **valence shell electron pair repulsion (VSEPR)** theory says that the relative positions of atoms bonded to an internal atom can be predicted if we know the number of *atoms and nonbonding pairs* (hereafter collectively referred to as **groups**) attached to that internal atom. The basic premise is this:

- The groups (i.e., the valence-electron pairs) around a second-period internal atom want to be as far from each other as possible.

For example, consider the molecule $BeCl_2$ (which is a covalent compound). To predict its most stable shape, the first step is to draw its line/dot structural formula and identify all internal atoms.

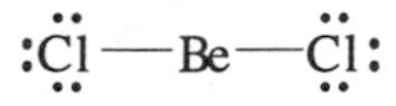

$$:\!\ddot{C}l\!-\!Be\!-\!\ddot{C}l\!:$$

The only internal atom is beryllium, which has just two groups (atoms plus nonbonding pairs) attached to it. What structure will allow these two atoms (and the associated bonding electron pairs) to be as far from each other as possible? Obviously, the *linear* structure above is just what the doctor ordered. Here, the structural formula accidentally *does* reflect the actual shape of the molecule.

When there are *three* groups around an internal atom, the structure that most effectively separates them is the *trigonal planar* arrangement we saw for BCl_3 in Example 3-3. And with *four* groups, the *tetrahedral* structure (as we saw for CH_4 in Section 3-1) is the most stable one. We can summarize these relationships as follows:

Number of groups attached to internal atom	Shape	Bond angle
2	linear	180°
3	trigonal planar	120°
4	tetrahedral	109.5°

note: Remember that the number of groups around an internal atom includes atoms plus nonbonding pairs (each nonbonding pair counts as one group).

EXAMPLE 3-4 Explain why water is *not* linear.

Solution The internal atom (oxygen) is surrounded by *four* groups—two hydrogens and two nonbonding pairs. So the predicted shape is tetrahedral, with an approximate bond angle of 109.5°. [The reason the actual bond angle is less than this (104.5°) is discussed in Section 3-4.]

EXAMPLE 3-5 Using VSEPR considerations, predict the shape and bond angles for each molecule below; then draw the shape of each molecule.

$$BeH_2, \quad BF_3, \quad NH_3$$

Solution

	BeH_2	BF_3	NH_3
Structural formula	H–Be–H	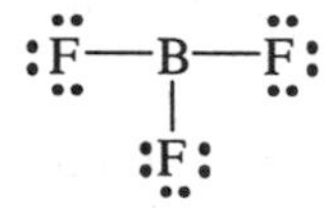 :F—B—F: with F below	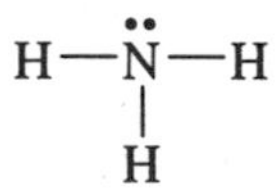 H—N—H with H below
Number of groups attached to central atom	2	3	4
Predicted shape	linear	trigonal planar	tetrahedral
Predicted bond angles	180°	120°	109.5°
Observed bond angles	180°	120°	107.1°
Drawing	H—Be—H	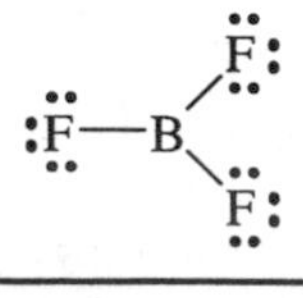 :F—B with F groups	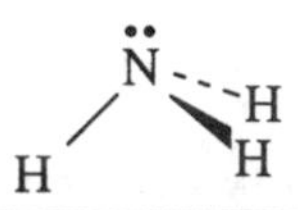 N with H groups

caution: When the internal atoms in a molecule are from the third period, our VSEPR theory isn't quite as reliable as it is for second-period atoms. For example, we'd probably predict that H_2S (H–$\overset{\cdot\cdot}{\underset{\cdot\cdot}{S}}$–H) would have an approximately tetrahedral structure, as we saw for water. Yet, the H–S–H bond angle is only 90°. We'll discuss the reasons for this in Section 3-4.

3-3. Hybrid Atomic Orbitals and the Necessity for Hybridization

Take a moment to look back at the shapes and orientations of *s* and *p* orbitals in Figure 1-1. What is the angle between two *p* orbitals, i.e., the **interorbital angle**? The answer is 90°, because they're orthogonal (see Section 1-7). Also, it should be obvious that there can be no interorbital angle when pure *s* orbitals are involved. But, looking back at the table of molecular shapes in Section 3-2, you can see that the common bond angles are 109.5°, 120°, and 180°. What AOs do you suppose are used to form the MOs that result in these bond angles? The answer is, a *mixture*.

A. Linear molecules: *sp* hybridization

Let's reconsider the linear molecule BeH_2 (Example 3-5). What atomic orbitals does beryllium use to make the bonds to the hydrogens, which are 180° apart? The valence-electron configuration of an uncombined beryllium atom is $(2s)^2$, with three empty $2p$ orbitals (see Section 1-6). Yet, to account for the linear shape, the beryllium atom needs two equivalent AOs 180° apart to form bonds to the hydrogens. Is there any way to pick out two equivalent orbitals from the *s* and *p* orbitals on beryllium that have the required 180° interorbital angle? Clearly, no pair of *s* and/or *p* orbitals has this property.

However, we *can* generate the two required beryllium orbitals by **hybridizing** (mixing) the $2s$ AO with *one* of the $2p$ AOs, then putting one beryllium electron in each of the two resulting **hybrid atomic orbitals (HAOs)**. Both of these equivalent HAOs will consist of half, or 50%, of an *s* orbital (fraction of *s* character = f_s = 0.50) and 50% of a *p* orbital (f_p = 0.50). Such a hybrid is labeled an *sp* HAO. The energy of an *sp* orbital is midway between that of a pure *s* orbital and a pure *p* orbital. The shape of an *sp* orbital can be deduced by considering the types of interference between an *s* orbital and a concentric *p* orbital:

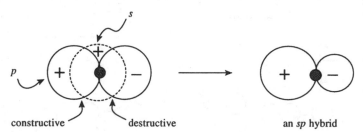

Just as the overlap of *n* atomic orbitals on *different* atoms gives *n* molecular orbitals (Section 2-7),

● The mixing of *n* AOs on the *same* atom generates *n* HAOs.

Therefore, the equal mixing of one $2s$ orbital and one $2p$ orbital on the beryllium generates two *equivalent* (same f_s and same f_p) *sp* HAOs with opposite orientations (i.e., a 180° interorbital angle).

We can summarize this discussion by means of an atomic orbital hybridization diagram. Notice that there is an energetic cost in promoting the electrons to the *sp* hybrids, but this cost is recovered by the energy released when the bonds are formed.

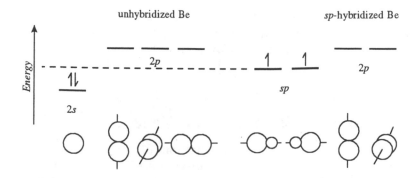

It is these *sp* HAOs that will be used to overlap with the hydrogen 1*s* orbitals to form the σ and σ* MOs of each Be–H bond. Also remember that there are two empty *p* orbitals left over on the beryllium, each perpendicular to the molecular axis and perpendicular to each other.

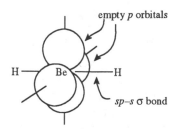

B. Trigonal molecules: *sp²* Hybridization

For a trigonal molecule such as BCl₃ we'll need *three* equivalent HAOs on boron to construct the σ bonds. These hybrids result from the equal mixing of the 2*s* AO with *two* 2*p* AOs; so they are labeled ***sp²*** hybrids (with $f_s = 0.33$ and $f_p = 0.67$). As expected, the interorbital angle between two *sp²* hybrids on the same atom is 120°.

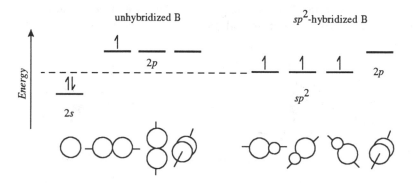

Notice that with *sp²* hybridization there is one empty 2*p* orbital (perpendicular to the plane of the molecule) left over on the boron. Each of the B–Cl σ (and σ*) MOs is the result of overlap between an *sp²* HAO on boron and an *sp³* AO on chlorine.

C. Tetrahedral molecules: *sp³* Hybridization

Let's see what happens with methane.

EXAMPLE 3-6 Describe the hybridization necessary for the tetrahedral shape of methane.

Solution Four equivalent hybrids are needed, requiring the equal mixing of the $2s$ and all three $2p$ orbitals on carbon. Such hybrids are labeled sp^3, with $f_s = 0.25$ and $f_p = 0.75$.

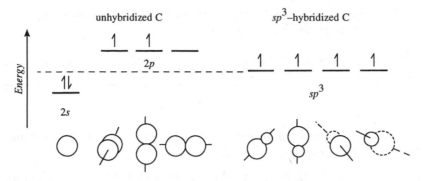

Each of the C–H σ (and σ^*) MOs is the result of overlap between an sp^3 HAO on carbon and a $1s$ AO on hydrogen.

D. The relationship of shape and hybridization

Let's add these facts to our table of "shapes," as in Table 3-1. The data in this table are *very* important, so be sure you understand and can reproduce these relationships.

TABLE 3-1. The Relationship of Shape and Hybridization

Attached groups*	Shape	Angle	Hybrid	f_s	f_p
2	linear	180°	sp	0.50	0.50
3	trigonal planar	120°	sp^2	0.33	0.67
4	tetrahedral	109.5°	sp^3	0.25	0.75

*Includes nonbonding pairs.

EXAMPLE 3-7 Consider the structure below:

$$
\begin{array}{c}
\quad\ \text{H}\ \ \text{H} \\
\quad\ |\ \ \ | \\
\text{H--O--C--C--N--H} \\
\quad\ |\ \ \ |\ \ \ | \\
\quad\ \text{H}\ \ \text{H}\ \ \text{H}
\end{array}
$$

(a) Give the predicted (VSEPR) hybridization at each internal atom
(b) Predict the magnitude of these bond angles: H–O–C, H–C–H, H–N–H.
(c) What type of orbital contains the nonbonding pair on nitrogen?

Solution
(a) First of all, don't forget to fill in (at least mentally) all nonbonding pairs, because they help determine the shape.

$$H-\overset{\displaystyle H}{\underset{\displaystyle H}{\overset{|}{\underset{|}{C}}}}-\overset{\displaystyle H}{\underset{\displaystyle H}{\overset{|}{\underset{|}{C}}}}-\ddot{N}-H$$

$$H-\ddot{\underset{\cdot\cdot}{O}}-\overset{\overset{\displaystyle H}{|}}{\underset{\underset{\displaystyle H}{|}}{C}}-\overset{\overset{\displaystyle H}{|}}{\underset{\underset{\displaystyle H}{|}}{C}}-\overset{\overset{\displaystyle H}{}}{\ddot{N}}-H$$

All four internal atoms (oxygen, nitrogen, and two carbons) have four groups around them, for tetrahedral (sp^3) hybridization at each one.

(b) All three angles are 109.5°.

(c) Based on VSEPR considerations, it's probably an sp^3 orbital that contains the nonbonding pair. (But see Section 3-4.)

Above all, remember that

- Hybridization is the chemist's attempt to explain the *observed* molecular shape by constructing HAOs with the appropriate interorbital angles.

A molecule adopts its most stable shape all by itself, then the chemist calculates what HAOs correspond to that shape. There is a one-to-one correspondence between shape (e.g., bond angles) and hybridization.

3-4. Departures from Nominal Hybridization

You've probably noticed that with molecules such as BeH_2, $BeCl_2$, BF_3, and CH_4 the bond angles predicted by VSEPR theory agree *exactly* with the observed values. However, with H_2O, NH_3, and H_2S the agreement is less exact. There are several reasons for this. First of all, the shape predicted by VSEPR theory assumes that all groups around an internal atom are equivalent, that is, electron pairs attached to the same type of atom. While this is true for a molecule like CH_4, the NH_3 molecule has only *three* equivalent groups (the three hydrogens); the fourth group is a nonbonding pair, which is clearly different from the bonding pairs. What can we infer from the fact that all the H–N–H bond angles in NH_3 are 107.1? The fact that all the H–N–H bond angles in NH_3 are less than 109.5°

must indicate that repulsions between bonding electron pairs are *less* severe than repulsions involving nonbonding pairs. Or, stated another way, nonbonding pairs are effectively larger (occupy more space) than the more directed bonding pairs. This causes a *contraction* of the H–N–H bond angles (from the nominal value of 109.5°) to allow for a larger-than-nominal interorbital angle between the nonbonding pair and any of the bonding pairs. The H atoms, being connected to the bonding pairs, are also forced closer together.

As implied in the previous section, the angle between any *equivalent* hybrid orbitals determines the fraction of s and p character of the hybrid, and vice versa. We already know the angles between equivalent nominal bonds from Table 3-1. But these represent just three points along a continuous curve, as shown in Figure 3-1. Notice how the interorbital angle (θ) between two equivalent HAOs increases monotonically (though not linearly) with the fraction of s character. Let's see how this relationship can be used to determine the *exact* type of hybrids used in the NH_3 molecule.

A. The hybridization index

To begin, we'll define the **hybridization index** i of a hybrid orbital as the superscript on p in the label. Thus, for sp^3, $i = 3$; for sp^2, $i = 2$; and for sp, $i = 1$. From this definition we can deduce the following relationships:

$$f_s = \frac{1}{i+1} \tag{3-1}$$

$$f_p = \frac{i}{i+1} \tag{3-2}$$

$$i = f_p / f_s \tag{3-3}$$

Also, when we add the f_s values for all the hybrids on a given atom, the total must equal 1, since we began the mixing process with just one s orbital:

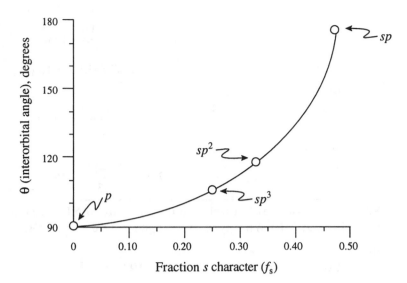

Figure 3-1. The relationship between hybridization (as expressed by f_s) and interorbital angle.

$$\sum f_s = 1.00 \tag{3-4}$$

And, for any hybrid generated from just *s* and *p* AOs,

$$f_s + f_p = 1.00 \tag{3-5}$$

Finally, the relationship that generates the curve in Figure 3-1 is

$$\cos\theta = \frac{-1}{i} \tag{3-6a}$$

or

$$i = \frac{-1}{\cos\theta} \tag{3-6b}$$

where θ is the *interorbital* angle between the two *equivalent* hybrids. This relationship can also be expressed in terms of f_s:

$$f_s = \frac{\cos\theta}{\cos\theta - 1} \tag{3-7}$$

EXAMPLE 3-8 I know this is not a math course, but just for practice see if you can derive Eq. (3-7) from Eqs. (3-1) and (3-6).

Solution
By Eq. (3-1),

$$f_s = \frac{1}{i+1}$$

Then, by Eq. (3-6),

$$f_s = \frac{1}{(-1/\cos\theta)+1}$$
$$= \frac{\cos\theta}{-1+\cos\theta}$$
$$= \frac{\cos\theta}{\cos\theta - 1}$$

Which is, of course, Eq. (3-7).

EXAMPLE 3-9 Show that Eq. (3-6) gives the correct interorbital angles for **(a)** *sp*, **(b)** sp^2, and **(c)** sp^3 HAOs.

Solution The appropriate *i* values are: *sp*, $i = 1$; sp^2, $i = 2$; sp^3, $i = 3$.
(a) Using Eq. (3-6), we calculate for *sp*

$$\cos \theta = \frac{-1}{i} = \frac{-1}{1} = -1.00$$

$$\theta = \cos^{-1}(-1.00) = 180°$$

(b) For sp^2

$$\theta = \cos^{-1}\left(\frac{-1}{2}\right) = 120°$$

(c) For sp^3

$$\theta = \cos^{-1}\left(\frac{-1}{3}\right) = 109.5°$$

EXAMPLE 3-10 **(a)** Returning to the structure of NH_3, what nitrogen hybrids are used to make the N–H bonds? **(b)** What type of nitrogen hybrid contains the nonbonding pair?

Solution

(a) Use Eq. (3-6) and the bond angle of 107.1°:

$$i = \frac{-1}{\cos \theta} = \frac{-1}{\cos(107.1°)} = 3.40$$

Thus, the three N–H bonds involve $sp^{3.40}$ hybrids on nitrogen. Using Eq. (3-1), we find that such an orbital has

$$f_s = \frac{1}{3.40 + 1} = 0.227$$

note: We could have gotten this same result (though less precisely) by using Figure 3-1 and finding that an interorbital angle of 107° corresponds to an f_s of 0.23 and [from Eq. (3-5)] an f_p of 0.77. Then, from Eq. (3-3),

$$i = \frac{f_p}{f_s} = \frac{0.77}{0.23} = 3.4$$

(b) We can use Eq. (3-4) to determine f_s of the nonbonding orbital:

$$\sum f_s = f_s(\text{nonbonding}) + 3(0.227) = 1.000$$

$$f_s(\text{nonbonding}) = 1.000 - 0.681 = 0.319$$

From this value, and Eqs. (3-5) and (3-3), we find that

$$f_p = 1 - f_s = 1 - 0.319 = 0.681$$

and

$$i = \frac{f_p}{f_s} = \frac{0.681}{0.319} = 2.13$$

So, the nonbonding hybrid is an $sp^{2.13}$ orbital.

It is usually true that the *internuclear* bond angle (as determined by, for example, X-ray crystallography) is equal to the *interorbital* angle between the hybrids that make up the bonds to those nuclei. However, there is no way to measure *directly* the angle between electron pairs; after all, they're just waves! In Problem 3-9 (at the end of this chapter) we'll see a case, however, where the internuclear angle is different from the interorbital angle.

B. Third-period elements

In the case of molecules with internal atoms from the third period, the atoms are large enough that VSEPR considerations sometimes take a back seat to other factors in determining the molecule's shape. The 90° H–S–H bond angle in H_2S (Section 3-2) indicates that pure (unhybridized) $3p$ orbitals on sulfur are used to bond the hydrogens. The two nonbonding pairs could be visualized as residing either in unhybridized $3s$ and $3p$ orbitals, or in two equivalent sp hybrids. Since we can't directly measure the angle between nonbonding orbitals, about the only way to determine their hybridization, if any, is to remove electrons (via ionization) and measure the relative energies of the electrons.

But take heart. The vast majority of molecules we'll be discussing involve internal atoms from the second period, where VSEPR theory works quite nicely, thank you!

3-5. Multiple Bonds Revisited

Let's take another look at *ethylene*, 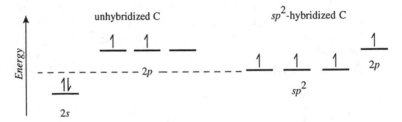, first discussed in Section 2-5. How many *groups* (atoms plus nonbonding pairs) are attached to each (internal) carbon atom? The answer is *three*.

- An atom connected to an internal atom by means of a multiple bond still counts as only *one* group.

Given this, what is the predicted shape and hybridization at each carbon? Answer: Trigonal planar and sp^2. What is the predicted H–C–H angle? Answer: 120°. Easy, right?

A. MO theory: Double bonds

Now let's look at this result in a bit more detail. An sp^2-hybridized carbon has an atomic orbital hybridization diagram that looks like this:

We place one electron in each orbital to provide for making four covalent bonds. Two of the three sp^2 HAOs on each carbon are used to make σ bonds with the hydrogens. The remaining sp^2 hybrids on both carbons overlap to form a σ bond between the carbons. This leaves one *unhybridized* 2p orbital on each carbon, as shown below (only σ molecular orbitals are shown; σ^* orbitals are not):

But where is the second shared electron pair of the double bond? It involves overlap of the two neighboring *parallel p* orbitals. And since the resulting MOs are *not* cylindrically symmetric around the bond axis, they are *not* σ MOs. Instead, these bonds are π **MOs** (with no node between the carbons) and π^* **MOs** (with a node between the carbons).

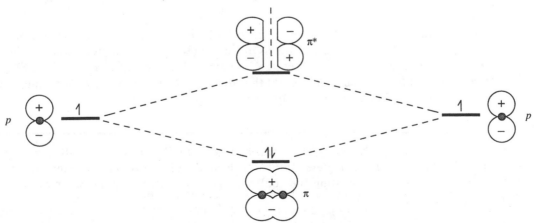

Note that a π^* orbital has four lobes, but it is still just *one* orbital. We can redraw the shape of ethylene with the σ bonds shown as lines, and the π bond as overlapping or parallel p orbitals (for clarity, the antibonding orbitals are not shown):

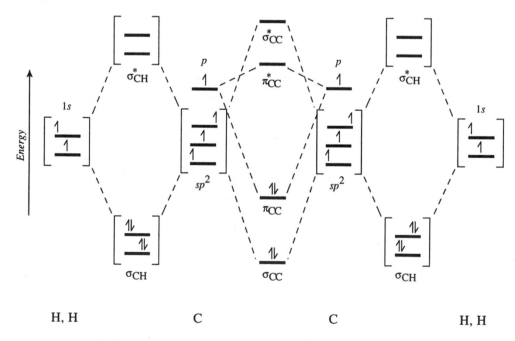

top view   side views   end view

A π bond can form only if the p orbitals are parallel. This fact, coupled with the trigonal planar shape around each carbon, requires that all six atoms in ethylene lie in the same plane (i.e., the molecule is planar).

- A double bond consists of one σ bond and one π bond.

The π-type overlap between two parallel p orbitals is less efficient than σ-type overlap. Therefore, the π–π^* energy separation is smaller than the σ–σ^* separation. As a direct consequence, the π component of a double bond is only 60% as strong as the σ component, as mentioned in Section 2-5.

The entire MOED for ethylene is shown below (square brackets [ ] indicate degenerate orbitals):

H, H   C   C   H, H

note: You may notice that the p orbitals drawn from here on will have somewhat elongated lobes, rather than the circular lobes used until now. There are two reasons for this convention. First, it's easier to draw the orbitals with these egg-shaped lobes. Second, the egg-shaped lobes actually represent the square of the p orbital wave function, which is directly proportional to electron density.

EXAMPLE 3-11   Describe the hybridization and bonding in CH_2O (Example 2-8). Predict the H–C–H bond angle, then draw the complete MOED for the molecule. Draw a picture that accurately shows the relative positions of all bonding and nonbonding MOs. (You may assume the nonbonding pairs occupy equivalent orbitals.)

Solution It's always useful to start with the line/dot structure to help count groups around the internal atom(s).

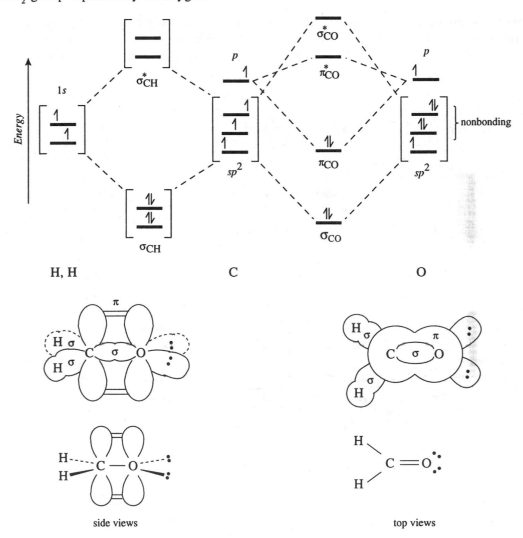

There are three groups around the carbon and three groups around the oxygen, though the oxygen is actually a terminal atom. So, we predict sp^2 hybridization for both carbon and oxygen, with H–C–H and H–C–O bond angles of 120°. The bonding scheme will resemble that of ethylene, with one CH_2 group replaced by an oxygen.

side views   top views

This formulation shows the oxygen nonbonding pairs residing in equivalent sp^2 hybrids, which, although perfectly reasonable from VSEPR considerations, is difficult to verify experimentally.

B. MO theory: Triple bonds

Next, we'll re-examine *acetylene*, C_2H_2 (Section 2-5), which has a triple bond between the carbons.

$$H-C\equiv C-H$$

How many groups are there around each (internal) carbon atom in C_2H_2? Two—remember, a multiply bonded atom is counted as only one group. So, the hybridization at carbon is sp, and the H–C–C bond angles are 180°. But what type of orbitals are used to make the triple bond? In order to hybridize a carbon *sp*, we'll leave two perpendicular (unhybridized) *p* orbitals on each carbon:

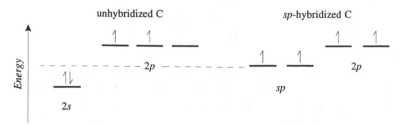

One of the *sp* orbitals on each carbon will be used to make the σ bonds to the hydrogens; the remaining *sp* orbitals on both carbons will overlap to form a σ bond between the carbons. The two sets of unhybridized *p* orbitals will form two mutually perpendicular π bonds.

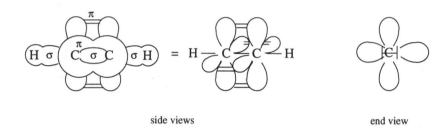

The complete MOED for acetylene looks like this:

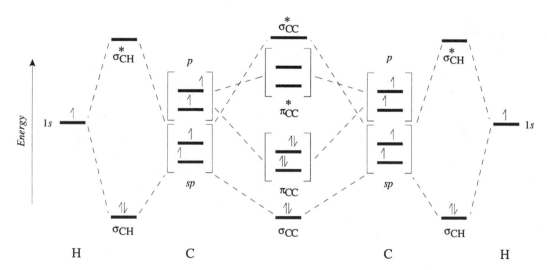

- A triple bond consists of one σ bond and two perpendicular π bonds.

It can be shown with a little trigonometry that the two mutually perpendicular π bonds are quantum-mechanically equivalent to a *torus* (cylinder) of π electrons.

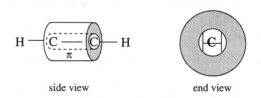

EXAMPLE 3-12   What is the bond order (Section 2-7) of the triple bond in acetylene?

Solution   Bond order = bonding pairs – antibonding pairs = 3 – 0 = 3.

C. Relative lengths and strengths of bonds

For a π bond to form between two atoms, the two nuclei must move closer together than they are in a single bond to allow for better *p–p* overlap. For this reason, bond lengths between any two bonded atoms follow the order: Triple < double < single. However, as stated in Section 2-5, a triple bond is only about twice as strong as a single bond. We now know why. Although a triple bond consists of one σ bond and two π bonds, a π bond is only about 60% as strong as a σ bond. Therefore, a triple bond is only about $1 + 2(0.6) = 2.2$ times as strong as a single (σ) bond.

EXAMPLE 3-13 Consider the structure below:

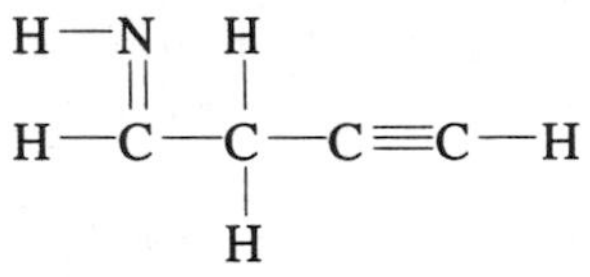

(a) From VSEPR considerations, predict the hybridization of each internal atom in the structure.
(b) Predict the magnitude of the following bond angles: H–C=N, H–C–C(H$_2$), H–N=C, H–C–H, C–C≡C, and C≡C–H.
(c) Redraw the molecule to show the bond angles as accurately as possible.

Solution Don't forget the nonbonding pair on the nitrogen.
(a) N, with a double bond to C and a single bond to H, is sp^2.
   C, with a double bond to N and single bonds to H and C, is sp^2.
   C, with single bonds to two H atoms and two C atoms, is sp^3.
   C, with a triple bond to C and a single bond to C, is sp.
   C, with a triple bond to C and a single bond to H, is sp.
(b) H–C=N̈, 120°; H–C–C(H$_2$), 120°; H–N̈=C, 120°; H–C–H, 109.5°; C–C≡C, 180°; C≡C–H, 180°.

(c)

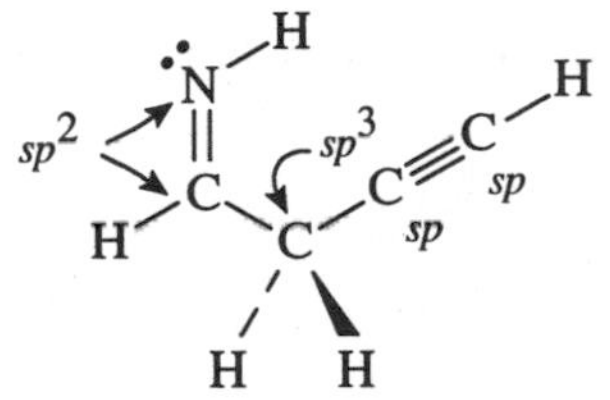

3-6. Nontraditional Models of Multiple Bonds

Some sets of molecular models are not capable of showing multiple bonds as combinations of σ and π bonds. Instead, multiple bonds are depicted by curved (so-called **bent**) bonds. For example, a model of ethylene might look something like this:

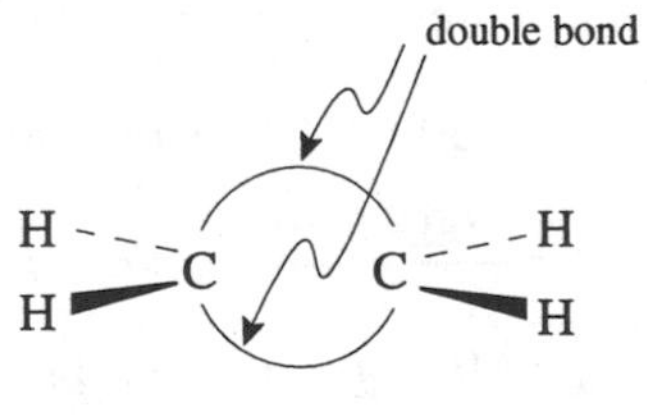

double bond

Does this mean that models like this one give an unrealistic picture of multiple bonds? No, not at all!

The σ/π picture is not the only way to describe a multiple bond, though it *is* the most common way. But let's consider an alternative model where the double bond is composed of two *equivalent* bent bonds that result from the overlap of equivalent carbon hybrids (labeled *sp^i* in the structure below). Of course, we'll still use sp^2 hybrids on carbon to form the bonds to the hydrogens, because we must maintain the 120° bond angle. Notice that these bent bonds are neither σ nor π type, but rather midway between.

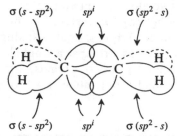

Here is the key question: Assuming that the two bent bond orbitals on each carbon are equivalent (same f_s and same f_p), what type of hybrids are they? (That is, what is the value of i in sp^i)? From Eq. (3-4) we know that the sum of f_s values for the four orbitals on either carbon must equal 1.00. Since f_s for the sp^2 hybrids is 0.33 (Table 3-1), we can write:

$$\sum f_s = 2(0.33) + 2f_s(sp^i) = 1.00$$

or

$$f_s(sp^i) = \frac{1.00 - 2(0.33)}{2} = 0.17$$

Using Eq. (3-5), we find that

$$f_p = 1.00 - f_s = 1.00 - 0.17 = 0.83$$

We can now calculate the value of i from Eq. (3-3):

$$i = \frac{f_p}{f_s} = \frac{0.83}{0.17} = 4.9 \cong 5$$

Thus, these bent bonds result from the overlap of sp^5 orbitals!

EXAMPLE 3-14   Instead of visualizing a triple bond as one σ bond and two π bonds, we can equally well view it as composed of three equivalent bent bonds:

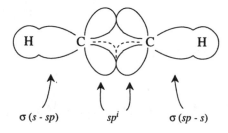

Can you figure out the type of equivalent hybrids needed to create these three bent bonds?

Solution   We must retain one sp hybrid on each carbon to form the linear bonds to the hydrogens. The remaining s character will be divided equally among the three bent bond hybrids.

$$\sum f_s = 0.50 + 3(f_s)(sp^i) = 1.00$$

$$(f_s)(sp^i) = \frac{1.00 - 0.50}{3} = 0.17$$

$$f_p = 1.00 - f_s = 1.00 - 0.17 = 0.83$$

and

$$i = \frac{f_p}{f_s} = \frac{0.83}{0.17} \cong 5$$

Surprise! These overlapping hybrids are also sp^5.

The whole point of this section is to acquaint you with the fact that the σ/π model is only one way to describe multiple bonds. The bent-bond model is equally valid in all respects, though less commonly discussed in undergraduate treatments of this topic. Because we can measure only the positions of nuclei and not the positions of electrons, it is impossible to say that one model is necessarily better than the other; they are equivalent representations. For example, it is just as easy to see why two bent bonds are not as strong as two σ bonds as it is to see that a π bond is not as strong as a σ bond. Nonetheless, throughout the rest of this book we'll use the σ/π model of multiple bonds, for it allows us to better visualize the effects of resonance (Chapter 4) on the structure and reactivity of molecules.

SUMMARY

1. The shape of a molecule is determined by the relative three-dimensional positions of its atoms. Of greatest interest to us are the bond lengths and bond angles that characterize the molecule.
2. The net dipole moment of a molecule is the (vector) sum of all the individual bond moments in the molecule.
3. Generally, structural formulas alone are not meant to convey the accurate shape of the molecules they represent.
4. VSEPR theory rationalizes the shape of molecules (whose internal atoms are from the second period) by letting the attached groups (connected atoms and nonbonding pairs) be as far from each other as possible.
5. Hybrid atomic orbitals (HAOs) are generated by chemists to provide orbitals with the appropriate interorbital angles to match the observed bond (internuclear) angles. These hybrids then overlap to form the σ and σ^* MOs.
6. The relationship between number of attached groups, shape, and hybridization is summarized in Table 3-1:

TABLE 3-1. The Relationship of Shape and Hybridization

Attached groups*	Shape	Angle	Hybrid	f_s	f_p
2	linear	180°	sp	0.50	0.50
3	trigonal planar	120°	sp^2	0.33	0.67
4	tetrahedral	109.5°	sp^3	0.25	0.75

7. Any sp^i hybrid is characterized by the following relationships:

$$f_s = \frac{1}{i+1} \qquad \textbf{(3-1)}$$

$$f_p = \frac{i}{i+1} \qquad \textbf{(3-2)}$$

$$i = \frac{f_p}{f_s} \qquad \textbf{(3-3)}$$

$$f_s + f_p = 1.00 \qquad \textbf{(3-5)}$$

All the s,p hybrids of a given atom must satisfy the condition

$$\sum f_s = 1.00 \qquad \textbf{(3-4)}$$

and the interorbital angle between two *equivalent* (same f_s and same f_p) hybrid orbitals is given by

$$\cos \theta = \frac{-1}{i} \qquad \textbf{(3-6)}$$

8. A π bond results from the overlap of two parallel p orbitals. Of the two resulting MOs (π and π^*), only the π MO normally contains electrons (two, with spins paired).

9. A double bond is usually formulated as one σ bond plus one π bond. This combination is shorter and about 1.6 times stronger than a σ bond. A triple bond consists of one σ bond and two perpendicular π bonds and is shorter than a double bond and about 2.2 times as strong as a σ bond. The two π bonds in a triple bond can be viewed as a "cylinder" of π electrons.

10. Alternatively, multiple bonds can be formulated as two (in the case of a double bond) or three (in the case of a triple bond) equivalent bent bonds, resulting from the overlap of sp^5 hybrids.

RAISE YOUR GRADES

Can you define...?

☑ shape (of a molecule)
☑ dipole moment
☑ bond (internuclear) angle
☑ terminal and internal atoms
☑ square planar
☑ tetrahedral
☑ bond length
☑ trigonal planar
☑ VSEPR theory
☑ bent bond

☑ groups (attached to internal atoms)
☑ linear
☑ hybridization
☑ hybrid atomic orbital (HAO)
☑ interorbital angle
☑ f_s and f_p of an HAO
☑ π bond
☑ π and π^* MOs
☑ planar (molecule)
☑ hybridization index i

Can you explain...?

☑ what information you need to specify the shape of a molecule
☑ how three-dimensional structures can be drawn in perspective
☑ how VSEPR theory can be used to predict approximate molecular shape
☑ the makeup and interorbital angles of sp, sp^2, and sp^3 hybrids
☑ why the shape of some molecules departs from VSEPR predictions
☑ how to relate exact bond angles with exact hybridization
☑ how π bonds are generated
☑ the σ/π model of multiple bonds
☑ why a triple bond is only about 2.2 times as strong as a single bond
☑ the bent bond model for multiple bonds

Can you draw the structures of...?

☑ methane
☑ ethylene
☑ acetylene

SOLVED PROBLEMS

PROBLEM 3-1 From memory, fill in the table below:

Groups	Shape	Angle	Hybrid	f_s	f_p
2					
3					
4					

Solution

Groups	Shape	Angle	Hybrid	f_s	f_p
2	linear	180°	sp	0.50	0.50
3	trigonal planar	120°	sp^2	0.33	0.67
4	tetrahedral	109.5°	sp^3	0.25	0.75

PROBLEM 3-2 (a) What shape would you predict for CHF_3? (b) Draw the shape of this molecule, showing the direction of all bond moments and the direction of the net dipole moment, if any.

Solution
(a) Begin with the line/dot structural formula:

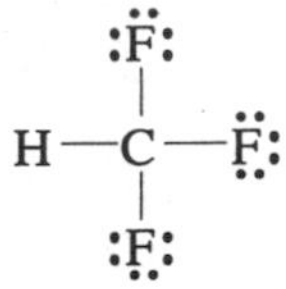

Thus there are four groups attached to the (internal) carbon atom, so we predict a tetrahedral shape with bond angles of 109.5°.

(b)

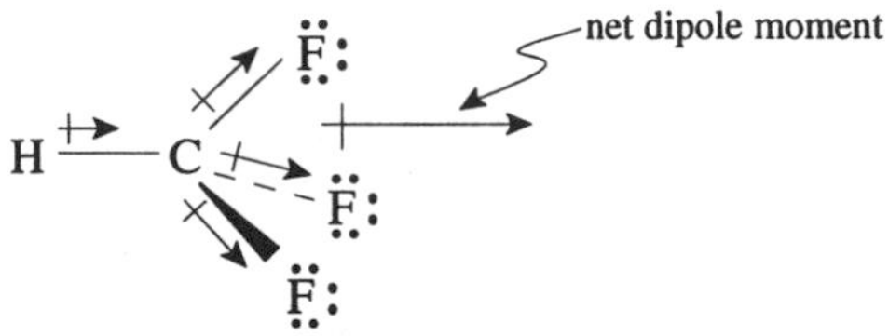

PROBLEM 3-3 For each structure below use VSEPR considerations to predict the hybridization at each internal atom and the magnitude of each indicated bond angle:

(a) H—O—B—O—H   with O—H above B

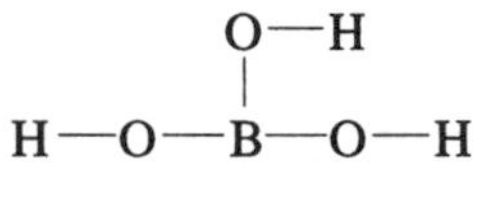

(b) Cl—C—Cl   with Cl above and Cl below

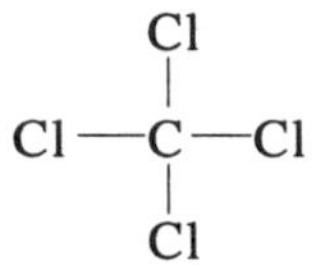

(c) H—C—C≡N   with H above and H below

∠ H–O–B, ∠ O–B–O   ∠ Cl–C–Cl   ∠ H–C–H, ∠ C-C≡N

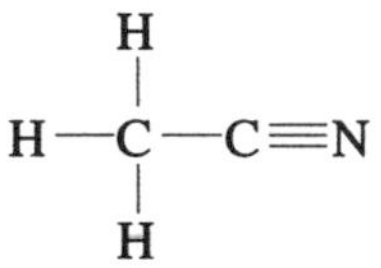

Solution Don't forget the nonbonding pairs.

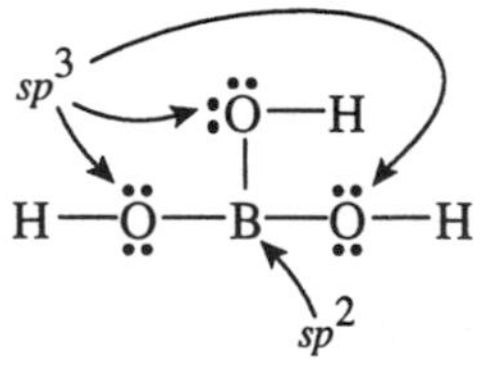

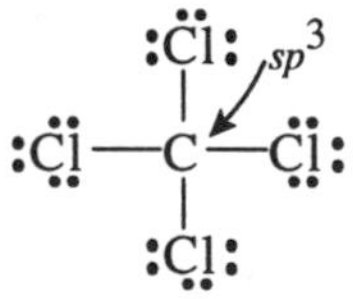

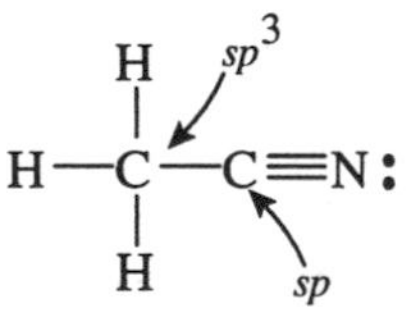

∠ H–O–B = 109.5°   ∠ Cl–C–Cl = 109.5°   ∠ H–C–H = 109.5°
∠ O–B–O = 120°   ∠ C-C≡N = 180°

PROBLEM 3-4 For the two structural isomers below (seen previously in Example 2-9), indicate the expected (VSEPR) hybridization at each carbon. Then redraw each structure to depict the shape of the molecule accurately. Show σ bonds as lines, and π bonds as parallel *p* orbitals.

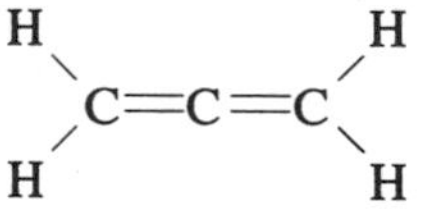

(A) 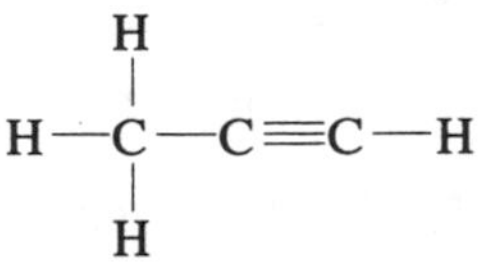   (B)

Solution The two outer C atoms in structure (**A**) are sp^2-hybridized, while the middle C is *sp* (why?). In structure (**B**) the first C atom, which is singly bonded to three H atoms and another C atom, is sp^3-hybridized while the other two C atoms are *sp*-hybridized.

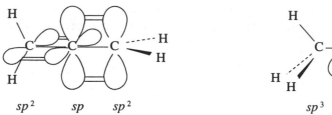

sp^2 sp sp^2 sp^3 sp s

PROBLEM 3-5 Draw a complete MOED for HCN, then complete the following statements:
(a) The H–C bond results from overlap of a(n) _____ orbital on hydrogen and a(n) _____ orbital on carbon.
(b) The C≡N σ bond results from the overlap of a(n) _____ orbital on carbon and a(n) _____ orbital on nitrogen.
(c) The C≡N triple bonds can either be formulated as consisting of one _____ bond and two _____ bonds, or as three _____ bonds.
(d) The nitrogen nonbonding pair resides in a(n) _____ orbital.
(e) The predicted H–C≡N bond angle is _____°.

Solution

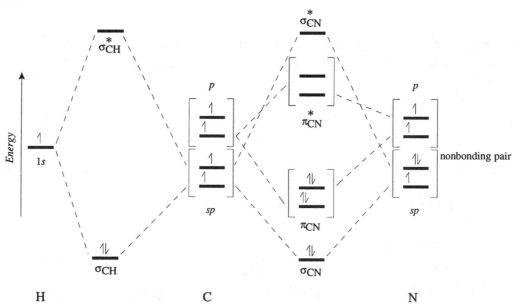

(a) $1s$, sp; **(b)** sp, sp; **(c)** σ, π, equivalent bent; **(d)** sp; **(e)** 180°.

PROBLEM 3-6 Although we can't measure interorbital angles directly, we can use VSEPR to guess what type of hybrid orbitals contain nonbonding pairs on terminal atoms. Using this approach, predict the hybridization at fluorine in CHF_3 (Problem 3-2).

Solution There are four groups around each fluorine, a carbon and three nonbonding pairs. So, we predict sp^3 hybridization, with the nonbonding pairs occupying sp^3 hybrids.

PROBLEM 3-7 Redraw each structure below to show the shape of the molecule as accurately as possible. Use VSEPR theory to predict the hybridization at each internal atom.

(a) $HOCNH_2$ **(b)** H_3CCHCH_2 **(c)** NCCN

Solution Don't forget nonbonding pairs.

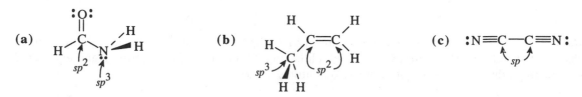

PROBLEM 3-8 The actual H–C–H bond angle in ethylene is 116.6°, rather than the 120° value predicted from VSEPR theory. **(a)** What is the magnitude of the H–C–C angle? **(b)** What actual carbon hybrids are used to make the C–H bonds?

Solution

(a)

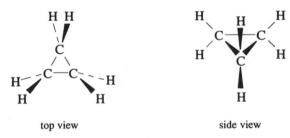

Because this is a planar molecule, we can see from the structure that

$$116.6° + \phi + \phi = 360°$$

$$\phi = \frac{360 - 116.6°}{2} = 121.7°$$

(b) Use Eq. (3-6):

$$i = \frac{-1}{\cos \theta} = \frac{-1}{\cos(116.6°)} = 2.23$$

Thus, the orbitals are $sp^{2.23}$ hybrids.

PROBLEM 3-9 This is a hard one! Cyclopropane is a cyclic molecule with the three carbons forming an equilateral triangle:

top view   side view

Although there are four groups around each internal carbon atom (and VSEPR theory would prefer sp^3 hybridization and 109.5° bond angles), the small ring requires C–C–C *internuclear* angles of 60°. The observed H–C–H angles are all 114°.
(a) What type of carbon hybrids are used to make the C–H bonds?
(b) What type of carbon hybrids are left to make the C–C bonds, and what is their *interorbital* angle?
(c) What type of bonds are the C–C bonds?

Solution
(a) Use Eq. (3-6):

$$i = \frac{-1}{\cos \theta} = \frac{-1}{\cos(114°)} = 2.46 \quad (sp^{2.46})$$

The f_s values for such hybrids are given by Eq. (3-1):

$$f_s = \frac{1}{i+1} = \frac{1}{2.46+1} = 0.29$$

(b) The f_s values for the C–C hybrids can be determined from Eq. (3-4):

$$\sum f_s = 1.00 = 2(0.29) + 2(f_s) \quad \text{(C–C hybrids)}$$

$$f_s(\text{C–C hybrids}) = \frac{1.00 - 2(0.29)}{2} = 0.21$$

And from Eq. (3-5):

$$f_p = 1.00 - f_s = 1.00 - 0.21 = 0.79$$

So using Eq. (3-3):

$$i = \frac{f_p}{f_s} = \frac{0.79}{0.21} = 3.76 \quad (sp^{3.76})$$

The interorbital angle is available from Eq. (3-6):

$$\theta = \cos^{-1}\left(\frac{-1}{3.76}\right) = 105°$$

(c) Because the C–C bonds have an *internuclear* angle (60°) that is different from the interorbital angle (105°), the C–C bonds are bent bonds. That is, the electron density in the C-C bonds lies somewhat outside the internuclear line.

4 RESONANCE AND THE DELOCALIZATION OF ELECTRONS

4-1. Formal Charge

We'll encounter many structures that are *polyatomic ions*, i.e., charged species with two or more atoms. It's not always obvious at first glance where the charge resides in such an ion. For example, the ammonium ion has the formula NH_4^+. Its structure is tetrahedral, as you can infer from the four groups around the nitrogen:

$$\left[\begin{array}{c} H \\ \quad N\!\!-\!\!-H \\ H \quad\quad H \\ H \end{array} \right]^+$$

Notice that the electron pair that was nonbonding in NH_3 (see Example 3-5) is bonded to a hydrogen in NH_4^+, and there is a positive charge resulting from the addition of a proton. But is the positive charge located on the nitrogen, the hydrogens, or both?

To answer this question, we'll need to define the **formal charge** of an atom as its number of valence electrons (in the uncombined state) minus the number of electrons it "owns" in the molecule. Electron ownership is determined in the following way. The electrons in any covalent bond (polar or not) are divided equally between the bonded atoms, while nonbonding electrons are assigned to the atom where they reside. If you would prefer an equation, you can write:

Formal charge = valence electrons – (half of bonding electrons + all nonbonding electrons)

or, when looking at a line–bond structure,

FORMAL CHARGE OF AN ATOM   Formal charge = valence electrons – (lines + dots)   (4-1)

Now, let's calculate the formal charges on the atoms in $\left[\begin{array}{c} H \\ | \\ H-N-H \\ | \\ H \end{array}\right]^+$:

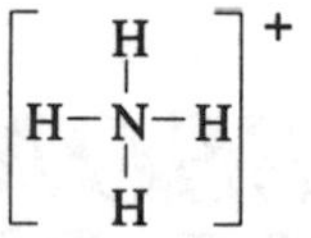

$$\text{Formal charge of H} = 1 - (1 + 0) = 0$$
$$\text{Formal charge of N} = 5 - (4 + 0) = +1$$

So, the + charge resides on the nitrogen. Further, note that the **net** (overall) **charge** on the ion is the sum of the formal charges of all the atoms in the structure:

$$\text{Net charge of NH}_4^+ = 4(\text{formal charge of H}) + \text{formal charge of N}$$
$$= 4(0) + 1 = +1$$

What may surprise you is that there are many neutral molecules (i.e., molecules with a net charge of zero) in which two or more of the atoms possess nonzero formal charges that cancel each other.

note: In this book, whenever you're asked to give formal charges of the atoms in a structure, all nonbonding pairs will be shown.

EXAMPLE 4-1   Give the formal charge of each atom in the structures below, then specify the net charge of each structure.

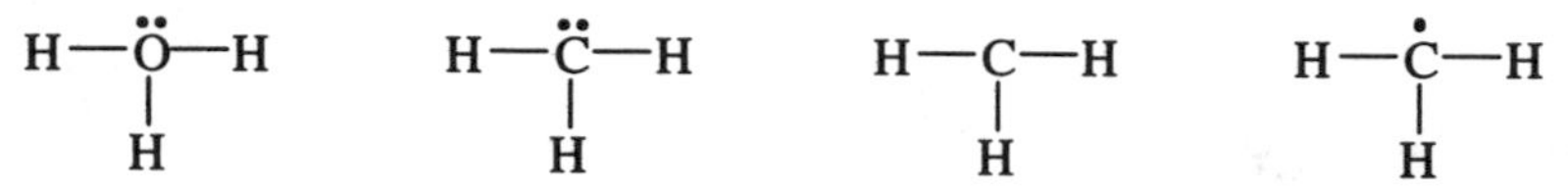

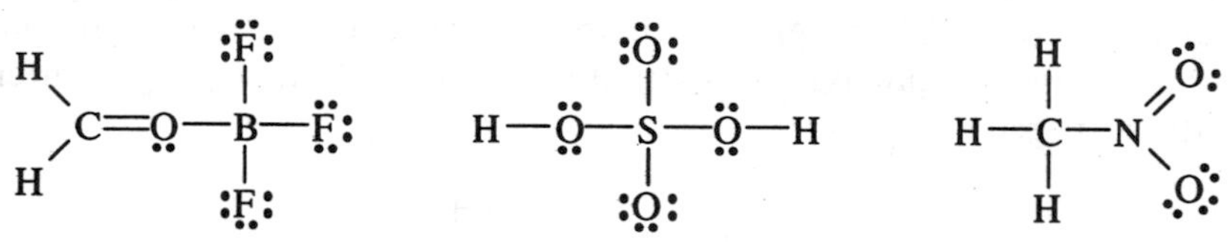

Solution

Formal charge	H: $1 - (1 + 0) = 0$ O: $6 - (3 + 2) = 1$	H: $1 - (1 + 0) = 0$ C: $4 - (3 + 2) = -1$	H: $1 - (1 + 0) = 0$ C: $4 - (3 + 0) = +1$	H: $1 - (1 + 0) = 0$ C: $4 - (3 + 1) = 0$
Net charge	$3(0) + 1 = +1$	$3(0) - 1 = -1$	$3(0) + 1 = +1$	$3(0) + 0 = 0$

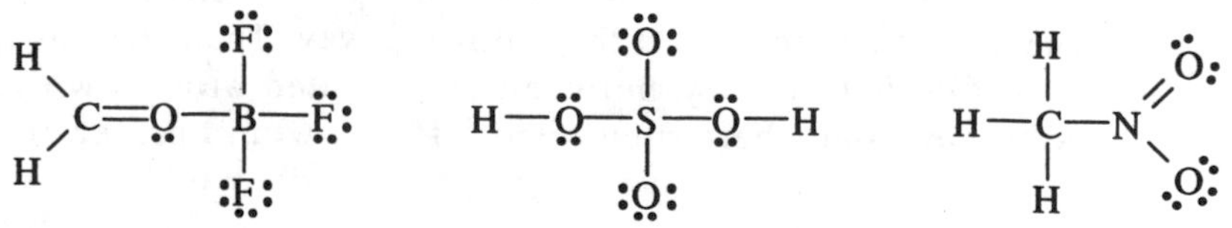

Formal charge	H: $1 - (1 + 0) = 0$ C: $4 - (4 + 0) = 0$ O: $6 - (3 + 2) = 1$ B: $3 - (4 + 0) = -1$ F: $7 - (1 + 6) = 0$	H: $1 - (1 + 0) = 0$ O(–H): $6 - (2 + 4) = 0$ O(–S): $6 - (1 + 6) = -1$ S: $6 - (4 + 0) = 2$	H: $1 - (1 + 0) = 0$ C: $4 - (4 + 0) = 0$ N: $5 - (4 + 0) = 1$ =O: $6 - (2 + 4) = 0$ –O: $6 - (1 + 6) = -1$
Net charge	$2(0) + 0 + 1 - 1 + 3(0) = 0$	$2(0) + 2(0) + 2(-1) + 2 = 0$	$3(0) + 0 + 1 + 0 - 1 = 0$

When there is a nonzero formal charge on an atom—and we choose to show the charge in a structural formula—we draw the charge as close as possible to the atom with the formal charge. Thus, we draw NH_4^+ like this, with the plus sign as close as possible to the N atom:

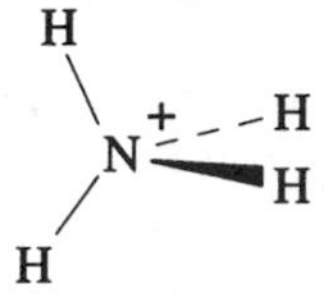

EXAMPLE 4-2   Redraw the structures in Example 4-1, showing all nonzero formal charges.

Solution

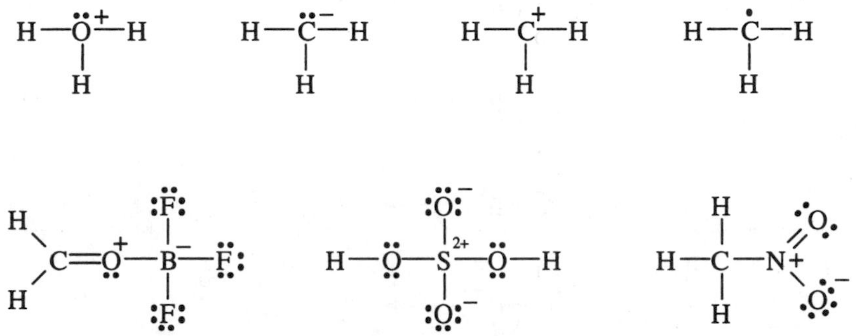

The three lower structures in Example 4-2, though neutral overall, are sometimes called **zwitterions,** which means they possess canceling positive and negative formal charges in the same molecule. Also, there are special names for ions in which carbon bears a formal charge. A positively charged ion with a carbon that bears the charge is called a **carbocation** (carbo-căt´-ion), while a negative ion with a carbon that bears the charge is called a **carbanion**. The neutral species ·CH₃, where the carbon has a zero formal charge and one unpaired electron, is called a carbon **free radical**.

Although they are not always shown explicitly in structural formulas, formal charges will prove to be very important when we investigate organic reactions.

4-2. An Introduction to Electron Pushing

Back in Chapter 2 we introduced you to the practice of using curved ("fish-hook") half-headed arrows to show how individual electrons are transferred or shared to make chemical bonds. Throughout the rest of your organic chemistry experience, you will continue to use curved arrows to keep track of electrons during chemical processes. This practice is known affectionately among organic chemists as **electron pushing**, or sometimes **arrow pushing.**

All chemical reactions involve the making and breaking of bonds, so let's see how we can show such processes with the aid of electron pushing. We've already seen (in Chapter 2) how the formation of a covalent bond can be depicted:

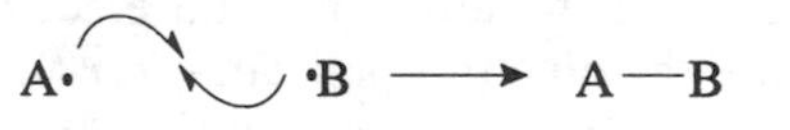

Such a bond-forming process, with each fragment donating one electron, is called **homogenic bond formation**. The reverse of this is the bond-breaking process called **homolytic bond cleavage**, in which each fragment accepts one electron from the bonding electron pair:

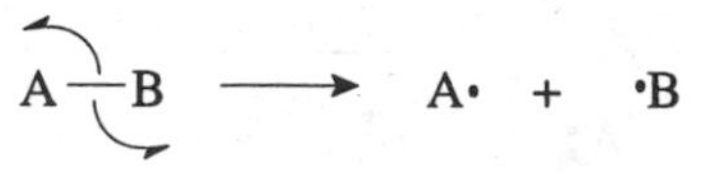

In many reactions bonds are broken by **heterolytic cleavage**, in which *both* electrons end up on *one* of the fragments. When a *pair* of electrons moves as a unit, we use a triangle-headed arrow:

$$A \overset{\frown}{-} B \longrightarrow A^+ + :B^-$$

Notice that when a bond is cleaved heterolytically, the formal charge on the atom *receiving* the electrons *decreases* (or becomes more negative) by 1, while the formal charge on the other atom *increases* (or becomes more positive) by 1. Another way of saying the same thing is that one fragment ends up with a filled orbital, while the other ends up with an empty orbital.

EXAMPLE 4-3   Show the formal charge remaining on each fragment after the indicated heterolytic bond cleavage occurs:

$$A \overset{}{-} B^+ \longrightarrow \quad ?$$

$$A^- \overset{}{-} B \longrightarrow \quad ?$$

$$A^- \overset{}{-} B^+ \longrightarrow \quad ?$$

$$A^- \overset{}{-} B \overset{}{-} C^+ \longrightarrow \quad ?$$

Solution

$$A \overset{}{-} B^+ \longrightarrow \quad A^+ + :B$$

$$A^- \overset{}{-} B \longrightarrow \quad A + :B^-$$

$$A^- \overset{}{-} B^+ \longrightarrow \quad A + :B$$

$$A^- \overset{}{-} B \overset{}{-} C^+ \longrightarrow \quad A + :B + :C$$

The reverse of heterolytic bond cleavage is called **heterogenic bond formation**, in which both electrons in the new bond come from one of the partners:

$$A: \overset{\frown}{} B \longrightarrow A^+ {-} B^-$$

(In older books a bond formed this way was sometimes called a **dative** or a **coordinate covalent bond,** and the bond was shown by an arrow directed from the donor toward the acceptor atoms: A → B.) Heterogenic bond formation requires that the electron-pair *donor* (A: in the above example) has a *filled* orbital, while the electron-pair *acceptor* (B) must have an empty orbital. An electron-pair donor is described as **nucleophilic** (nucleus-loving), while an electron-pair acceptor is described as **electrophilic** (electron-loving).

EXAMPLE 4-4   Show the formal charge remaining on each atom after the indicated heterogenic bond formation takes place:

$$A:^- \overset{\smile}{} B^+ \longrightarrow \quad ?$$

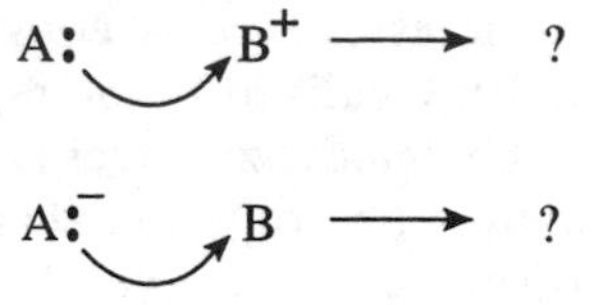

$$A: \quad \curvearrowright \quad B^+ \longrightarrow \quad ?$$

$$A:^- \quad \curvearrowright \quad B \longrightarrow \quad ?$$

Solution

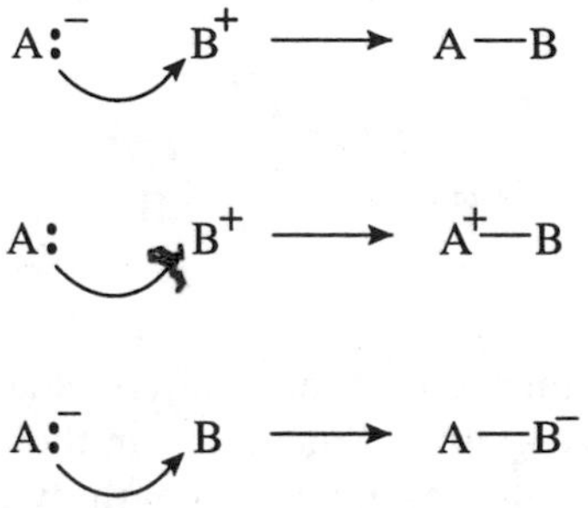

$$A:^- \quad \curvearrowright \quad B^+ \longrightarrow \quad A\!-\!B$$

$$A: \quad \curvearrowright\!\!\!\times \quad B^+ \longrightarrow \quad A^+\!\!-\!B$$

$$A:^- \quad \curvearrowright \quad B \longrightarrow \quad A\!-\!B^-$$

EXAMPLE 4-5 Complete the reactions below by showing arrows and formal charges; identify all nucleophiles and electrophiles.

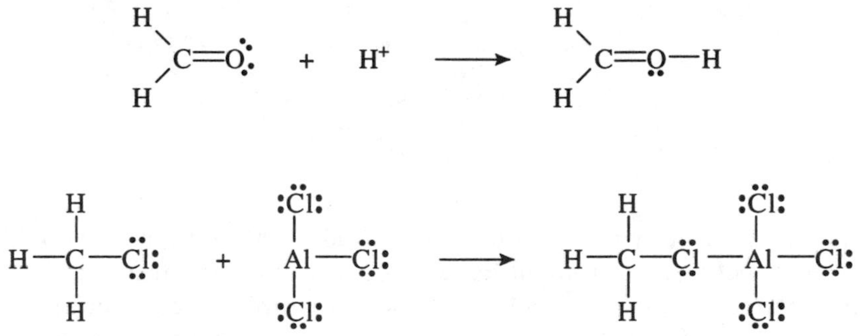

Solution

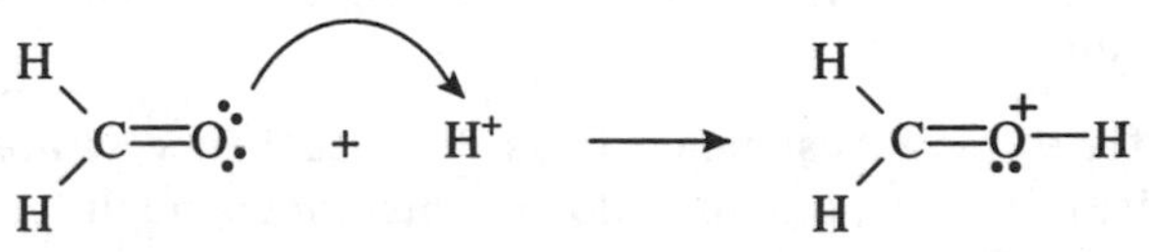

nucleophile   electrophile

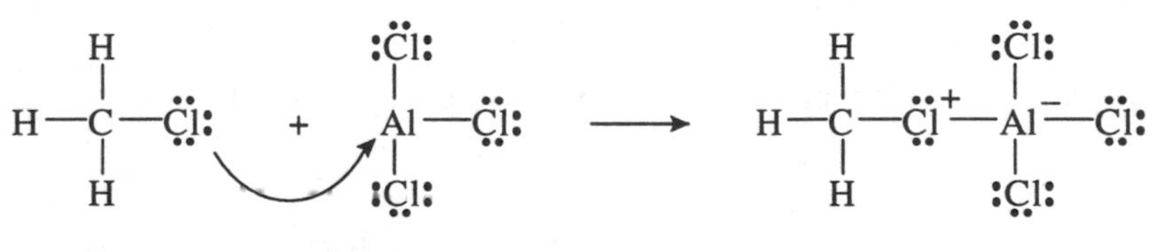

nucleophile   electrophile

Remember that electron pushing is nothing more than electron bookkeeping, designed to help us keep track of electron reorganization and changes in formal charge.

4-3. Electron Delocalization: The Valence-Bond Approach

In our discussion of covalent bonds in Chapter 2 we saw how electrons that were originally **localized** on (or confined to) individual atoms became **delocalized** (spread out) when shared between two nuclei. It may surprise you to learn that in some circumstances electrons can be delocalized over three, four, or more atoms, even entire molecules! This delocalization of electrons over three or more atoms can be depicted by so-called **resonance** structures.

Because of the wave properties of electrons, we can "locate" them only in the sense of probability or electron density. Nonetheless, there is considerable evidence that, while most bonding electron pairs remain essentially localized in the region between two nuclei, certain π-bonding and nonbonding electrons can be delocalized over three or more atoms. For example, the **allyl carbocation** has several very interesting properties that are not obvious from its usual (so-called **valence bond**) structural formula. Here, we show the allyl carbocation with its positive charge located on the third carbon, which is reasonable in light of our formal charge calculations:

But we know from experimental evidence that the positive charge is actually found to be divided equally between the first and third carbons, though *not* on the middle carbon. Furthermore, each of the carbon–carbon bonds is equal in length and strength, with a bond order (Section 2-7,F) of 1.5! Now what?

To help explain these facts, let's draw a second structure for the allyl carbocation. In this new structure the carbon that we'd previously shown as bearing the positive charge is highlighted in boldface:

This structure is also a perfectly reasonable one for the allyl carbocation. What is the relationship between these two structures? You might be tempted to call them "structural isomers" because, although they have the same positions of all atoms, there is a different sequence of bonds (electrons). But neither structure *by itself* adequately represents the allyl carbocation's charge distribution. The actual structure of the allyl carbocation is a *hybrid* of *both* forms. And since the interconversion of these two forms involves only the *resonance* (relocation) of the π-bond electrons, they are called **resonance forms** of the allyl cation.

note: Most chemists refer to these resonance forms as "resonance *structures*" or "valence-bond *structures*," in spite of the fact that they do not exist independently. The idea of "resonance" is really a confession that we need more than one structure to represent the actual molecule because we are unable to account for all of the molecule's known properties with a single structure.

The relationship between resonance structures is denoted by a double-headed arrow:

The curved arrow indicates how the electrons are relocated during the conversion of the first structure to the second. In one sense, the π bond "pivots" around the middle carbon. Notice how these two structures, when taken together, show how the charge is divided between the first and third carbons. They also explain the carbon–carbon bond order of 1.5, because each carbon–carbon bond is half single bond and half double bond.

You might be wondering how long it takes for one resonance structure to "convert" to another. There is no answer to this question—because there is no actual interconversion! Neither resonance form has an independent existence. Rather, the actual structure of the molecule is a hybrid of all contributing resonance forms, and at any instant it has the "averaged" properties of all resonance contributors.

- We invoke resonance structures when no single valence-bond structure adequately portrays the properties of the species.

    Sometimes you'll see the allyl carbocation drawn this way:

This is a somewhat feeble attempt to arrive at a single formula to account for all of the chemical properties. The dotted bond represents a "partial" bond, since the electron pair is delocalized over three carbons. The placement of the plus sign indicates that the charge is also delocalized.

EXAMPLE 4-6 The **allyl carbanion** is similar to the allyl carbocation, except that the nonbonding pair and resulting negative charge are delocalized between the first and third carbons.

(**a**) Draw a second resonance structure for the allyl carbanion, using curved arrows to show how the electrons are relocated between the first and second structures. (**b**) Draw a dotted-bond representation of the allyl carbanion. (**c**) What average bond order do you predict for the carbon–carbon bonds?

Solution

(**a**)

(**b**)

(**c**) As in the allyl carbocation, each carbon–carbon bond is half single and half double, for a bond order of 1.5.

Example 4-6 shows how a nonbonding pair can be involved in resonance with a π bond located at the next atom. We'll practice drawing more resonance structures in the next section.

    Now, how does resonance affect energy and stability? The answer is clear and definite:

- Resonance normally *lowers* the energy and *increases* the stability of a molecule or ion.

Thus, the greater the number of contributing resonance structures for a molecule, the more stable it is expected to be.

4-4. Rules for Drawing Valence-Bond Resonance Structures

We have already said that resonance effects are encountered mainly in molecules with multiple (i.e., π) bonds. Except in a few special molecules, the σ-bond framework is not involved in significant resonance interactions. This fact can be rationalized by noting that electrons in π bonds are higher in energy than those in σ bonds (see the MOEDs in Section 3-5), making the π-bonding electrons more reactive and also more **polarizable** (easily displaced) than σ-bonding electrons.

When a species contains multiple bond(s), it may have several resonance contributors. For example, even a molecule as simple as ethylene (Section 3-5) has several resonance contributors. But not all resonance structures contribute equally to the hybrid's properties. For example, here are four structures we might write for ethylene:

The first structure (**I**) is by far the main contributor. Structures **II** and **III** are minor contributors for three reasons: The π bond has been broken, there are charge separations not present in the major contributor, and the valence of the positively charged carbon is unsatisfied. Structure **IV** does not contribute at all—not only are both carbons' valences unsatisfied, but previously paired electrons have become unpaired. From considerations such as these we can devise a list of rules that will help us determine if a given resonance contributor has a valid structure, and how important its relative contribution is.

1. In going from one resonance structure to another *no nuclei can change position*; only electrons (bonds) can move.
2. No atom in any resonance structure can ever have more electrons around it than will fill its valence shell; that is, there should be no more than eight electrons around any second-period element.
3. Most organic structures have no unpaired electrons. But if there *are* unpaired electrons, the number of unpaired electrons cannot change from one structure to the next.
4. The most important resonance contributors are those with the maximum number of bonds and the minimum number of charge separations.
5. When zwitterionic charges are present, negative charges prefer to be on the electronegative atoms, with positive charges on the more electropositive atoms.

Rules 1, 2, and 3 determine whether a structure is "legal," while Rules 4 and 5 determine the relative contribution by a resonance structure.

EXAMPLE 4-7 For each pair of structures below, state whether they are valid resonance contributors and, if they are both valid, identify which structure is the more important contributor. Which rule applies to each pair?

(a)

(b)

(c)

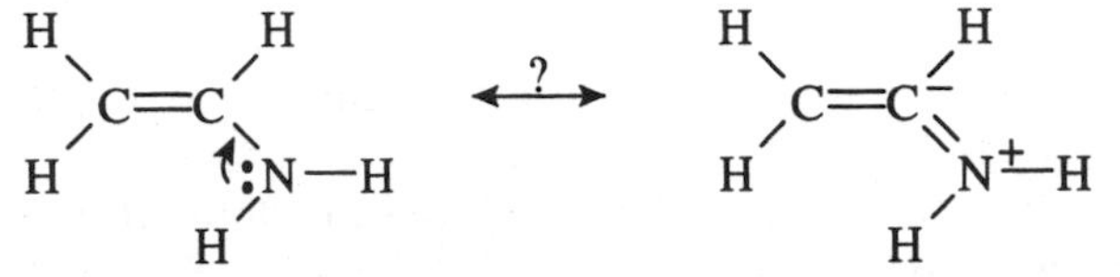

(d)

(e)

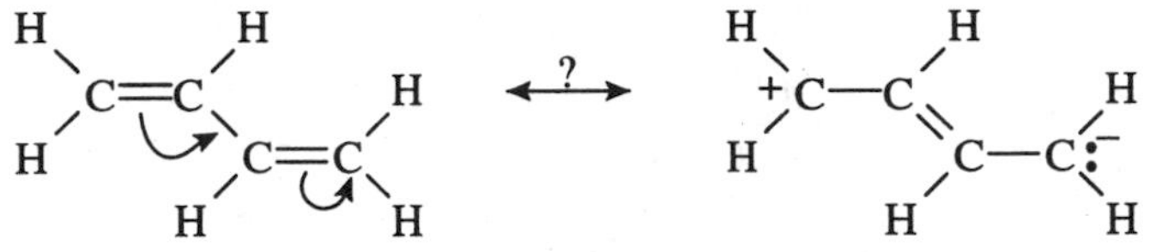

Solution

(a) The second structure is "illegal" because previously paired electrons have become unpaired (Rule 3). However, notice that Rule 3 does *not* preclude resonance in the **allyl radical** shown below because each structure has one unpaired electron.

Here again the carbon–carbon bond order is 1.5.

(b) Because the chlorine atom moves, these are structural isomers, not resonance structures (Rule 1).

(c) The second structure is "illegal" because the valence of the negatively charged carbon is exceeded with ten electrons (Rule 2).

(d) Both structures are valid, but the first structure is the more important contributor because the negative charge is on the more electronegative O atom and the positive charge is on the less electronegative C atom (Rule 5). However, the *most* important contributor for this molecule is the one shown here, which has a maximum number of bonds and a minimum number of charge separations (Rule 4):

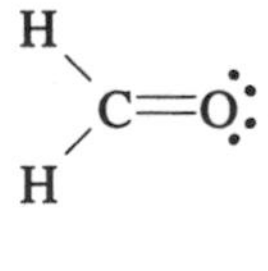

(e) Both are valid, but the first structure—which has more bonds and fewer charge separations—is the major contributor (Rule 4).

Example 4-7e shows the interaction between two neighboring π bonds. Notice that for two π bonds to engage in resonance they must be separated by *one* other bond. Two (or more) π bonds that bear this relationship are said to be **in conjugation** or **conjugated**.

EXAMPLE 4-8 For each structure below indicate which π bonds, if any, are in conjugation:

(a)   (b) 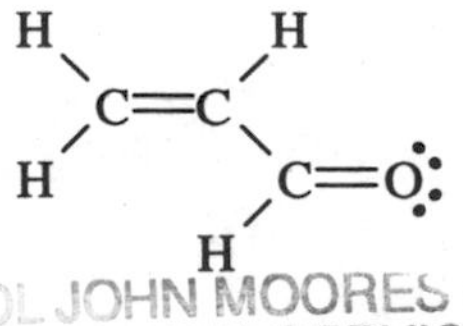   (c)

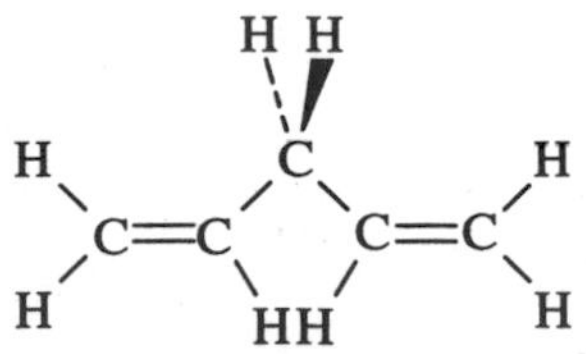

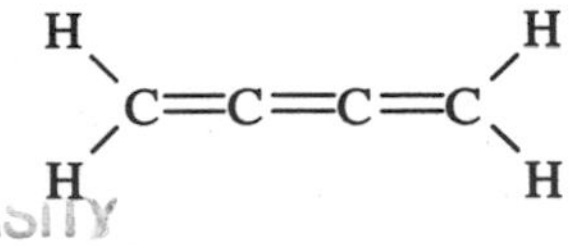

Solution

(a) Because the π bonds are separated by *two* carbon–carbon bonds, they are *not* in conjugation.

(b) The two π bonds *are* conjugated.

(c) This is a tricky one. The first and third π bonds *are* in conjugation because they are separated by the middle carbon–carbon (double) bond. But the middle π bond is *not* in conjugation with either of the other two π bonds because it shares a carbon with each one and is therefore not separated from either one. Such a sequence of double bonds is often called *cumulated*.

EXAMPLE 4-9   Now you can recognize situations in which resonance is likely to be important, and you know the rules for drawing valid resonance structures. Draw the resonance structure that would result from the electron reorganization indicated by the curved arrows in each case below:

(a)

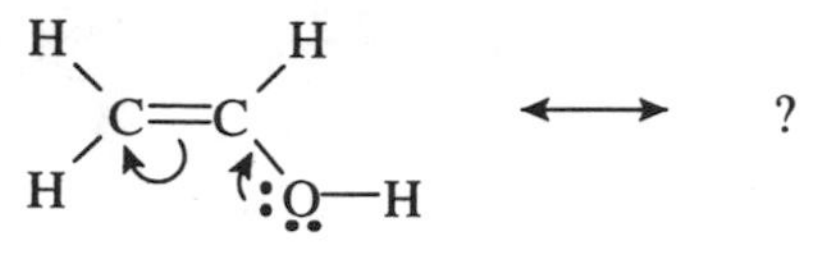

(b)

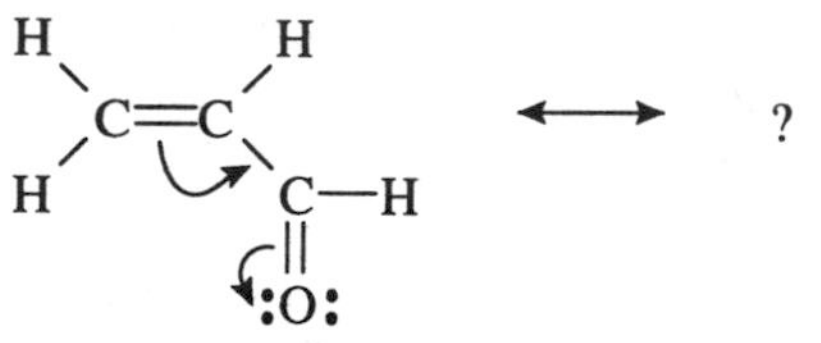

(c)

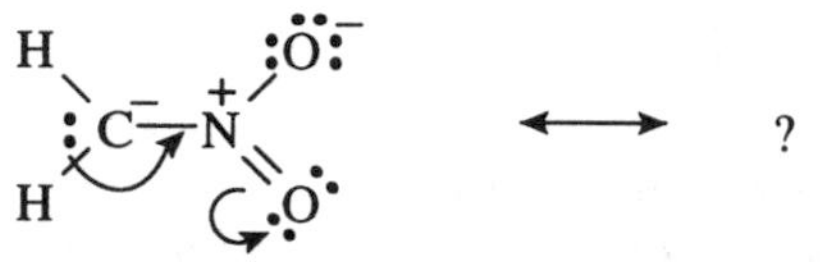

Solution

(a)

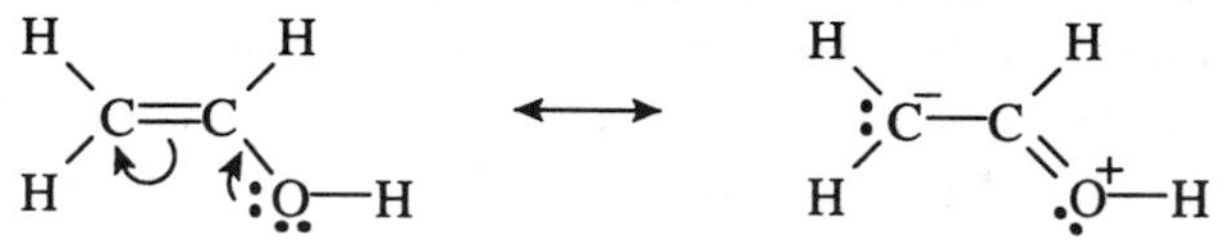

(b)

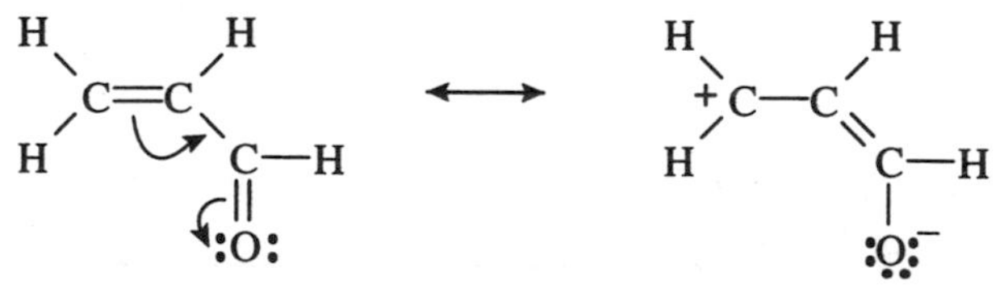

(c)

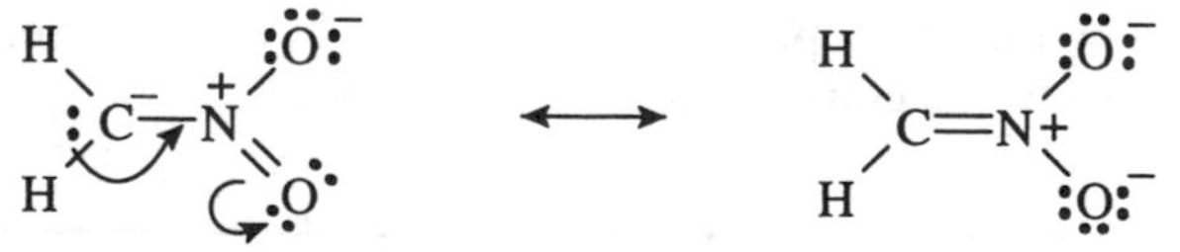

EXAMPLE 4-10   The carbonate ion, CO_3^{2-}, can be written like this:

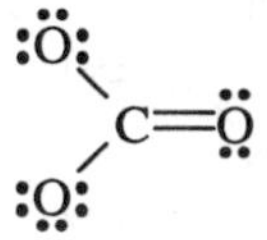

(a) Complete the structure by adding in any formal charges.
(b) Write as many resonance structures of carbonate as you can, using curved arrows to show how the resonance structures are interconverted. Which of these is (are) the major contributor(s)?
(c) What is the hybridization at carbon and the magnitude of the O–C–O bond angles?
(d) Predict the bond order of the C–O bonds, and the average charge on each oxygen.

Solution

(a)

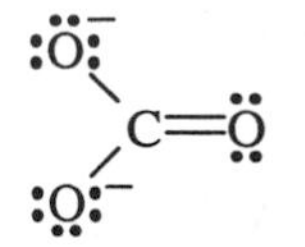

(b)

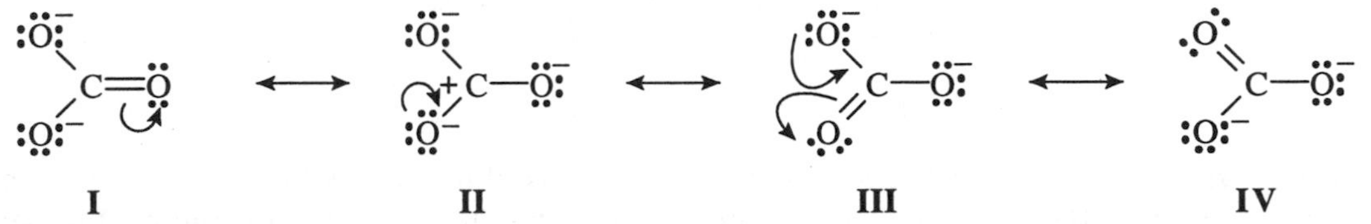

| I | II | III | IV |

Of these, structures **I**, **III**, and **IV** are equivalent in energy and will be equal contributors. Structure **II** has fewer bonds and more charge separations, so it will be a less important contributor.

(c) Regardless of which of the above resonance structures you examine, the carbon always has three groups around it. Therefore we predict sp^2 hybridization and O–C–O bond angles of 120°.

(d) The carbonate ion is essentially a hybrid of three equally important resonance contributors (structures **I**, **III**, and **IV**). Each of the C–O bonds is a single bond in two of the structures and a double bond in the third structure. So, the average C–O bond order is $(2/3)(1) + (1/3)(2) = 4/3$, i.e., one σ bond and one-third of a π bond. By similar reasoning, each oxygen has a charge of $-2/3$.

4-5. Electron Delocalization: The Molecular-Orbital Approach

Let's see if we can now show why π bonds, rather than σ bonds, are most readily involved in resonance. Remember that a π bond results from the overlap of two neighboring parallel p orbitals (Section 3-5).

Suppose we redraw the allyl carbocation, but this time we'll show all σ bonds as lines while drawing in the lobes of the p orbitals. Recall that the carbon bearing the positive charge is sp^2 hybridized (with an empty p orbital), for it has only three groups attached.

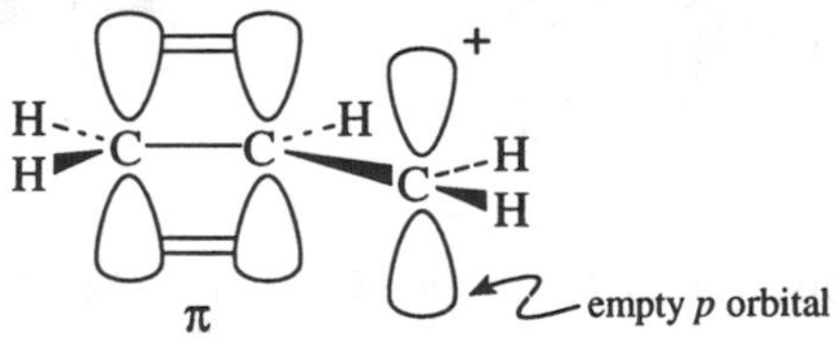

From this structure we see that it is possible to have all three p orbitals *parallel*, a fact that is critical to overlap between them. Further, the two π electrons can be shared between either the first and second carbons, or between the second and third carbons:

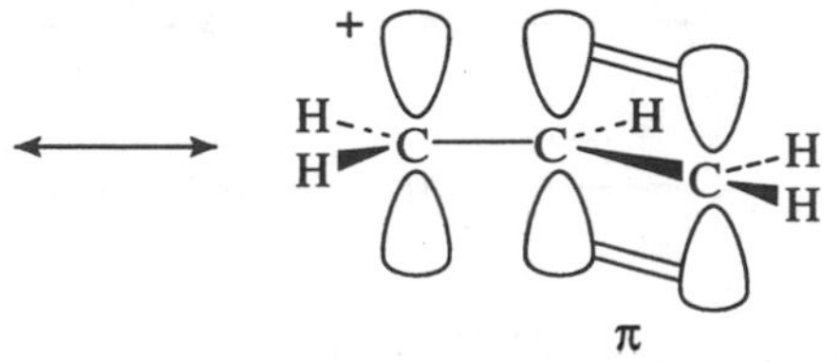

But these two structures are still nothing more than glorified valence-bond resonance structures of the allyl carbocation. The important question now is, what type of π-molecular orbitals can be constructed with these *three* parallel *p* orbitals?

In Section 3-5 we saw how *two* parallel *p* orbitals on neighboring carbons overlap to generate a π MO (no node between the atoms) and a π^* MO (with a node between the atoms). We also know (Section 2-7) that when there are *n* AOs in the basis set, *n* MOs will be generated. Therefore, if we combine *three* parallel *p* orbitals on three neighboring atoms, we will generate *three* π-type MOs, *each one of which will be spread out over all three atoms.* Of these, the lowest-energy MO, which will be labeled π_1, has no nodes between neighboring atoms. We can represent this orbital in any of three ways (remember that the signs inside the orbital lobes indicate orbital phase, *not* charge):

Because there are no nodes between neighboring atoms, this is a *bonding* MO. Most importantly, even though this MO is spread over *three* atoms, it can still contain only a maximum of *two* electrons (with spins paired).

The next higher energy MO, π_2, will have one node placed as symmetrically as possible, through the central carbon:

Although this orbital has four lobes spread out over three atoms, it is nonetheless just *one* MO. And since there are neither bonding nor antibonding phases between *neighboring* atoms, this MO is *non*bonding (not to be confused with nonbonding *atomic* orbitals).

The highest-energy MO, π_3, will have two symmetrically placed nodes (between each pair of neighboring atoms)—and will thus be *antibonding:*

We can summarize this discussion with a π MOED for the allyl system:

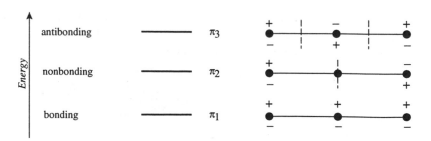

Now, let's put the *p* electrons in these MOs. In the allyl carbocation there is only one π bond, that is, just two *p* electrons. These are placed (paired) in the lowest energy MO, π_1.

$$
\begin{array}{ll}
\rule{3em}{0.4pt} & \pi_3 \\[1.2em]
\rule{3em}{0.4pt} & \pi_2 \\[1.2em]
\underline{\uparrow\!\downarrow} & \pi_1
\end{array}
$$

Allyl carbocation

This pair of electrons in a bonding MO constitutes *one* π bond distributed over *three* atoms. Therefore, the π bond order between each pair of neighboring carbons is 1/2, giving a total bond order for each carbon–carbon bond of 1.5 (one σ plus one-half π). This is the same result we deduced on the basis of valence-bond resonance structures in Section 4-3. Further, notice that π_2 has empty orbital character only at the first and third carbons, with a node at the central carbon. This, too, is consistent with our two valence-bond resonance structures for the allyl carbocation, which placed the positive charge only on the first and third carbons, but not the middle one.

EXAMPLE 4-11   The allyl radical has *three* π-bonding electrons, while the allyl carbanion has four. Using a π MOED for each, show where these electrons reside. What is the total carbon–carbon bond order in each species? [*Hint:* You can assume that the carbon atom bearing the odd electron (in the radical) or the "nonbonding pair" (in the carbanion) is sp^2-hybridized.]

Solution

Allyl radical   Allyl carbanion

Because π_2 is nonbonding, these additional electrons will neither increase nor decrease the C–C bond orders (they remain 1.5). And again, the unpaired electron (in the radical) and the "nonbonding pair" (in the carbanion) are found only at the first and third carbons, not the middle one, just as we saw in their valence-bond resonance structures.

If you're concerned about forcing the third carbon to be sp^2-hybridized, you may find the following rationalization helpful. VSEPR considerations would have predicted sp^3 hybridization for the third carbon of the carbanion, and something between sp^2 and sp^3 for the radical. But in order for there to be a resonance interaction between the π bond and the third carbon, the latter's hybridization must provide a parallel p orbital (i.e., sp^2 hybridization). There is an energetic cost of this rehybridization, but it is more than offset by the benefit of the resulting resonance interactions.

You might be wondering why we even need to worry about these molecular-orbital representations if we can get the same answers from the valence-bond picture. The reasons are threefold. First, MO theory allows us to get more quantitative answers to such questions as "What is the electron density at a given atom?" and "What is the exact bond order between these two atoms?" Valence-bond resonance theory usually gives only qualitative answers to these questions because it's hard to know exactly how much contribution each resonance structure makes to the hybrid. Second, there are many reactions that are most easily understood in terms of orbital-phase relationships. And finally, there are a few special molecules (such as O_2; see Problem 4-8) where the valence-bond structures simply don't adequately represent the true structure of the molecule, while MO theory does.

4-6. Rules for Drawing π-Molecular Orbitals of a Linear System

We're now ready to generalize the rules for drawing π MOs of a *linear* (i.e., noncyclic) array of parallel p orbitals. [*Note: Linear* in this context does not indicate molecular shape (Section 3-1) but rather *connectivity*, i.e., no rings or branch points.]

1. There must be a *p* orbital on each atom in the array, and the *p* orbitals must all be parallel.
2. If there are *n* *p* orbitals in the array, these will generate *n* MOs labeled π_1 through π_n. In terms of energy, these MOs will be distributed like the rungs of a ladder, symmetrically above and below the energy of a *p* orbital, with π_1 being the lowest and π_n the highest.
3. The number of nodes along the array in each MO equals its label number minus one (i.e., no nodes for π_1, one node for π_2, etc.). These nodes are placed symmetrically in such a way as to have as many bonding relationships as possible.
4. If *n* is *odd*, there will be one nonbonding MO, and this one will be at the same energy as a *p* orbital. Half of the remaining MOs will be at lower energy (bonding) and half will be higher (antibonding). If *n* is *even*, there will be no nonbonding MO.
5. The π electrons (the number of which doesn't necessarily equal *n*) are grouped in pairs and placed in the lowest available π MOs.

Let's apply these rules to the construction of the π MOs of the molecule below (seen previously in Example 4-7e).

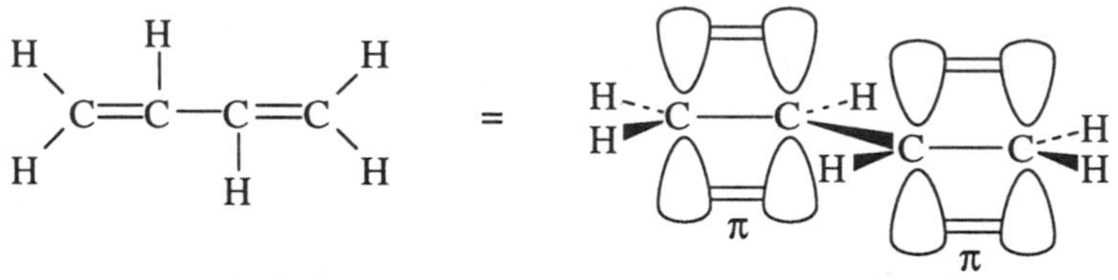

In this case *n* equals four, and there are also four *p* electrons in the system. The lowest-energy π MO is always easy to draw, because there are no nodes between atoms:

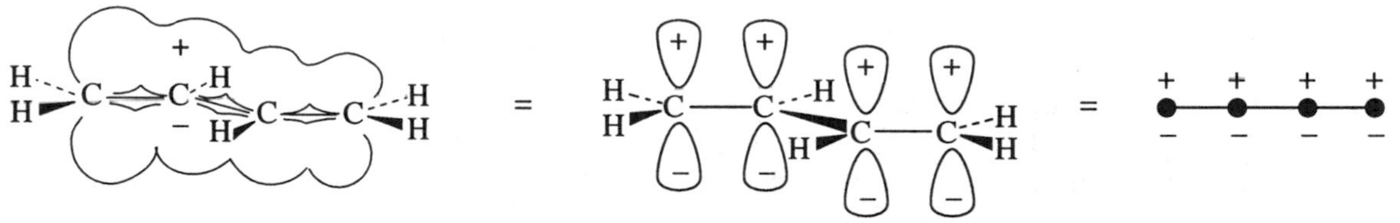

π_1

The next MO, π_2, will have one node, midway between the second and third carbons:

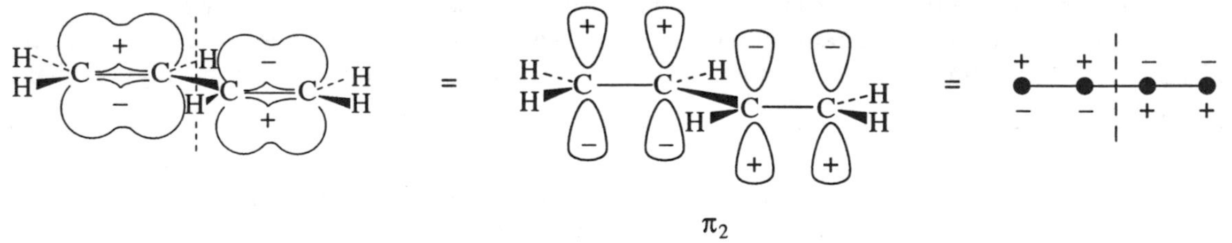

π_2

Because this MO has *two* bonding interactions (in-phase overlap) and only one antibonding interaction (out-of-phase overlap) along the array, it is still considered to be a bonding MO, though it has a higher energy than π_1.

The third MO, π_3, has two nodes:

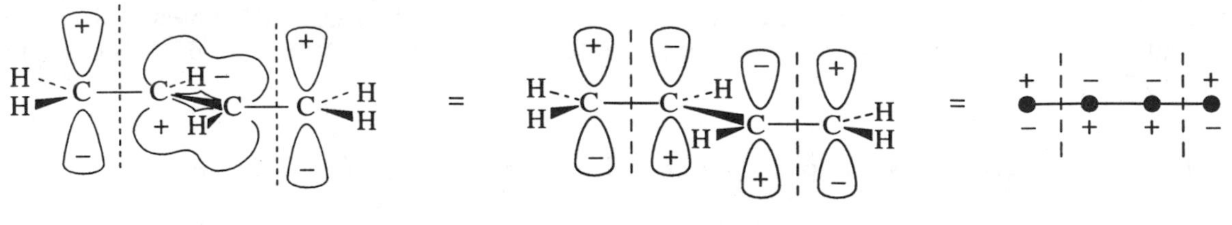

π_3

Here we have *two* antibonding relationships and one bonding relationship along the array, so this MO is considered an antibonding MO.

Finally, π_4 has three nodes, one between each pair of neighboring carbons; it is totally antibonding.

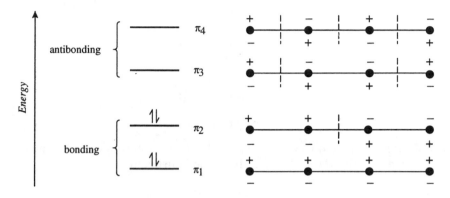

π_4

The complete π MOED for this system, including the π electrons, looks like this:

EXAMPLE 4-12   Draw a complete π MOED for the molecule below:

Use the abbreviated way of showing the relative phase relationships within each MO.

Solution   For this system n is six, and there are six π electrons.

4-7. The Special Place of Aromatic Molecules

There is a special group of organic compounds that holds the prize for resonance interactions. The unique chemical behavior of these compounds is a subject of much interest, but for now we'll limit our discussion to their special structural characteristics. These compounds are called **aromatic,** a term that is something of a misnomer. Membership in this class of compounds has nothing to do with a molecule's fragrance; rather, the term is descriptive of a molecule's electronic structure.

A. Structure of benzene

The prototype aromatic molecule is **benzene, C_6H_6**. The carbons form a six-membered ring (loop), so the molecule is described as **cyclic** (having one or more rings). Each carbon is also attached to one hydrogen. Therefore, in order to satisfy the valence of each carbon, there must be three double bonds (six π electrons) located in alternate positions around the ring. Each carbon in benzene must be sp^2-hybridized (three groups around each one, right?), so all twelve atoms are forced to lie in the same plane. Thus, benzene (like ethylene in Section 3-5) is a planar molecule.

To avoid drawing all the C's and H's, and to focus attention on the bonds, we'll usually draw benzene rings using the following structural shorthand:

In such a shorthand representation, it's implied that there's a carbon at each corner of the hexagon, and enough hydrogens (one in the case of benzene) attached to each carbon to satisfy the valence of that carbon.

note: When using these shorthand formulas, be very careful to mentally fill in all hydrogens. Notice also that the double bonds are normally drawn inside the ring, rather than outside like this:

EXAMPLE 4-13   What is the molecular formula that corresponds to the shorthand representation below?

Solution $C_{10}H_8$. If you had trouble with this question, draw out the complete structural formula, being careful to satisfy the valence of each carbon by adding enough hydrogens to bring the valence of each carbon to four.

B. Resonance in benzene

Looking at the structure of benzene, we might be tempted to predict that the carbon–carbon bond lengths around the ring would alternate from short (double bond) to long (single bond) to short, etc. But in fact all six carbon–carbon bonds are *exactly* the same length, roughly midway between the length of a single bond and a double bond. To explain this, consider the two structures below (two carbons are highlighted to maintain the viewer's orientation):

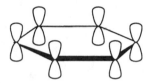

I	**II**

These two structures are equivalent resonance forms of benzene because only the double bonds (or π bonds, to be exact) have exchanged positions. The reason this "bond-switching" is so easily accomplished is readily appreciated from a side-view picture showing the *p* orbitals involved:

As you can see, the six *parallel p* orbitals in this *cyclic* array (as opposed to the *linear* arrays we saw in the previous section) are ideally situated for efficient resonance interactions involving no zwitterionic charges, no violated valences, no unpairing of electrons, and 120° bond angles throughout. The π-bonding electrons in benzene thus occupy doughnut-shaped regions above and below the plane of the ring. Because of this extremely efficient resonance,

- Benzene is one of the most stable organic molecules known.

note: To draw attention to this very special type of resonance interaction, benzene rings are often drawn with a circle (or a dotted circle) in the ring, instead of three π bonds:

Suppose you were asked to draw some additional resonance structures for benzene. How about these?

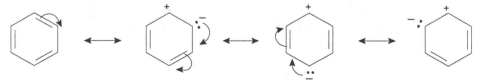

These three zwitterionic structures, while "legal" according to Rules 1, 2, and 3 of Section 4-4, are not expected to contribute significantly to the hybrid structure of benzene because of Rule 4. So, benzene is primarily a hybrid of valence-bond structures **I** and **II**, which contribute equally to its overall structure. By comparing these two resonance forms you can see why all six carbon–carbon bonds, each of which is a double bond in one form and a single bond in the other, are equal in length and all six carbons are indistinguishable.

C. The benzene ring attached to other groups

Not only is the "aromatic" benzene ring stable by itself, but it can also engage in resonance interactions with other atoms and groups of atoms connected to it. For example, the **benzyl cation** is a benzene ring in which one of the hydrogens has been replaced by a CH_2^+ group. The carbon atom in the CH_2^+ group is, like the ring carbons, sp^2-hybridized and has a *p* orbital parallel to the ring *p* orbitals. There are six π-bonding electrons distributed throughout *seven* parallel *p* orbitals. Here are five resonance structures of the benzyl cation:

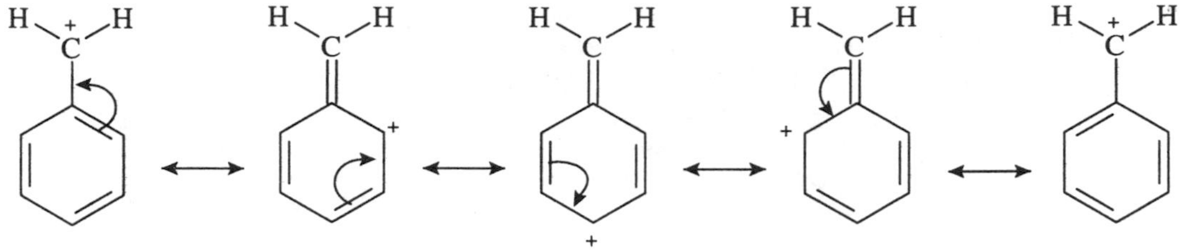

Notice how the positive charge is delocalized over three ring carbons, as well as the original CH_2 carbon. This delocalization of charge makes the benzyl cation much more stable than, for example, CH_3^+ (where there is no resonance stabilization). But most importantly, notice how the charge can be placed only at the carbons that are **ortho** and **para** to the CH_2 group, but not at the carbons that are **meta** to the CH_2 group.

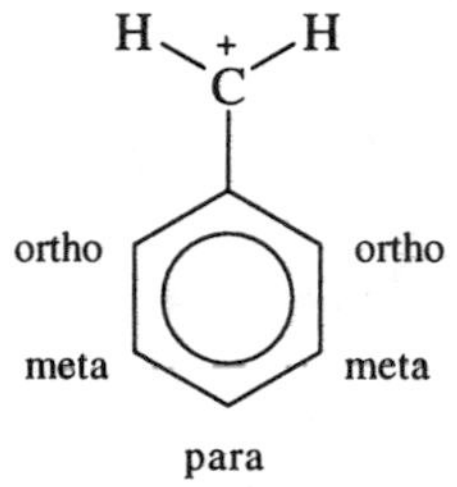

note: In the context of benzene derivatives, *ortho* is defined as "next to," *para* means "across from," and *meta* means "next to the next" or "once removed from." (See Section 5-9.)

EXAMPLE 4-14 Draw as many *valid* resonance structures as you can for the phenoxide ion below:

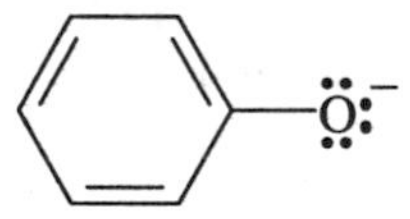

Use curved arrows to show how the electrons rearrange themselves in going from one structure to the next.

Solution Here are the resonance structures with just one negative charge:

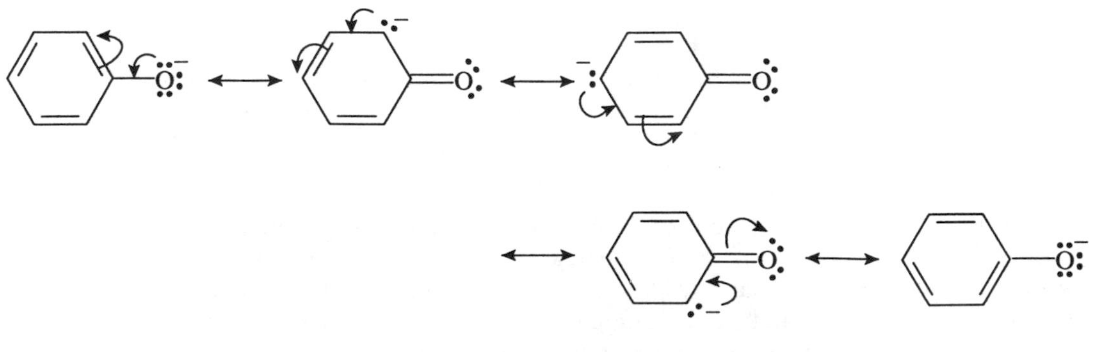

EXAMPLE 4-15 What, if anything, is wrong with these resonance structures for the ion in Example 4-14?

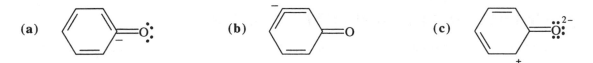

Solution All three structures violate Rule 2 by exceeding the valence of an atom:
(a) There are five bonds (ten electrons) around the C=O carbon.
(b) There are five bonds around the negatively charged carbon (don't forget the hydrogen!).
(c) There are ten electrons around oxygen.

D. Other aromatic rings

There are thousands of organic compounds whose molecules contain one or more benzene rings. These are referred to as **benzenoid aromatic compounds.** If one of the carbons *in* the ring is replaced by a **heteroatom** (an atom other than carbon or hydrogen), the resulting structure is said to be **heteroaromatic.** A good example of such a compound is **pyridine,** which contains one nitrogen in place of a ring carbon:

EXAMPLE 4-16 What is the molecular formula for pyridine? (Careful!)

Solution The molecular formula for pyridine is C_5H_5N. Remember that nitrogen is trivalent (i.e., it forms *three* bonds); there is no hydrogen attached to N in pyridine, though there *is* a nonbonding pair.

Alas, things are not as simple as they might appear. Here's a molecule whose structure is quite similar to that of benzene. It is called **cyclooctatetraene** (COT).

COT

This molecule has *none* of the properties (e.g., resonance stabilization, bond switching, equal bond lengths) usually associated with aromatic compounds. Nonetheless, it does appear to possess a cyclic array of eight *p* orbitals, containing eight π-bonding electrons (four π bonds).

But as you read the next section, you'll learn why this molecule should not be expected to be aromatic.

4-8. Criteria for Aromaticity: The $4n + 2$ Rule

There are two major requirements a molecule must meet to possess aromatic character:

1. It must have a planar cyclic array of parallel p orbitals, one at each atom in the ring.
2. The number of electrons in the π system must equal **$4n + 2$ (Hückel's rule),** where n is any integer, zero to infinity.

note: This n is not to be confused with the principal quantum numbers we used in Chapter 1.

EXAMPLE 4-17   What is the value of $4n + 2$ when $n = 0, 1, 2,$ and 3?

Solution   2, 6, 10, and 14. These are the first few "magic" numbers of π electrons in a cyclic array of p orbitals for the system to exhibit aromatic properties. Now we can begin to see why benzene (with *six* π-bonding electrons) *is* aromatic, while COT (with *eight*) is *not*.

EXAMPLE 4-18

CBD

(a) Draw another resonance structure of **cyclobutadiene** (CBD).
(b) Is CBD likely to show aromatic properties?

Solution
(a) We'll darken two of the carbons to help keep our perspective. So, another resonance structure looks like this:

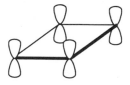

(b) This molecule does meet the first criterion for aromaticity—a planar cyclic array of parallel p orbitals, one at each ring atom:

However, the number of π electrons is four (two π bonds), so this molecule will not be aromatic.

To understand the basis of the $4n + 2$ rule, we need to look at the energies of the π MOs for cyclic arrays. Fortunately, there is an easy mnemonic device to help you prepare the π MOED for a monocyclic (single ring) aromatic system. Begin by representing the ring as a polygon of the appropriate size, drawn with one corner *down*. Then circumscribe the polygon with a circle. The center of the circle represents the energy of an isolated p orbital. At each intersection of the polygon with the circle there will be a π MO with that relative energy. All MOs *below* the center of the circle will be bonding, and all those above the center of the circle will be antibonding. Those MOs at the same level as the center of the circle are nonbonding.

Here's how the π MOED for benzene looks. Notice that π_2 and π_3 are degenerate, as are π_4 and π_5.

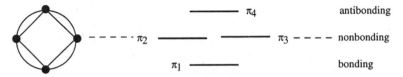

There are six π-bonding electrons to go into these π MOs, and they will be arranged in pairs and placed in π_1, π_2, and π_3, all of which are *bonding* MOs.

Compare the MOED diagram for benzene with the MOED for CBD:

How will the four π-bonding electrons be distributed among these four MOs? The first two will clearly go (with spins paired) into π_1, which is bonding. But Hund's rule (Section 1-6) says that the remaining two electrons will be placed, one each, in π_2 and π_3 (which are degenerate), with spins *parallel*.

Thus, half of the π electrons in CBD remain *unpaired*—in nonbonding MOs. This configuration greatly decreases the stability of the π bonds, making CBD an extremely reactive molecule. By contrast, we know from Section 4-4 that a *linear* array of two π bonds enjoys some measure of resonance stabilization. Thus, CBD is an example of a molecule that is actually *destabilized* by cyclic resonance. Such a molecule is often referred to as **antiaromatic.**

EXAMPLE 4-19 Construct the π MOED for COT, and show how the π-bonding electrons are distributed. Explain why COT is not expected to be aromatic.

Solution Draw an octagon, corner down, then circumscribe it with a circle. The first six π-bonding electrons, in pairs, go into π_1, π_2, and π_3. But the last two electrons (left unpaired) go, one each, into π_4 and π_5.

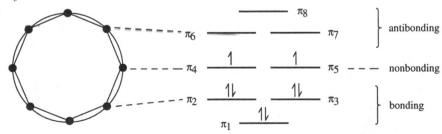

Therefore, planar COT would be predicted to have two unpaired electrons in nonbonding orbitals, making it unstable like CBD. However, the COT molecule can do something that the CBD molecule can't. Because of the increased flexibility of the eight-membered ring (compared to the four-membered ring), the COT molecule contorts itself away from planarity into a tub shape where the C–C–C bonds can be closer to the trigonal angle of 120°. The p orbitals are thus no longer parallel and can no longer engage in resonance. In this structure of COT, each of the double bonds behaves as if it were isolated. Cyclobutadiene can also relieve some of its antiaromaticity, by stretching to form a rectangular, rather than squre, geometry.

Remember: For all the *p* orbitals in a cyclic array to be parallel, the ring must be *planar*.

The reason behind the $4n + 2$ rule may now be apparent to you. Whenever there are $4n + 2$ π electrons in a cyclic array, they will generally be paired and reside in bonding MOs. This will contribute to lowering the energy and increasing the stability of the molecule. If the number of π electrons is other than $4n + 2$, there will be unpaired electrons in nonbonding orbitals, which creates instability in the molecule.

We can summarize much of our discussion in this chapter by saying that resonance interactions are usually associated with a significant stabilization of a molecule. In future chapters we'll often use our knowledge about resonance to predict and explain patterns in reactivity.

SUMMARY

1. The formal charge on an atom equals the number of valence electrons on that atom (in the uncombined state) minus the number of electrons it "owns" in the bonded state. "Owned" electrons include *half* of all shared electrons and *all* nonbonding electrons on the atom. From the line–bond Lewis structure of a molecule the formal charge of each atom is given by:

$$\text{Formal charge} = \text{valence electrons} - (\text{lines} + \text{dots}) \qquad \textbf{(4-1)}$$

2. Electron (arrow) pushing describes the use of curved arrows to keep track of electron reorganizations in a molecule during bond-making and bond-breaking processes.
3. Bond-breaking processes are labeled on the basis of how the electrons in the bond are divided:

   homolytic cleavage—each fragment retains one of the electrons from the bond;
   heterolytic cleavage—one fragment receives *both* electrons from the bond.

   Bond-making processes are labeled similarly:

   homogenic formation—one electron in the bond comes from each fragment;
   heterogenic formation—both electrons in the bond come from one of the fragments.

4. A nucleophile is an electron-pair donor; an electrophile is an electron-pair acceptor.
5. In certain molecules, π and nonbonding electrons can be delocalized (spread out) over more than just one or two atoms. This extended electron delocalization is called resonance or conjugation.
6. Certain molecules have properties that are not adequately represented by just one (valence-bond) structural formula. Such molecules are often better represented as hybrids of two or more so-called resonance structures (structural formulas differing only in the positions of electrons).
7. To be valid, a resonance structure must conform to the following rules:
   (a) All resonance structures of a given molecule must have all nuclei in exactly the same position; only electrons can change position.
   (b) No atom can have more electrons around it than will fill its valence shell (i.e., no second-row atom may have more than eight electrons around it).
   (c) All resonance structures for a given molecule must have the same number (usually zero) of unpaired elecntrons.
   (d) The relative contribution of a given resonance structure to the hybrid's overall structure increases when the number of bonds in the structure is maximized and the number of charge separations is minimized.

(e) When formal charges in a resonance structure are unavoidable, like charges should be as far from each other as possible, with negative charges on electronegative atoms and positive charges on electropositive atoms.

8. In most cases resonance interactions lower the energy of a molecule and increase its stability.

9. For two π bonds, or a π bond and a neighboring nonbonding pair (in a p orbital), to engage in a resonance interaction they must be separated by one bond, and the p orbitals must be parallel.

10. The rules for drawing π MOs for a linear system of p orbitals are:

    (a) There must be a p orbital on each atom in the array, and they must all be parallel.

    (b) If there are n p orbitals in the array, there will be n π MOs labeled π_1 through π_n. In terms of energy, these will be distributed symmetrically above and below the energy of a p orbital, with π_1 being the lowest and π_n the highest.

    (c) The number of nodes along the array in each MO equals the label number minus one. These nodes are placed symmetrically so as to have as many bonding relationships as possible.

    (d) If n is *odd* there will be one nonbonding MO, and this one will be at the same energy as a p orbital. Half of the remaining MOs will be at lower energy (bonding) and half will be higher (antibonding).

    (e) The π electrons (the number of which doesn't necessarily equal n) are grouped in pairs and placed in the lowest available π MO.

11. In principle, the valence bond/resonance approach to molecular structure usually gives the same predictions as the MO approach does, but the MO approach affords more quantitative predictions.

12. Certain cyclic molecules, such as benzene, have exceptional stability because of unusually efficient resonance interactions. Such molecules are called aromatic. For a molecule to have aromatic properties it must have a planar cyclic structure with a parallel p orbital at each ring atom, and it must have $4n + 2$ (where n is any integer) electrons in the π MOs.

13. To draw the π MOED for an aromatic molecule, begin by representing the ring as a polygon of the appropriate size, drawn with one corner *down*. Then circumscribe the polygon with a circle. The center of the circle represents the energy of an isolated p orbital. At each intersection of the polygon with the circle there will be a π MO with that relative energy. All π MOs *below* the center of the circle will be bonding, while those above the center of the circle will be antibonding. Those MOs at the same level as the center of the circle are nonbonding.

RAISE YOUR GRADES

Can you define...?

- ☑ formal charge
- ☑ carbocation
- ☑ carbanion
- ☑ free radical
- ☑ electron pushing
- ☑ homolytic/heterolytic bond cleavage
- ☑ homogenic/heterogenic bond formation
- ☑ $4n + 2$ rule
- ☑ nucleophile

- ☑ electrophile
- ☑ delocalization (of electrons)
- ☑ resonance
- ☑ conjugation
- ☑ resonance structure
- ☑ aromatic
- ☑ antiaromatic
- ☑ heteroatom
- ☑ heteroaromatic

Can you explain...?

- ☑ how formal charges for individual atoms are determined
- ☑ how electron pushing can be used to keep track of electron reorganization
- ☑ why some molecules have structures that are best regarded as hybrids of several resonance forms
- ☑ how to draw valid resonance structures and assess their relative contribution
- ☑ the structural requirements for resonance interactions to take place
- ☑ how to draw the π MOs for a linear array of p orbitals
- ☑ how to construct the π MOED for linear and cyclic arrays of p orbitals
- ☑ how to draw shorthand structures for aromatic molecules and their resonance forms
- ☑ the criteria for aromaticity and the $4n + 2$ rule
- ☑ how to find the relative energy of each π MO in an aromatic molecule

Can you draw a structure for...?

☑ allyl carbocation, carbanion, and radical
☑ ammonium ion
☑ carbonate ion
☑ benzene

☑ pyridine
☑ cyclooctatetraene
☑ cyclobutadiene

SOLVED PROBLEMS

PROBLEM 4-1 Write in any nonzero formal charges in the structures below:

(a)  :F̈:

(b) H—N̈—H

(c)
H—C
with =Ö: and :Ö:

(d)
:Ö:
H—C—Ö—H
:Ö—H

(e)
H
C=C—H
H

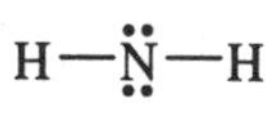

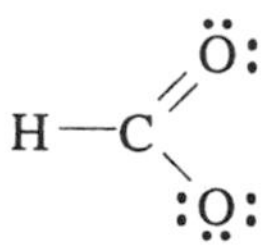

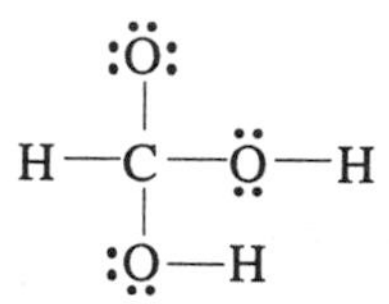

(f) H—C≡O:

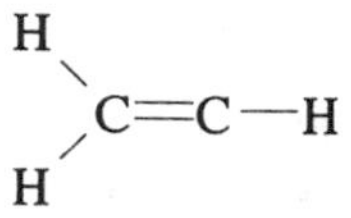

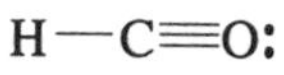

Solution Remember to suspect a nonzero formal charge whenever an atom has a number of bonds other than its normal valence. As a reminder, let's list the normal valences of common second-period elements: C, 4; N, 3; O, 2; F, 1. Now, supply the formal charges that account for "discrepancies":

(a)  :F̈:⁻

(b) H—N̈⁻—H

(c)
H—C
with =Ö: and :Ö:⁻

(d)
:Ö:⁻
H—C—Ö—H
:Ö—H

(e)
H
C=C⁺—H
H

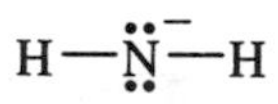

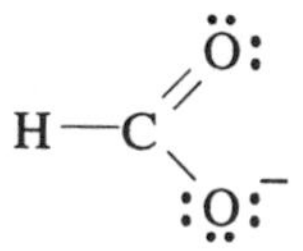

(f) H—C≡O⁺:

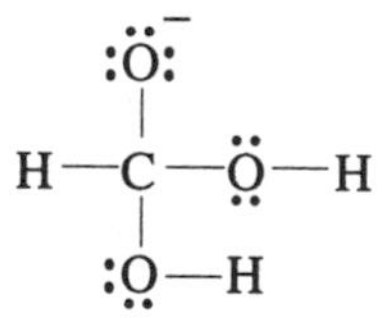

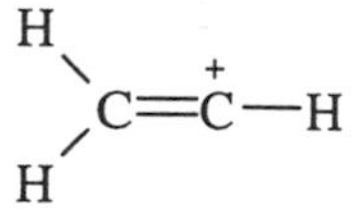

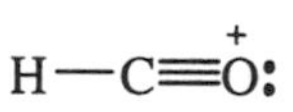

PROBLEM 4-2 Draw the structure(s) of the product(s) for each set of bond cleavages and formations indicated by the curved arrows below. State the type of bond cleavage (heterolytic or homolytic) and the type of bond formation (heterogenic or homogenic).

(a) 
H
H—C—H   ·Cl̈:   ⟶   ?
H

(b)
H
H—C—Cl̈:   ⟶   ?
H

(c)
:Ö:
H—C—Ö—H   ⟶   ?
:Cl̈:

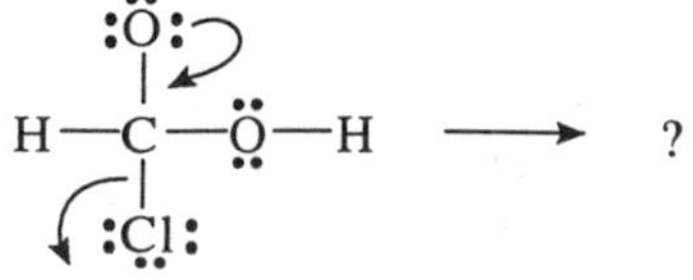

(d) 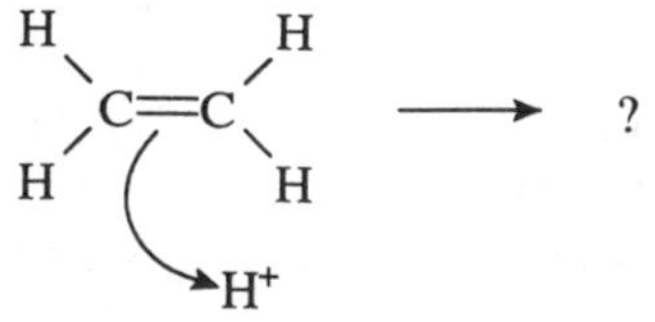
H   H
C=C   ⟶   ?
H   H
H⁺

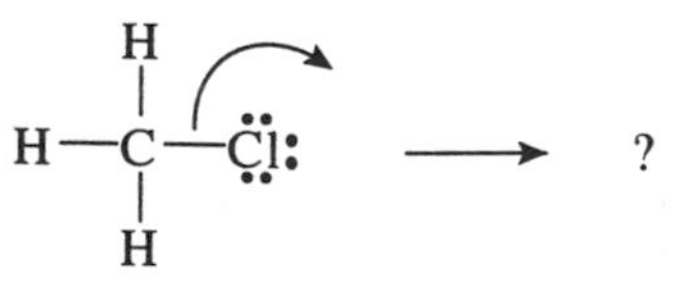

(e) 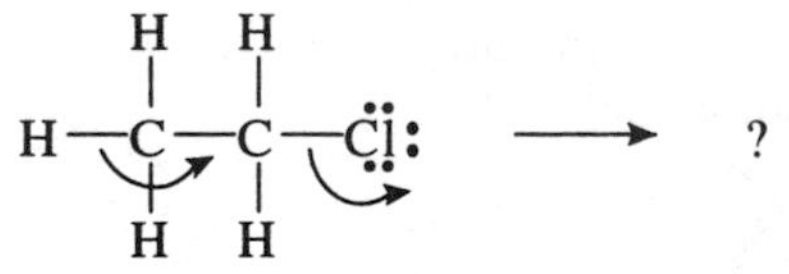   H—C—C—Cl: $\longrightarrow$?

(structure: H₃C—CH₂—Cl with curved arrows)

Solution

(a) 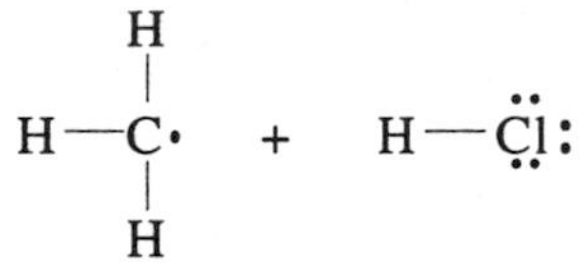   H—C· + H—Cl: homolytic cleavage, homogenic formation

(b) 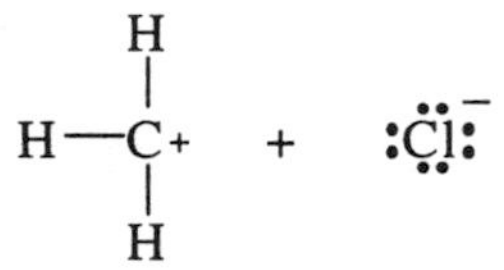   H—C+ + :Cl:⁻   heterolytic cleavage

(c) 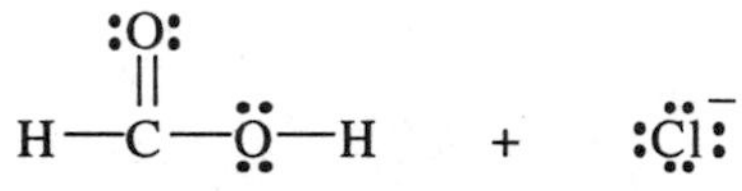   H—C—O—H (with :O: double bonded) + :Cl:⁻   heterolytic cleavage, heterogenic formation

(d) 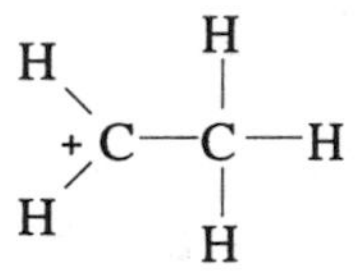   ⁺C—C—H (ethyl cation structure)   heterogenic formation

(e) 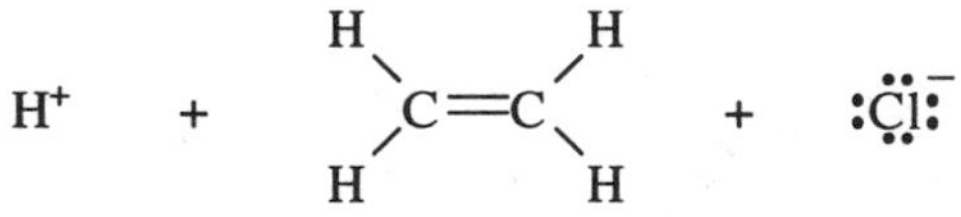   H⁺ + C=C (ethylene) + :Cl:⁻   two heterolytic cleavages and one heterogenic formation

PROBLEM 4-3   Given the structure of diazomethane (**I**) below,

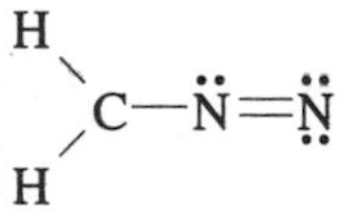

$$\text{H}_2\text{C}-\ddot{\text{N}}=\ddot{\text{N}}$$

I

(a) Complete structure **I** by adding in all nonzero formal charges.

(b) Below are five additional resonance structures for molecule **I**. Use curved arrows to show how each structure is converted to the next, after first drawing in all nonzero formal charges.

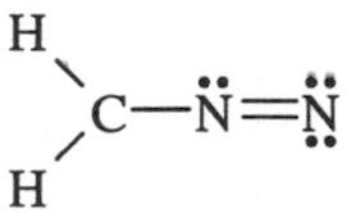

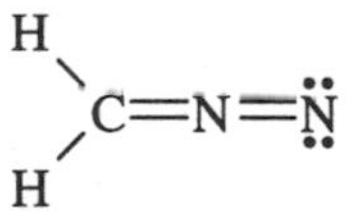

H₂C—N̈=N̈ $\longleftrightarrow$ H₂C=N=N̈

Ia   **Ib**

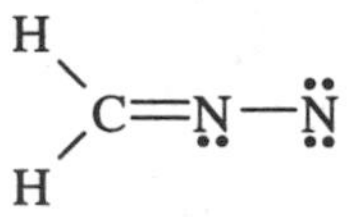

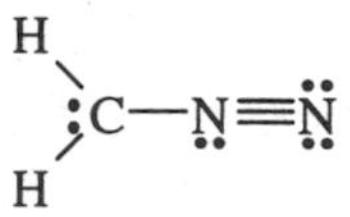

$\longleftrightarrow$ H₂C=N—N̈ $\longleftrightarrow$ H₂C—N≡N

Ic   **Id**

$\longleftrightarrow$ H₂C—N=N $\longleftrightarrow$ H₂C—N≡N:

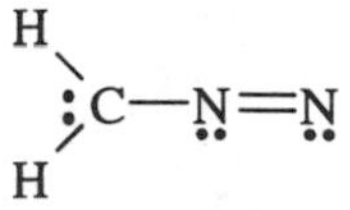

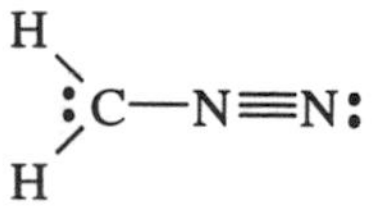

Ie   **If**

(c) Which of the above resonance structures is (are) likely to be the major contributor(s)? Why?

(d) What, if anything, is wrong with the resonance structures below?

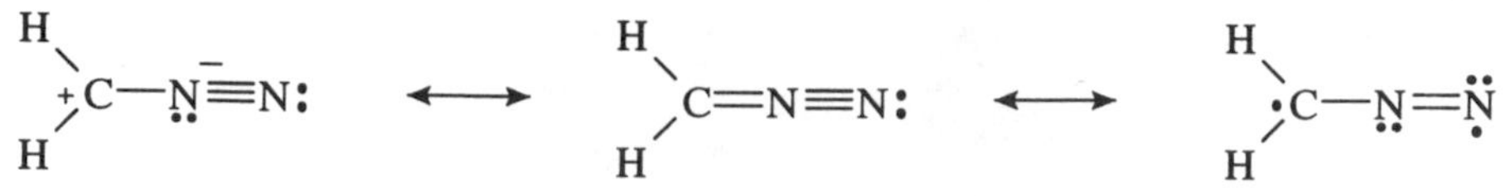

Ig   **Ih**   **Ii**

Solution

(a)

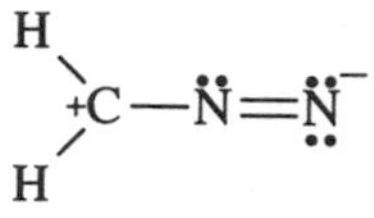

I

(b)

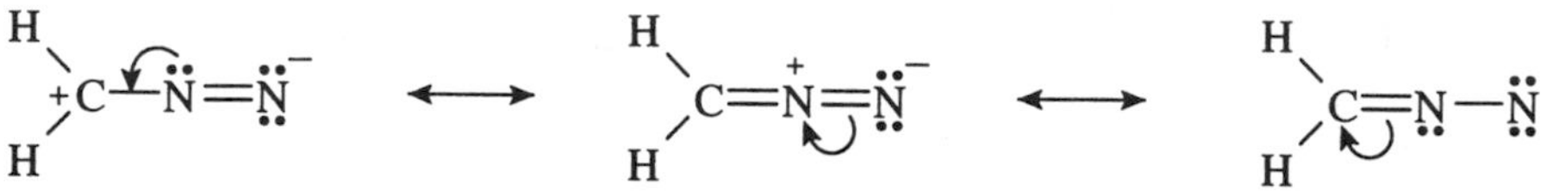

Ia   **Ib**   **Ic**

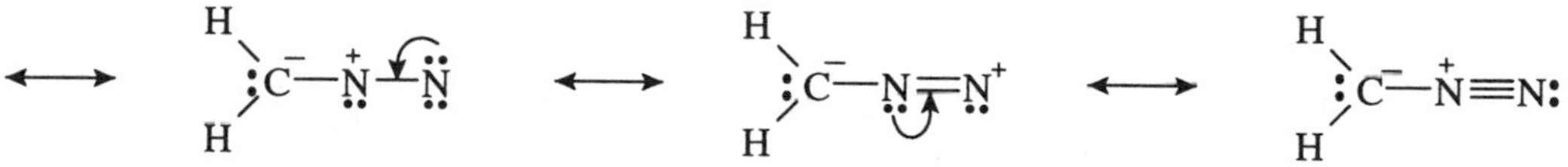

Id   **Ie**   **If**

(c) Structures **Ib** and **If** are the major contributors, since they have the greatest number of bonds (six), and each atom has a satisfied valence. Structure **Ib** is probably somewhat better because in **If** there is a negative charge on carbon, rather than on the more electronegative nitrogen.

(d) The first two structures, **Ig** and **Ih**, have too many electrons (ten) around the internal nitrogen. The third structure, **Ii**, has unpaired electrons not present in the other resonance structures.

PROBLEM 4-4   Circle the π bonds and nonbonding pairs in the structures below that are likely to engage in conjugation with each other.

(a)

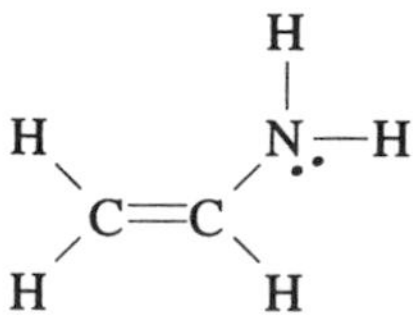

(b)

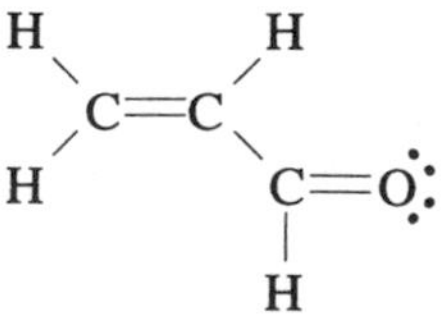

(c)

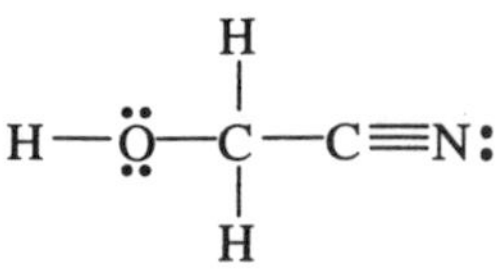

(d)

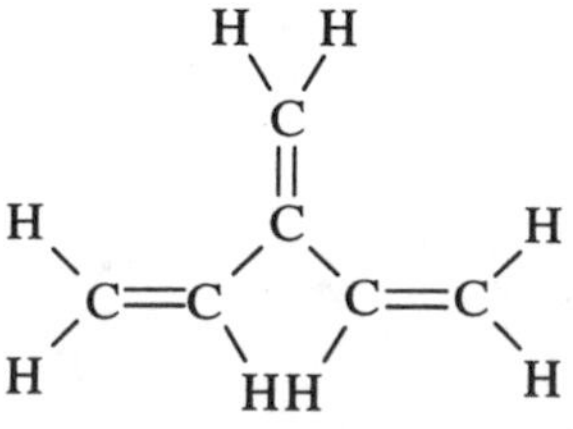

Solution

(a)

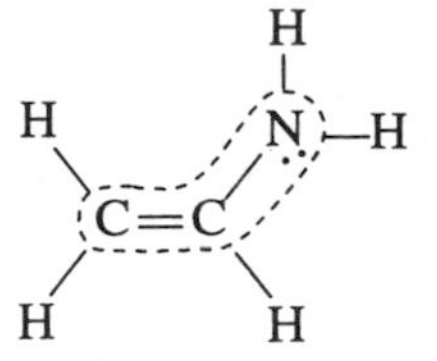

(b)

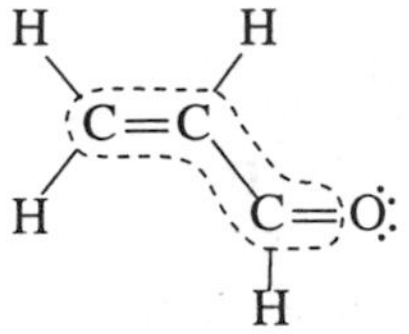

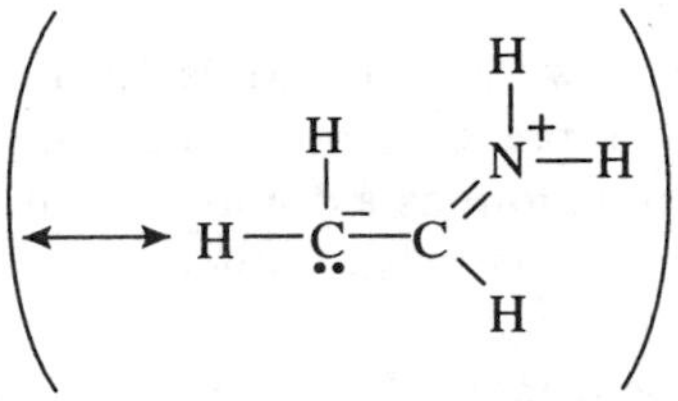

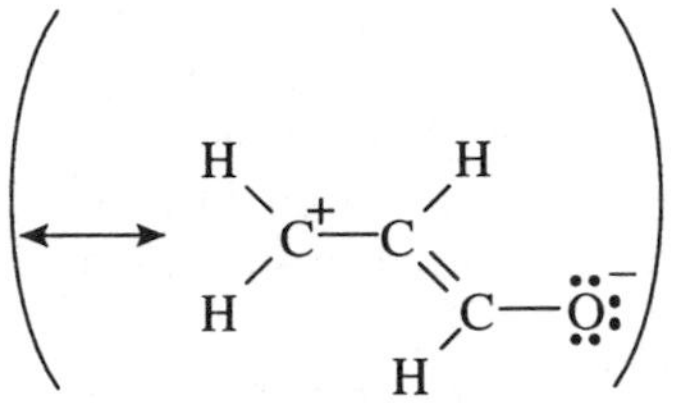

(c) No conjugation is possible because the distance between π bonds and nonbonding pairs exceeds one bond.

(d)

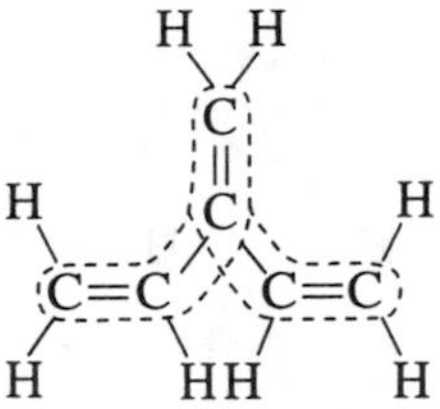

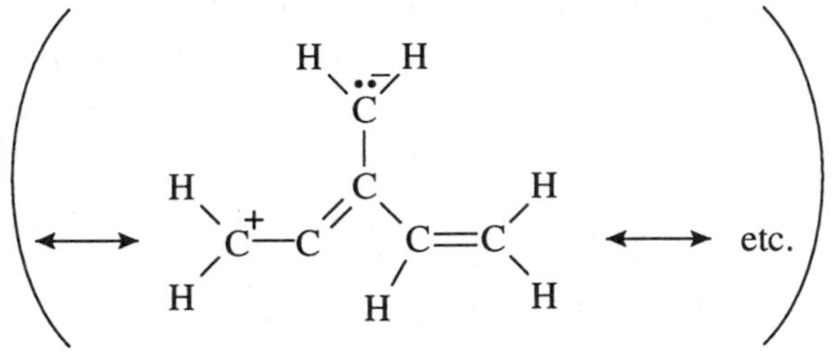

Notice that in this case the middle π bond is conjugated with both other π bonds, but the latter π bonds are *not* conjugated with each other. This molecule is referred to as **cross-conjugated.**

PROBLEM 4-5 Draw the resonance structure that results from the electron reorganization indicated by the curved arrow(s).

(a)

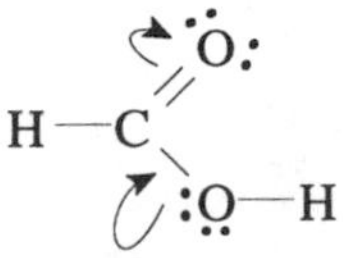

(b)

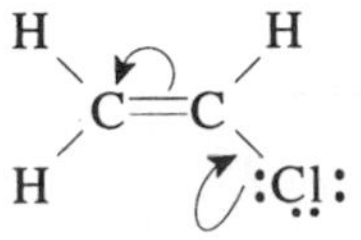

(c)

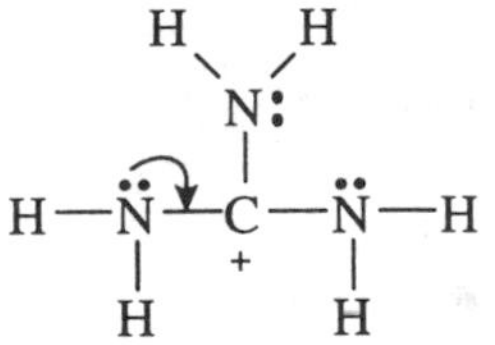

(d)

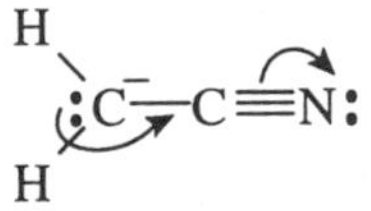

Solution

(a)

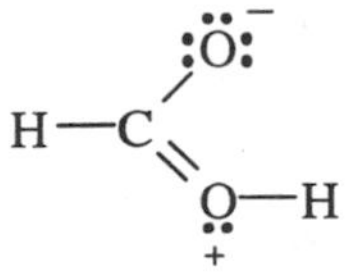

(b)

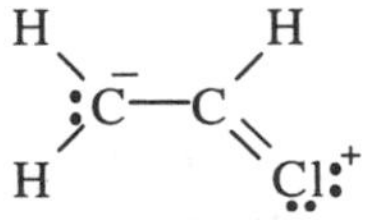

(c)

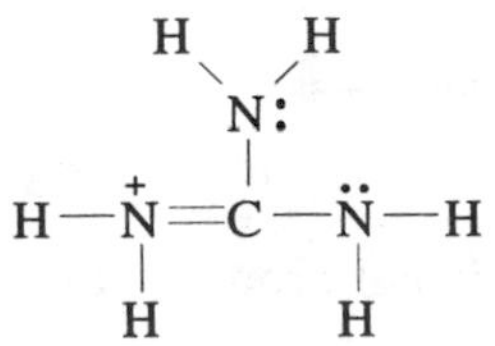

(d)

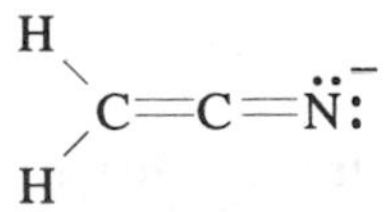

PROBLEM 4-6 Given

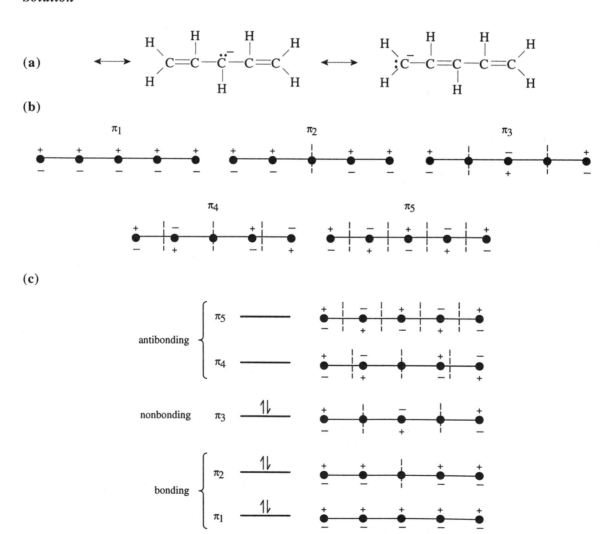

(a) Draw two additional resonance structures for this carbanion, showing how the charge is delocalized.

(b) This carbanion consists of a linear array of five *p* orbitals. Draw the five π MOs for this system, labeling them π_1 through π_5. You may use the shorthand representation, where the σ-bond backbone of the molecule is indicated by a line, each carbon atom as a dot along the line, and the orbital phases are shown as +/− signs above and below the dots. Also show nodes as vertical dashed lines.

(c) Draw a π MOED for the system, showing the relative energy of each π MO and indicating which are bonding, nonbonding, or antibonding.

(d) How many π-bonding electrons are there in this system? Using the above MOED, show how they are distributed among the π MOs. Is this carbanion aromatic?

(e) Describe how the valence-bond resonance structures and the MO approach give the same predictions as to the location of the negative charge.

Solution

(a)

(b)

(c)

(d) There are six π-bonding electrons in the array. But the molecule is not aromatic because it is not a *cyclic* array.

(e) In the MO approach the nonbonding pair (i.e., negative charge) resides in π_3, which has electron density on the first, third, and fifth carbons, but nodes at the second and fourth carbons. This is in accord with the three resonance structures in part (a).

PROBLEM 4-7 Carbon monoxide (CO) has two principal resonance contributors.
(a) Draw both resonance contributors and indicate any formal charges.
(b) Predict the direction of the bond dipole in each structure.
(c) Can you predict which of the two structures would be the major contributor? Why?
(d) The actual dipole moment of CO is small, with carbon being the negative end. How does this fact affect your answer to part (c)?

Solution

(a)   :C꞊Ö: ⟷ ⁻:C≡O⁺:

(b) The dipole in the first structure will have oxygen (the more electronegative element) as the negative end. But the formal charges in the second structure (being larger in magnitude than the partial charges in the first structure) cause a reversal in the direction of the dipole.

$$\overset{\delta+}{}\ \overset{\delta-}{}$$
:C꞊Ö: ⟷ ⁻:C≡O⁺:

(c) The first structure is better from the standpoint of having no formal charges, but the second one is better from the standpoint of number of bonds. Judging from structures alone, the choice seems to be a toss-up.

(d) The small but reversed dipole of CO is consistent with the second structure's being the main contributor, but with a substantial contribution from the first structure.

PROBLEM 4-8 Consider the oxygen molecule (O_2).
(a) Draw a line/dot structure of O_2 consistent with the fact that O_2 has two unpaired electrons.
(b) The O–O bond in O_2 has the strength of a double bond. Draw a line/dot structure consistent with this fact.
(c) Are the structures you drew for parts (a) and (b) resonance structures? Why?
(d) The valence-level MOED for O_2 is reproduced below. Show where the valence electrons reside.

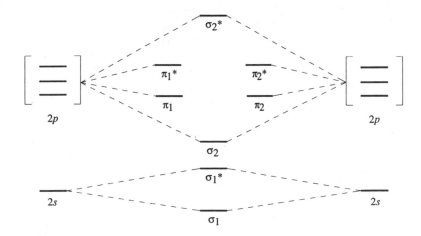

(e) Calculate the O–O bond order in O_2.
(f) Does the MO approach correlate with the observed properties of O_2?

Solution

(a) ·Ö—Ö·

(b) :O꞊O:

(c) No. The number of unpaired electrons (two in the first structure and none in the second) differs in the two structures.

(d) There are $2 \times 6 = 12$ total valence electrons in O_2. These are distributed as follows:

$$\overline{}\ \sigma_2^*$$

$$\underline{\uparrow}\ \ \ \ \ \ \underline{\uparrow}$$
$$\pi_1^*\ \ \ \ \ \pi_2^*$$

$$\underline{\uparrow\downarrow}\ \ \ \ \ \underline{\uparrow\downarrow}$$
$$\pi_1\ \ \ \ \ \pi_2$$

$$\underline{\uparrow\downarrow}$$
$$\sigma_2$$

$$\underline{\uparrow\downarrow}$$
$$\sigma_1^*$$

$$\underline{\uparrow\downarrow}$$
$$\sigma_1$$

(e) Bond order = bonding pairs – antibonding pairs = $4 - (1 + 1/2 + 1/2) = 2$.

(f) Yes. The MO picture shows why there are two unpaired electrons (one each in π_1^* and π_2^*, owing to Hund's rule), as well as why the net bond order is 2.

PROBLEM 4-9 Consider

(a) What is the molecular formula of this structure?

(b) Does the π system in the above structure constitute a linear conjugated system, an aromatic system, or neither?

Solution

(a) The molecular formula is C_6H_8. Don't forget to mentally fill in enough hydrogens at each carbon to satisfy its valence:

(b) This is not an aromatic system because two of the carbons in the ring are sp^3-hybridized and therefore do not have p orbitals to complete the cyclic array. Still, the two π bonds *are* conjugated (separated by just one bond), so the π system is considered to be linear even though the molecule itself is cyclic.

PROBLEM 4-10 Consider

(a) Draw four additional resonance structures (lacking formal charges) for this radical. Use curved arrows to show how one structure is converted to the next.

(b) How many *p* orbitals and *p* electrons are involved in this π system?
(c) Draw π_1, showing the orbital phase of each lobe in the MO.

Solution

(a)

(b) For the unpaired electron to be delocalized as indicated in the resonance structures, the CH_2 carbon must be *sp^2*-hybridized, with the unpaired electron in a *p* orbital parallel to the aromatic π-system. Therefore, there are seven *p* orbitals and seven electrons in the π system.

(c)

PROBLEM 4-11 **(a)** Is the structure

(Example 4-13) expected to be aromatic? Why? **(b)** Draw as many additional resonance structures (lacking formal charges) for this structure as you can. Use curved arrows to show how one structure converts to the next.

Solution
(a) Yes. It is a cyclic array of *p* orbitals, one at each ring atom (each is *sp^2*-hybridized), with ten π electrons in the system. (The Hückel *n* is 2.)

(b)

PROBLEM 4-12 Draw as many resonance structures as you can for the carbocation below, showing how the positive charge can be delocalized. Each structure should have only one formal charge. Use curved arrows to show how one structure is converted to the next.

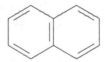

Solution

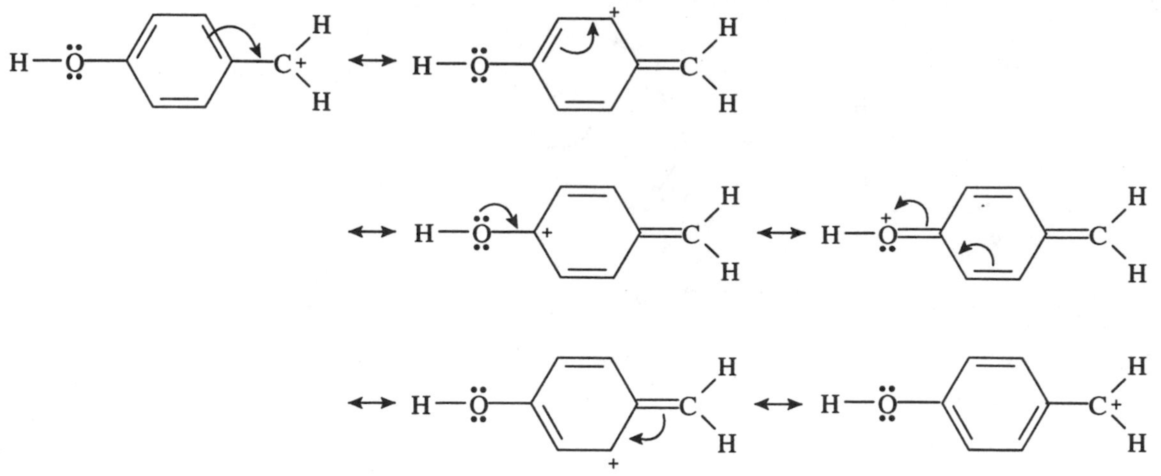

PROBLEM 4-13   The molecule represented below shows a resonance interaction between the two aromatic rings when R = H. However, when R is changed to CH_3, the resonance interaction *between* the rings is lost (though there is still resonance *within* each ring). Explain. (*Hint:* Make a molecular model of both compounds.)

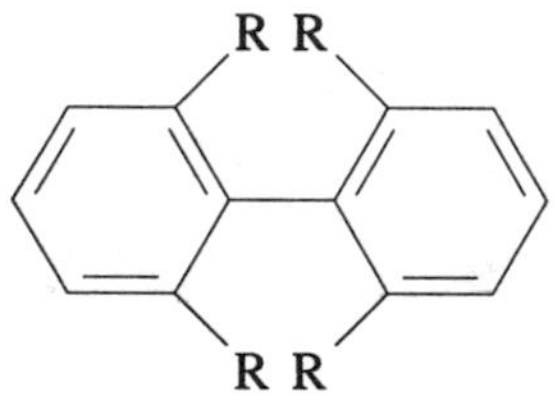

Solution   When R = H, the molecule can adopt a fully planar structure, which is required for all twelve *p* orbitals (six in each ring) to be parallel. But in the second case, the two rings are prevented from occupying the same plane because of interference between the four CH_3 groups. So, the molecule adopts a structure in which the two aromatic rings are perpendicular to each other:

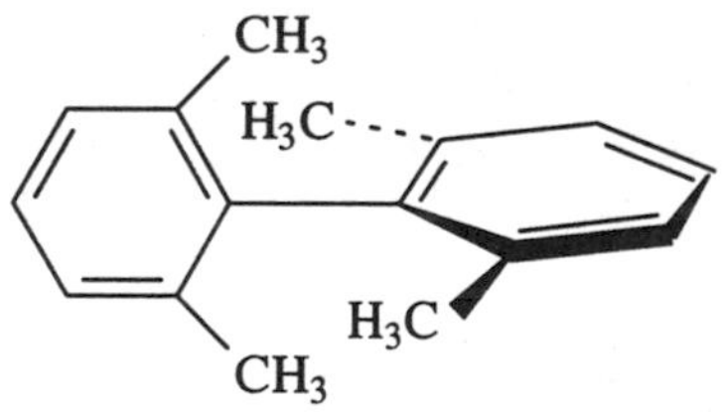

PROBLEM 4-14   For each pair of structures below, one exhibits aromatic properties (e.g., exceptional stability, charge delocalization, equal bond lengths, etc.), while the other shows antiaromatic characteristics.

(a) Select the aromatic structure in each pair, and explain why you made your selection.

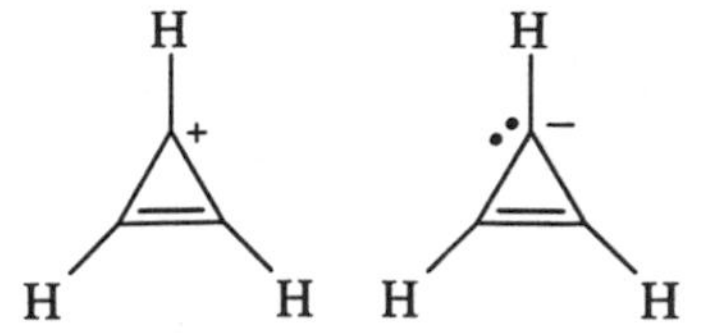

I 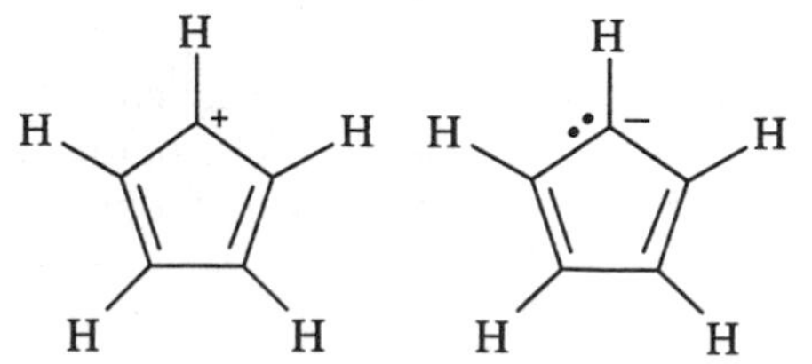   II

(b) Draw additional resonance structures for each of the aromatic ions, showing how the charge is delocalized.

(c) Draw the π MOEDs for the aromatic ions and show how the π-bonding electrons are distributed among the MOs.

Solution

(a) In **I** the first (carbocation) structure will be aromatic because it has two π electrons ($n = 0$) in a cyclic array. The carbanion has four π electrons, so it will be antiaromatic. In **II** the second (carbanion) structure has six π electrons and will be aromatic. The carbocation has only four π electrons.

(b)

I

II

In both cases, note how the charge is distributed equally to each atom in the ring.

(c)

I

II

EXAM 1
(Chapters 1–4)

1. Complete each sentence below with the appropriate term or number:

   (a) The main-group elements Li through Ne make up the _____ period. Their valence electrons are in the $n =$ _____ principal quantum level. The maximum number of valence electrons that can be accommodated in this principal quantum level is _____.

   (b) An atom of Si has atomic number _____ and belongs to group _____. Its electronic configuration is _____, which includes _____ unpaired electrons. An atom of Si has _____ valence electrons, _____ subvalence electrons, a kernel charge of _____, a nuclear charge of _____, and _____ empty $3d$ orbitals.

   (c) An electron in a $3p$ orbital has $n =$ _____, $l =$ _____, $m =$ _____, and $s =$ _____. Each of these quantum numbers is related to a property of the orbital or the electron in it: n is related to the _____ and _____ of the orbital, l is related to its _____, m to its _____, and s to the electron's _____.

   (d) The normal valence of a main-group atom (i.e., the number of _____ it forms) equals its group number for elements in groups _____, and eight minus its group number for elements in groups _____.

   (e) A(n) _____ bond forms between two atoms of equal _____, and involves _____ the electrons.

   (f) The structural formula for a molecule shows the _____ of atoms and bonds but not necessarily the _____ of the molecule.

   (g) The overlap of five parallel neighboring p orbitals in a linear array will generate _____ bonding π MO(s), _____ antibonding π MO(s), and _____ nonbonding π MO(s).

   (h) One s AO and one p AO on the same atom are mixed to form two *non*equivalent hybrids. One of these hybrids has $f_s = 0.4$; it has $f_p =$ _____ and is a(n) _____ hybrid. The other hybrid has $f_s =$ _____.

   (i) An MO that is cylindrically symmetric around the bond axis is called a _____ MO, while one that arises from the overlap of two or more parallel p orbitals is called a _____ MO.

   (j) As the fraction of s character (f_s) of two *equivalent* hybrids on the same atom *increases*, their energy _____, and the interorbital angle between them _____.

2. Give a brief (one-phrase or -sentence) answer to each question below:

   (a) How is an atomic orbital different from a molecular orbital?

   (b) Arrange the following isoelectronic species from smallest radius to largest: O^{2-}, Ne, Mg^{2+}.

   (c) Arrange the following atoms from least electronegative to most: O, F, S.

   (d) What are the criteria for aromaticity?

   (e) What hybridization and bond angle do we associate with each of the following shapes? linear, trigonal planar, and tetrahedral.

   (f) From VSEPR considerations, what shape and Cl–P–Cl bond angles do you predict for PCl_3?

3. **(a)** For each description below, draw the indicated number of different line/dot structural formulas with the molecular formula C_3H_4O:

   (1) one structure with conjugated π bonds;

   (2) three structures, each with a triple bond;

   (3) six cyclic structures.

   (b) What term describes the relationship of all the structures in (a)?

4. **(a)** Draw in all nonbonding electron pairs in the structure below and indicate the hybridization at each carbon:

$$\overset{\displaystyle O}{\underset{\displaystyle H}{H-C}}\overset{\displaystyle H}{\underset{\displaystyle H}{-C}}-Cl$$

(b) How many of each of the following are there in the above molecule? π bond; σ bonds; π* orbitals; and σ* orbitals.

(c) Based on VSEPR considerations, estimate the following bond angles: H–C=O, H–C–H, and C–C–Cl in the molecule represented by the structure in part (a).

(d) Redraw the molecule, showing bond angles and molecular shape as accurately as you can.

(e) Based on electronegativities (H, 2.2; C, 2.5; O, 3.4; Cl, 3.1), list all the bonds in the above molecule from least polar to most.

5. In certain compounds of third-period elements it is believed that a *p* orbital on one atom can overlap with a *d* orbital on a neighboring atom to form a *p–d* π bond. Draw both a bonding and an antibonding *p–d* π MO, being careful to show the relative phase of each lobe in the MOs.

6. In the structure below there is a resonance interaction between the nonbonding pair on nitrogen and the C–O π bond:

$$\overset{\displaystyle :O:}{\underset{\displaystyle H-C}{\|}}\overset{\displaystyle H}{\underset{\displaystyle \overset{..}{N}}{|}}-H$$

(a) Draw two additional resonance structures of the molecule, showing all nonzero formal charges. Use curved arrows to show how one structure is converted to the next.

(b) Based on your answer to (a), what is the expected hybridization at N? Is this molecule expected to be planar? Why?

(c) Based on VSEPR considerations, what is the magnitude of each bond angle in this molecule?

(d) Redraw the structure as viewed along the molecular plane, showing each *p* orbital involved in π bonding.

(e) Draw the π MOED for this molecule, labeling all MOs and showing where the π electrons reside.

(f) What is the expected C–N bond order?

7. **(a)** Draw six additional resonance structures of the molecule below. Show all formal charges and use curved arrows to show how one structure is converted to the next.

(b) Is the above molecule likely to be planar? Why?

8. **(a)** Which of the ions below is expected to exhibit aromatic properties?

(Remember that there is one hydrogen attached to each ring carbon.)

(b) Draw six additional resonance structures for the aromatic ion, showing how the charge can be delocalized to each carbon atom in the ring. Use curved arrows to show how each structure is converted to the next.

(c) Using MOEDs for each of the ions in part (a), explain why one is so stable, while the other is so unstable.

(d) Draw π_1 for the aromatic ion, showing the relative phase of each lobe of the MO.

Answers to Exam 1

1. (a) second; 2; 8

   (b) 14; IVA; $(1s)^2(2s)^2(2p)^6(3s)^2(3p)^2$; 2; 4; 10; +4; +14; 5

   (c) 3; 1; −1, 0, or +1; $\pm\frac{1}{2}$; energy and size; shape; orientation; spin

   (d) bonds; IA to IVA; VA to VIIIA

   (e) covalent; electronegativity; sharing

   (f) sequence; shape

   (g) 2; 2; 1

   (h) 0.6; $sp^{1.5}$; 0.6

   (i) σ; π

   (j) decreases; increases

2. (a) An atomic orbital is centered at a single nucleus, while a molecular orbital is spread over two or more nuclei.

   (b) $Mg^{2+} < Ne < O^{2-}$

   (c) $S < O < F$

   (d) An uninterrupted cyclic array of parallel p orbitals containing $4n + 2$ (where n is any integer) π-bonding electrons.

   (e) sp, 180°; sp^2, 120°; sp^3, 109.5°

   (f) $:PCl_3$ (with four groups around the phosphorus) is predicted to be approximately tetrahedral, with Cl–P–Cl bond angles of 109.5°.

3. (a) *(1)*

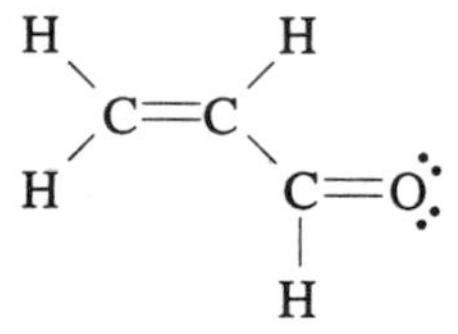

   (2)

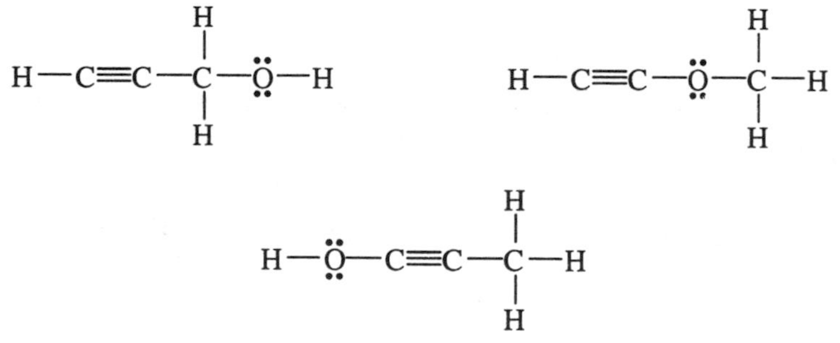

   (3)

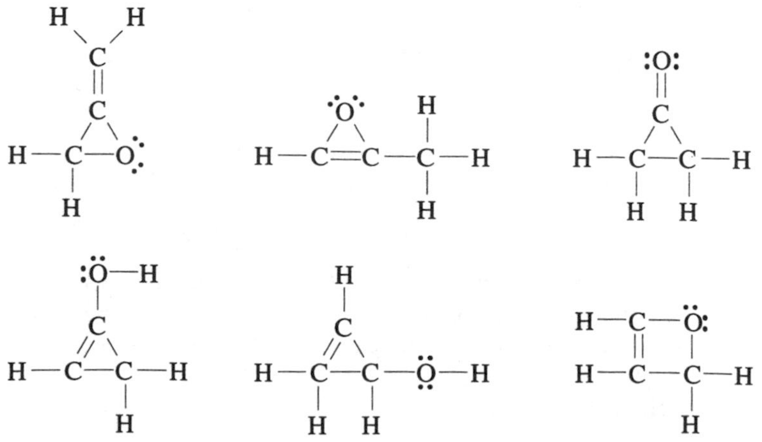

(b) structural isomers

4. (a)

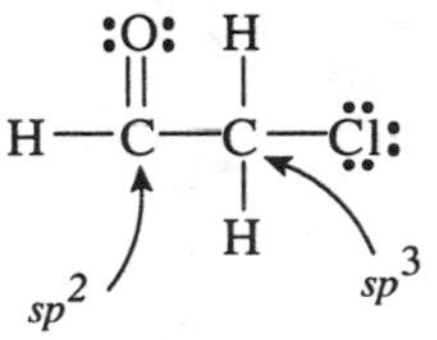

(b) π bonds, 1;   σ bonds, 6;   π* orbitals, 1;   σ* orbitals, 6
(c) 120°;   109.5°;   109.5°

(d)

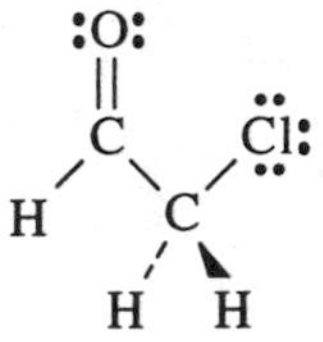

(e)

bond polarity	C–C	<	C–H	<	C–Cl	<	C–O
electronegativity difference	0		0.3		0.6		0.9

5.

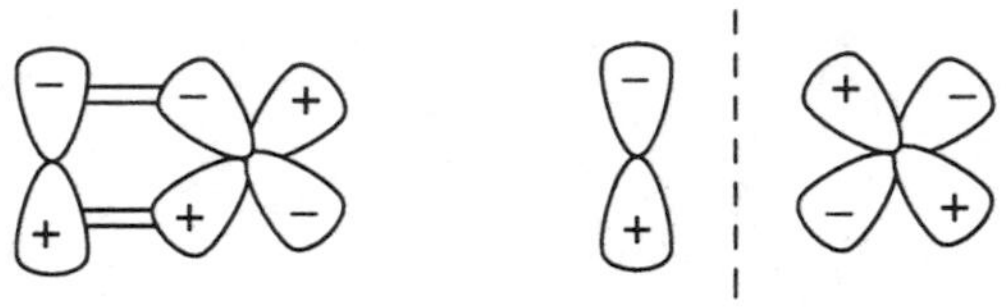

π   π*

6. (a)

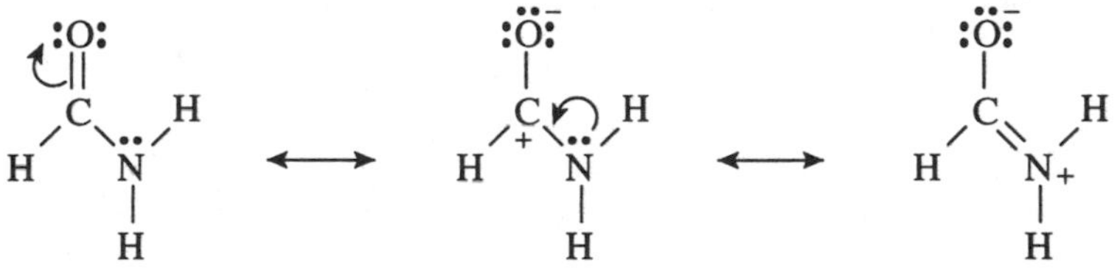

(b) The hybridization at N is *sp²*, in order to provide a *p* orbital for the nonbonding pair. Yes; both internal atoms are *sp²*-hybridized with parallel *p* orbitals.
(c) 120°

(d)

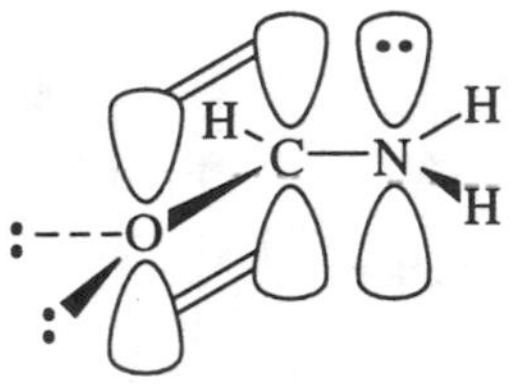

(e) The system in this molecule is isoelectronic with the allyl carbanion:

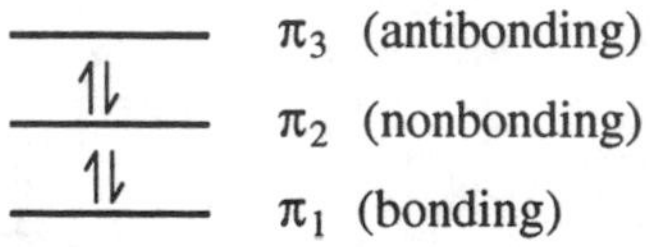

———	π₃ (antibonding)
↑↓ ———	π₂ (nonbonding)
↑↓ ———	π₁ (bonding)

(f) one σ bond plus a half π bond, for a total bond order of 1.5

7. (a)

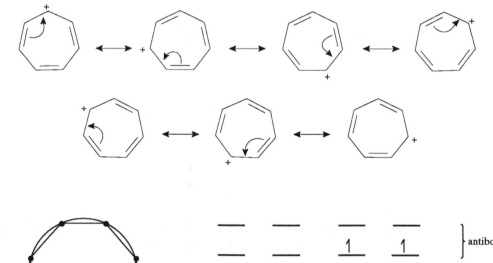

(b) Yes; each internal atom is sp^2-hybridized with a parallel p orbital.

8. (a) The carbocation, which has six $(4n + 2)$ π-bonding electrons, is aromatic. The carbanion has eight, so it isn't aromatic.

(b)

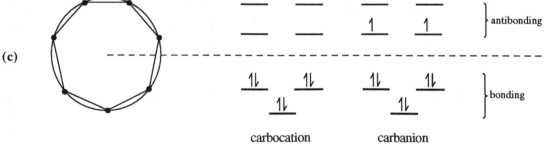

(c)

carbocation   carbanion

The carbocation has its six π electrons paired in bonding orbitals. By contrast, the carbanion has two unpaired π electrons in antibonding orbitals.

(d)

5 HYDROCARBONS: STRUCTURE AND NOMENCLATURE

THIS CHAPTER IS ABOUT

☑ **Alkanes**
☑ **An Introduction to the IUPAC Nomenclature System**
☑ **The Index of Hydrogen Deficiency**
☑ **Cycloalkanes**
☑ **Stereoisomers**
☑ **Alkenes**
☑ **Alkynes**
☑ **Polyenes, Cycloalkenes, Polyynes, Enynes, and Cycloalkynes**
☑ **Aromatic Hydrocarbons**

5-1. Alkanes

You probably know that organic chemistry owes its existence to carbon's unique properties—its valence of four and its capacity to form strong bonds to a wide variety of elements (see Section 2-1). It might surprise you to learn, however, that there are literally thousands of organic compounds that contain only carbon and hydrogen. Such compounds are called **hydrocarbons** (not to be confused with *carbohydrates*, to be discussed later). The reason there are so many different hydrocarbons is because carbon atoms also form strong bonds to each other, thereby making chains and rings of varying size and complexity.

We'll first discuss **alkanes**, those hydrocarbons that contain *no multiple bonds*. That is, alkane molecules consist only of carbon–carbon and carbon–hydrogen single (σ) bonds.

A. The first twenty alkanes

The simplest alkane, which has just one carbon, is **methane**, CH_4, seen previously in Chapters 2 and 3. Recall that we can represent the structure of methane in several different ways, depending on whether we want to focus attention mainly on the sequence of atoms and bonds (as in **1**), or on the shape of the molecule (as in **2**).

1 **2**

Always remember that, regardless of how we draw it, the structure of methane is tetrahedral. Indeed, in any alkane each carbon is nominally sp^3 hybridized, with bond angles of approximately 109.5°.

The next simplest alkane, with two carbons, is **ethane**, C_2H_6, which can be represented in several different ways:

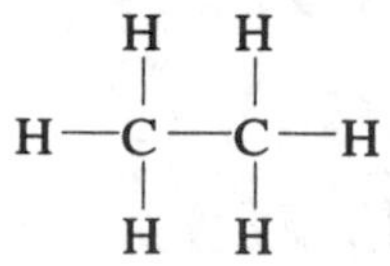

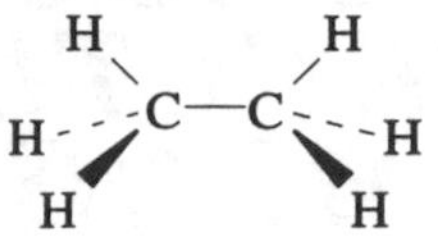

Sometimes the structure of ethane will be represented by the condensed structural formulas $H_3C–CH_3$ or CH_3CH_3, just to save space.

EXAMPLE 5-1   Draw three different representations each for the structures of (a) propane (C_3H_8) and (b) butane (C_4H_{10}).

Solution

(a) $H_3C–CH_2–CH_3$

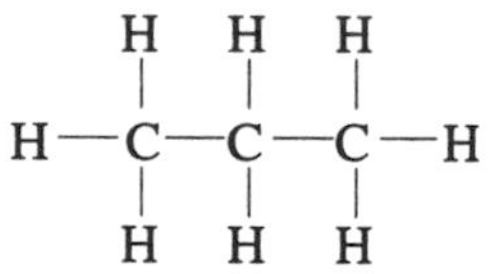

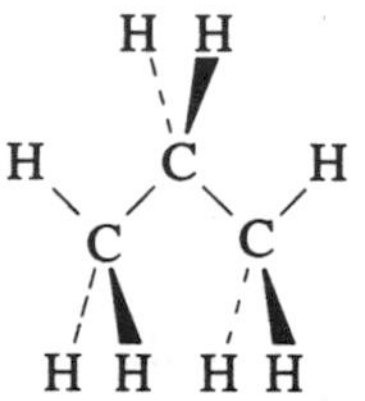

(b) $H_3C–CH_2–CH_2–CH_3$

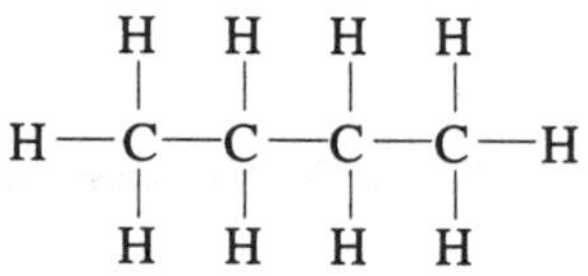

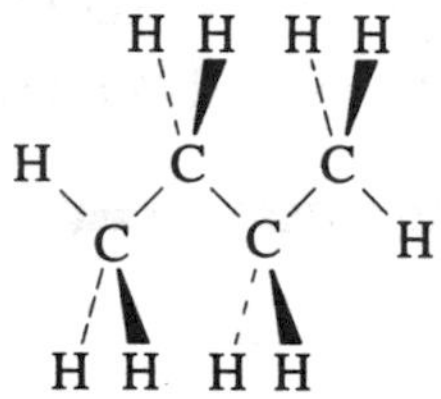

One way to view alkanes is as chains of singly bonded carbons, with enough hydrogens to satisfy the valences of the carbons in the chain. Table 5-1 gives the names and molecular formulas of the first twenty alkanes, i.e., the alkanes having from one to twenty carbons in the chain. Notice how each of these names combines the suffix *"ane"* (from alk*ane*) with a prefix that indicates the number of carbons (meth = 1; eth = 2; prop = 3; and so on).

Take a few minutes now to memorize the first ten names.

TABLE 5-1. The First Twenty Alkanes

Name	Molecular formula	Name	Molecular formula
Methane	CH_4	Undecane	$C_{11}H_{24}$
Ethane	C_2H_6	Dodecane	$C_{12}H_{26}$
Propane	C_3H_8	Tridecane	$C_{13}H_{28}$
Butane	C_4H_{10}	Tetradecane	$C_{14}H_{30}$
Pentane	C_5H_{12}	Pentadecane	$C_{15}H_{32}$
Hexane	C_6H_{14}	Hexadecane	$C_{16}H_{34}$
Heptane	C_7H_{16}	Heptadecane	$C_{17}H_{36}$
Octane	C_8H_{18}	Octadecane	$C_{18}H_{38}$
Nonane	C_9H_{20}	Nonadecane	$C_{19}H_{40}$
Decane	$C_{10}H_{22}$	Icosane	$C_{20}H_{42}$

B. The alkane molecular formula

Another way to view a typical alkane structure is as a chain of CH_2 groups (called **methylene** groups), with one additional hydrogen at each end. Thus, the "generic" structure of an alkane might be drawn like this:

$$H-(CH_2)_n-H$$

This structure corresponds to a molecular formula of C_nH_{2n+2}.

- Any compound that fits the molecular formula C_nH_{2n+2} is an alkane.

EXAMPLE 5-2 **(a)** A certain alkane molecule has 41 carbons. How many hydrogens does it have? **(b)** Another hydrocarbon has the molecular formula $C_{30}H_{60}$. Is it an alkane?

Solution

(a) For $n = 41$, $2n + 2 = 2(41) + 2 = 84$, so the molecule must have 84 hydrogens.

(b) No, $C_{30}H_{60}$ can't be an alkane because $2(30) + 2 \neq 60$.

C. Branched-chain alkanes: A complication

All the alkanes discussed so far in this section are said to have **unbranched** (or **linear**) carbon backbones. This means that all the carbons form a *single* chain. By contrast, consider the structure below:

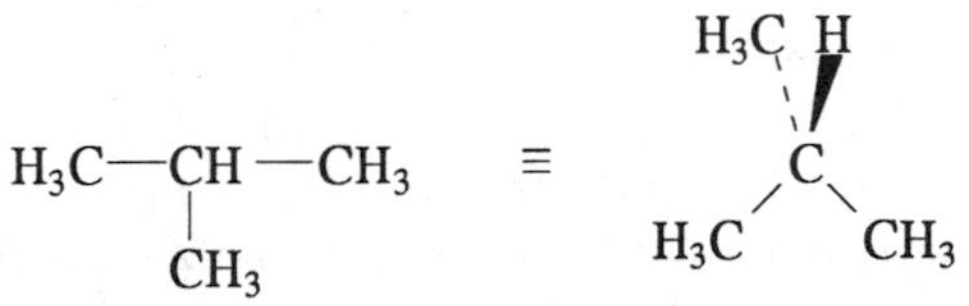

$$H_3C-\underset{\underset{CH_3}{|}}{CH}-CH_3 \equiv$$

(The structure on the right is partially condensed, showing only the shape around the middle carbon.) The molecular formula for this compound is C_4H_{10}, which should make it a butane, yet its structure is clearly different from the structure of butane shown in Example 5-1. While butane has an unbranched carbon chain, this structure has a **branched chain**: the fourth carbon is connected to the middle of the chain formed by the other three. Alternatively, this structure can be described as having three equivalent **methyl** (CH_3) groups attached to a **methine** (CH) group.

Regardless of how we choose to draw or describe the branched structure, it is clear that it is a structural isomer (Section 2-4) of butane. We are therefore left with a problem: How can we name this compound in order to distinguish it from butane? Originally, this problem was solved by adding the prefix "*n-*" (for "normal") to the *unbranched* structure, and "iso" (for "isomer") to the *branched* structure. Thus the two structures of C_4H_{10} would be called *n*-butane and isobutane:

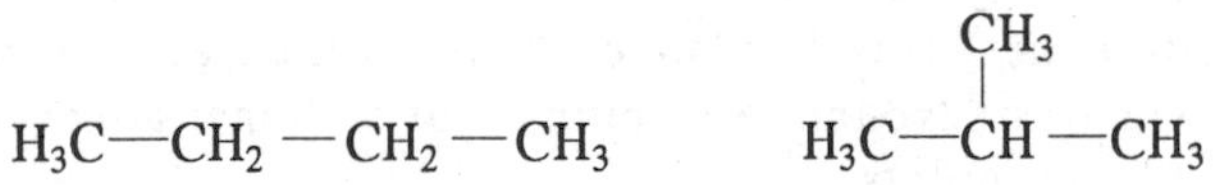

$$H_3C-CH_2-CH_2-CH_3 \qquad \qquad H_3C-\underset{\underset{CH_3}{|}}{CH}-CH_3$$

n-butane   isobutane

This prefix solution to the naming problem may seem like a good idea; but, as Example 5-3 shows, this approach to nomenclature leads to more problems than it solves.

EXAMPLE 5-3 Draw all possible structural isomers of C_5H_{12}. Indicate the number of methyl, methylene, and methine groups in each structure.

Solution Whenever you're asked to draw all the isomers of a molecule with a given structural formula, begin by writing only the carbon backbone, filling in the hydrogens later on. Write the first structure with all the carbons in an unbranched chain. Next, disconnect one carbon from the chain and connect it to the second carbon of the new (shorter) chain. Then move the disconnected carbon to the third carbon, and so on. Be careful not to "double count" any isomers: A carbon attached to the second carbon gives the same structure as a carbon attached to the second-to-last carbon. When you've drawn all the possible (different) methyl-substituted isomers, shorten the longest chain by another carbon, and begin writing all the possible isomers with two one-carbon branches or one two-carbon branch, and so on. Using this system, we can come up with just three different isomers of C_5H_{12}:

	3	**4**	**5**
Methyls	2	3	4
Methylenes	3	1	0
Methines	0	1	0

Structures:

$H_3C-CH_2-CH_2-CH_2-CH_3$ (**3**)

$H_3C-\overset{\overset{\displaystyle CH_3}{|}}{CH}-CH_2-CH_3$ (**4**)

$H_3C-\overset{\overset{\displaystyle CH_3}{|}}{\underset{\underset{\displaystyle CH_3}{|}}{C}}-CH_3$ (**5**)

If you came up with more than these three structures, look carefully at yours. Some of them must either be different representations of the above structures, or they are incorrectly drawn.

Structure **3** in Example 5-3 is *n*-pentane, but which of the other two isomers should receive the name isopentane? We can agree to use the "iso" prefix for the isomer with just one branch (as in structure **4**), but we'll need some other prefix for the doubly branched structure **5**. We could call **5** neopentane (which was its original name), but using yet another prefix suggests that we'd have to think up special prefixes for every structural isomer. And, unfortunately, the number of structural isomers of a given molecular formula increases approximately exponentially with the number of carbons: There are only 3 pentane isomers, but 75 decane isomers, 366,319 icosane isomers, and so on. It would be most impractical to have so many prefixes to memorize. What we need is a nomenclature system that has as few new terms to memorize as possible. And, sure enough, there is such a system.

5-2. An Introduction to the IUPAC Nomenclature System

A. IUPAC nomenclature rules

There were many problems and ambiguities with what came to be called the *common* (or *trivial*) *names* of organic compounds. So, the **International Union of Pure and Applied Chemistry (IUPAC)** charged a committee with the task of establishing a systematic way to name organic compounds. The resulting system, which has been refined over the years, now fills a good-sized book. However, we'll focus only on the most important aspects of these nomenclature rules.

Rule 1: Identify the *longest* carbon chain, the one with the greatest number of carbons, neglecting branches. This determines the root (or base) name of the compound (Table 5-1). If two chains have the same number of carbons, select the one with the greater number of branch points.

Rule 2: Number the carbons in this longest chain, starting at the end nearest the first branch point.

Rule 3: Identify the location and type of each **substituent** (branching group). Add this information (with the substituent names in alphabetical order) as a prefix to the root name.

B. Naming methyl-substituted alkanes

Let's apply these rules to the structures in Example 5-3. Structure **3** has an unbranched five-carbon chain, so it's simply *pentane*—the prefix *n*- is no longer needed because the name "pentane" tells us that all five carbons are in the same chain. Structure **4** has a *four*-carbon chain, so it is named as a derivative of *butane*, not pentane. Numbering the carbons from the end nearest the branch point finds the methyl group attached to carbon 2.

$$H_3C-\underset{2}{\overset{\overset{\displaystyle CH_3}{|}}{CH}}-\underset{3}{CH_2}-\underset{4}{CH_3}$$
$$\underset{1}{}$$

So, the complete name of this structure is 2-methylbutane.

note: There is no space between the substituent name and the root name, and a hyphen is used to separate a numerical locator from the substituent name.

Structure **5** has a *three*-carbon chain, making it a derivative of *propane*.

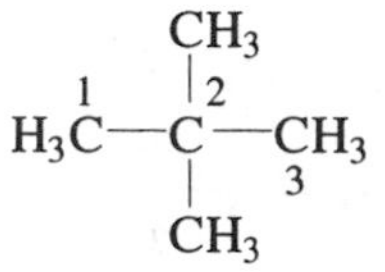

$$H_3C \overset{1}{\underset{\underset{\displaystyle CH_3}{|}}{C}} \overset{\overset{\displaystyle CH_3}{|}}{\underset{2}{C}} \overset{}{\underset{3}{CH_3}}$$

The two methyl groups are connected to carbon number 2, so you might be tempted to name it 2-methyl-2-methylpropane. Instead, we group like substituent names using the prefixes di (two), tri (three), and tetra (four). However, we retain all the numerical locators of substituents so that there is no question where each of the substituents is attached. Thus structure **5** is named 2,2-dimethylpropane. (Notice that when two or more numerical locators are used in sequence, they are separated by commas.)

EXAMPLE 5-4 Give the complete IUPAC name for the structure below. Be careful to identify the *longest* carbon chain.

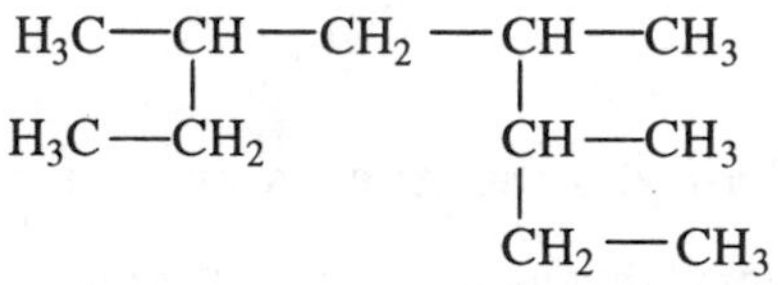

$$\begin{array}{l} H_3C - CH - CH_2 - CH - CH_3 \\ \qquad\quad | \qquad\qquad\quad | \\ \quad\ H_3C - CH_2 \qquad\ CH - CH_3 \\ \qquad\qquad\qquad\qquad\ | \\ \qquad\qquad\qquad\ CH_2 - CH_3 \end{array}$$

Solution Don't be misled by the twisted way the structure is drawn. The longest chain in the structure has *eight* carbons, so the root name is octane. Here are two possible ways we might number the longest chain. Which is better?

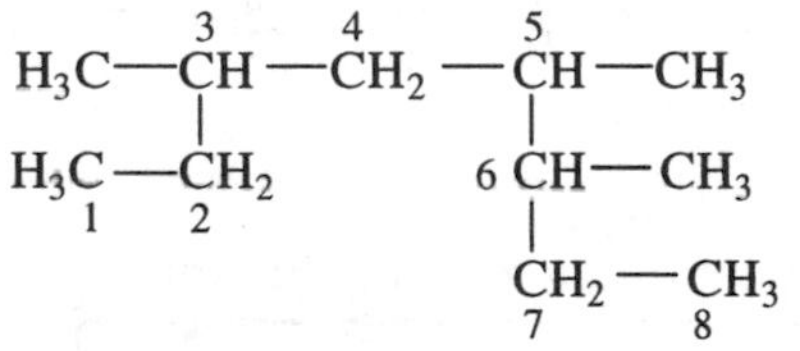

$$\begin{array}{l} \overset{3}{H_3C} - \overset{4}{CH} - \overset{5}{CH_2} - \overset{5}{CH} - CH_3 \\ \qquad\qquad | \qquad\qquad\qquad | \\ \quad\ H_3C - CH_2 \qquad 6\ CH - CH_3 \\ \ \ \ 1 \quad\ \ 2 \qquad\qquad\qquad | \\ \qquad\qquad\qquad\qquad CH_2 - CH_3 \\ \qquad\qquad\qquad\qquad\ 7 \qquad 8 \end{array} \quad \text{or} \quad \begin{array}{l} \overset{6}{H_3C} - \overset{5}{CH} - \overset{5}{CH_2} - \overset{4}{CH} - CH_3 \\ \qquad\qquad | \qquad\qquad\qquad | \\ \quad\ H_3C - CH_2 \qquad 3\ CH - CH_3 \\ \ \ \ 8 \quad\ \ 7 \qquad\qquad\qquad | \\ \qquad\qquad\qquad\qquad CH_2 - CH_3 \\ \qquad\qquad\qquad\qquad\ 2 \qquad 1 \end{array}$$

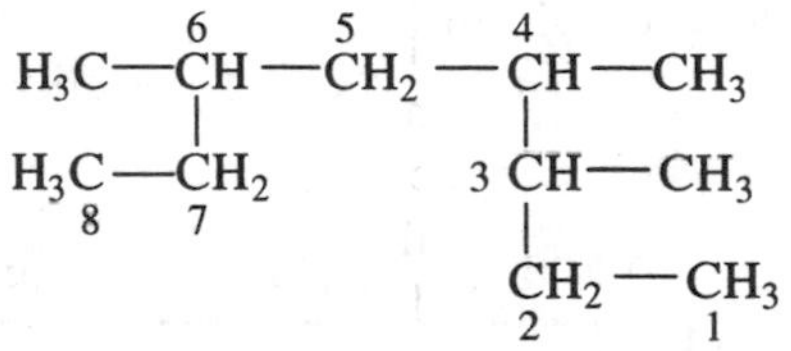

Both numbering sequences have a methyl branch at carbon 3. But the second branch occurs at carbon 5 in the left sequence, and on carbon 4 in the right sequence. We pick the latter sequence because it provides lower numbers for the numerical locators. Thus, the name of the compound is 3,4,6-trimethyloctane, not 3,5,6-trimethyloctane.

C. Naming other alkyl-substituted alkanes

Of course, not all substituent groups are methyls. For example, a CH_3CH_2 group branching off the longest chain is called an **ethyl** group, $CH_3CH_2CH_2$ is a **propyl** group, and so on. In each case, the "ane" suffix of the alkane root name (Table 5-1) is replaced by "yl" when labeling the derived substituent group. Such groups are collectively referred to as **alkyl groups**, or **alkyl substituents**. Notice that an alkyl group has one less hydrogen than the corresponding alkane.

But what if the substituent itself is branched, as in the structure below (where the zigzag line represents the backbone of an unspecified molecule):

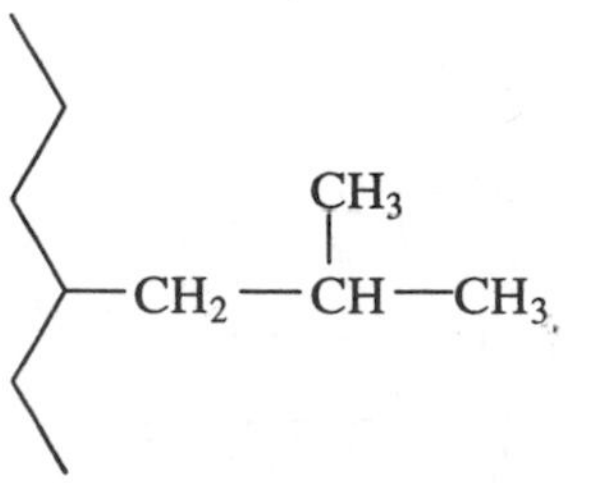

$$\begin{array}{l} \qquad\qquad\qquad\ CH_3 \\ \qquad\qquad\qquad\ | \\ - CH_2 - CH - CH_3 \end{array}$$

In this case the longest chain of the alkyl group is identified (and numbering begun) *from the point of attachment:*

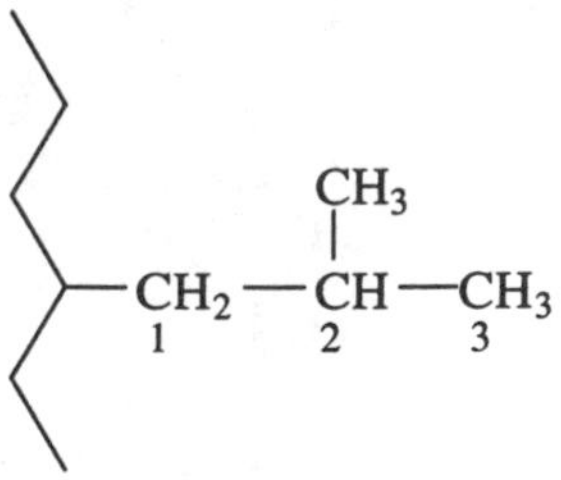

$$CH_2 \overset{CH_3}{\underset{1 \quad 2 \quad 3}{-CH-CH_3}}$$

So, the name of this *substituent group* is 2-methylpropyl. When adding this complex substituent name as a prefix, we set it off with parentheses to avoid confusion with the root name. For example, the complete name for the structure below is 5-(2-methylpropyl)nonane.

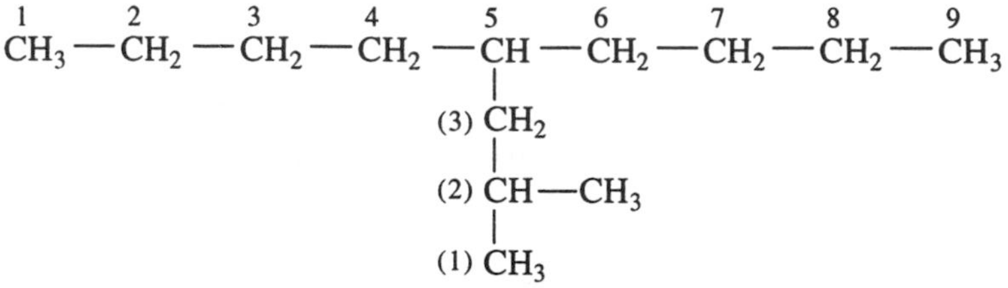

$$\overset{1}{CH_3}-\overset{2}{CH_2}-\overset{3}{CH_2}-\overset{4}{CH_2}-\overset{5}{CH}-\overset{6}{CH_2}-\overset{7}{CH_2}-\overset{8}{CH_2}-\overset{9}{CH_3}$$

$$(3)\ CH_2$$
$$(2)\ CH-CH_3$$
$$(1)\ CH_3$$

D. Nonsystematic names of alkyl substituents and carbon substitution patterns

There are a few alkyl substituents that have "common" or nonsystematic names, which can be—and often are—used in place of their systematic equivalents. These names are listed in Table 5-2 and are worth memorizing. Notice that, when using one of these common substituent names, you need not include the parentheses. Thus, for example, you might call 5-(2-methypropyl)nonane by its nonsystematic name 5-isobutylnonane. Notice also the prefixes *sec* and *tert*, which stand for "secondary" and "tertiary," respectively; these prefixes are used to describe the substitution pattern of the first carbon of the alkyl substituent.

TABLE 5-2. Names of Commonly Encountered Alkyl Substituents

Structure	Systematic name	Common substituent name
$-\overset{CH_3}{\underset{}{CH}}-CH_3$	(1-Methylethyl)	Isopropyl
$-\overset{CH_3}{\underset{}{CH}}-CH_2CH_3$	(1-Methylpropyl)	*sec*-Butyl
$-CH_2-\overset{CH_3}{\underset{}{CH}}-CH_3$	(2-Methylpropyl)	Isobutyl
$-\overset{CH_3}{\underset{CH_3}{C}}-CH_3$	(1,1-Dimethylethyl)	*tert*-Butyl

The terms "secondary" and "tertiary" can also be used to describe the carbons in the molecule. A **secondary (2°) carbon** is directly attached to two other carbons, while a **tertiary (3°) carbon** has three other carbons directly attached to it. You'll also see the terms **primary (1°) carbon** (which has one carbon directly attached to it) and **quaternary (4°) carbon** (which has four carbons directly attached to it). And, just to add to the confusion, a primary *hydrogen* is one attached to a primary carbon, a secondary hydrogen is one attached to a secondary carbon, and so on.

EXAMPLE 5-5 **(a)** Identify the 1°, 2°, 3°, and 4° carbons and hydrogens in the structure below. **(b)** Provide the complete IUPAC name for the compound.

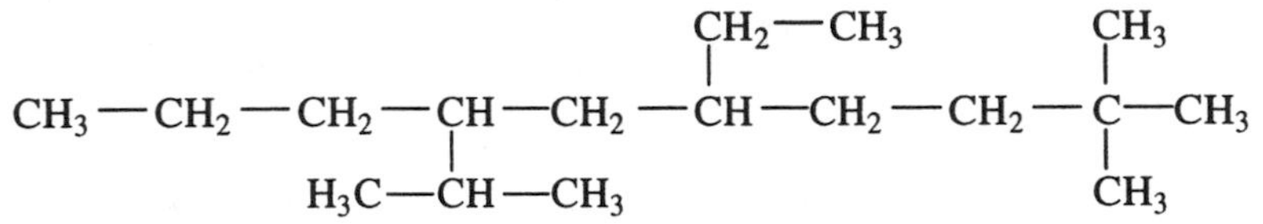

Solution
(a) 1° carbons and hydrogens:

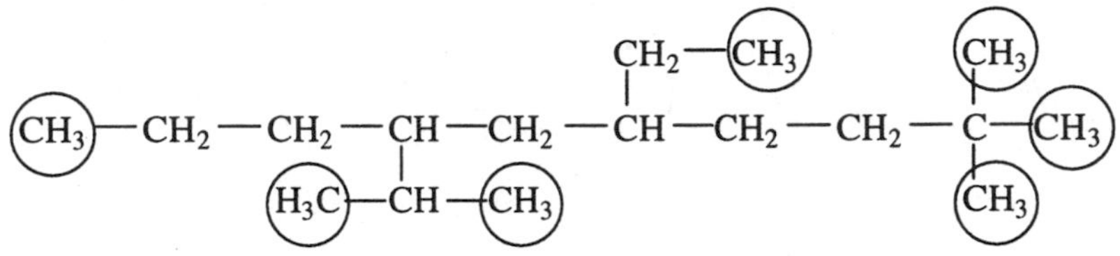

2° carbons and hydrogens:

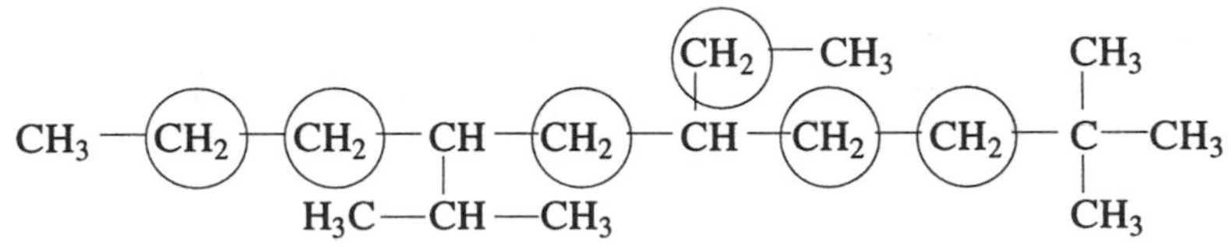

3° carbons and hydrogens:

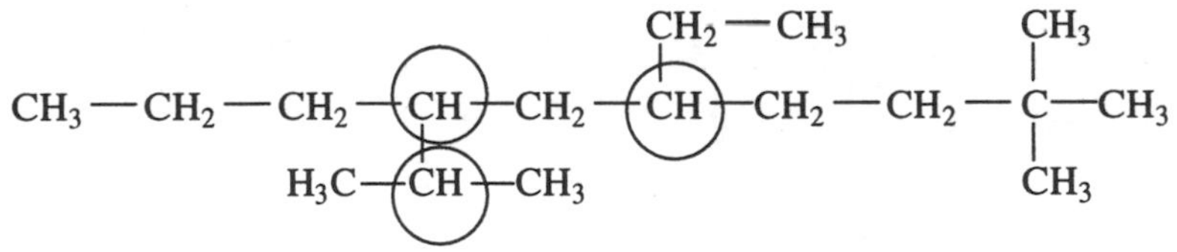

4° carbons:

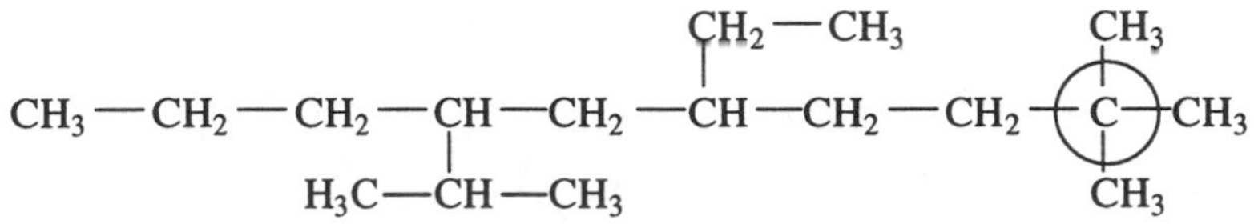

> *note:* There is no such thing as a "quaternary hydrogen," because a quaternary carbon already has its valence satisfied with four carbons, leaving no room for any hydrogens to be attached.

(b) The name of the compound is 5-ethyl-7-isopropyl-2,2-dimethyldecane. Notice that *ethyl* appears before *dimethyl* because *e* takes alphabetical precedence over *m*. Prefixes such as di, tri, and tetra are *not* considered when placing group names in alphabetical order, but the prefixes "iso," "sec," and "tert" *are* included.

E. Representing alkanes: Skeletal structures

You may remember from our discussion of aromatic rings (Section 4-7) that we have a shorthand way to represent organic structures. With this method, only the carbon–carbon bonds are shown (as lines), hence the name "skeletal" structures. No carbons or hydrogens are drawn in unless we want to draw particular attention to them. Rather, each end and intersection of each line represents a carbon, and we are left to add in the missing hydrogens mentally. In the case of alkanes, we draw the longest chain as a zigzag line of the appropriate length, then add lines for the substituent groups. Thus the structure in Example 5-4 would be drawn like this:

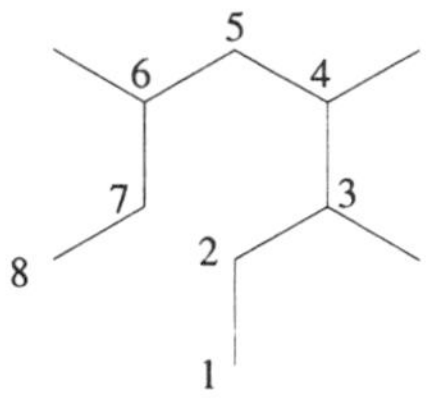

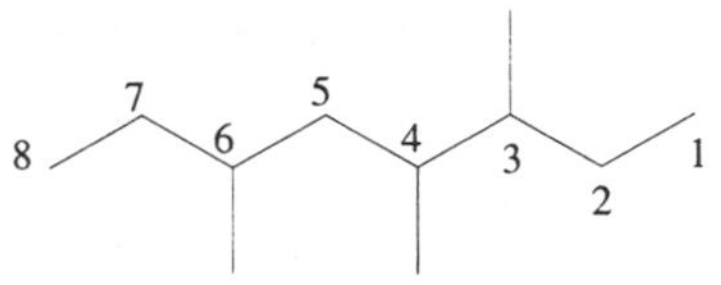

Similarly, the structure in Example 5-5 would appear as shown below.

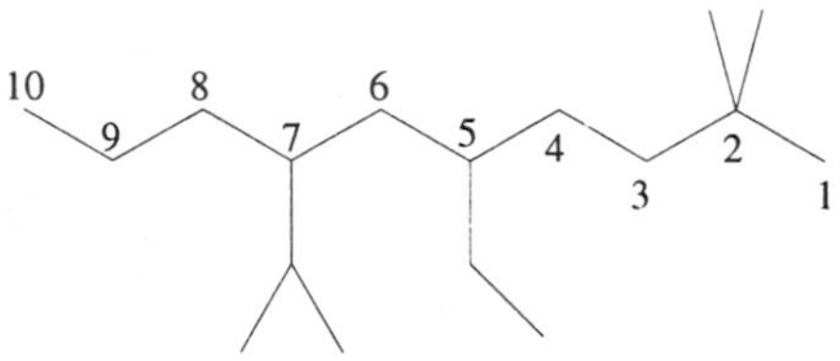

EXAMPLE 5-6 Use the skeletal method to draw structures for all five isomers of C_6H_{14}. Provide the IUPAC name for each structure. Indicate the number of 1°, 2°, 3°, and 4° carbons and hydrogens in each structure.

Solution

Structure	Name	1°		2°		3°		4°
		C	H	C	H	C	H	C
	Hexane	2	6	4	8	0	0	0
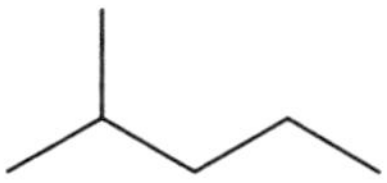	2-Methylpentane	3	9	2	4	1	1	0
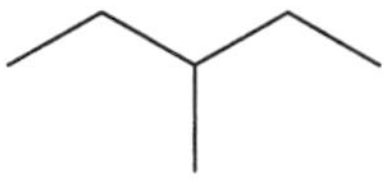	3-Methylpentane	3	9	2	4	1	1	0
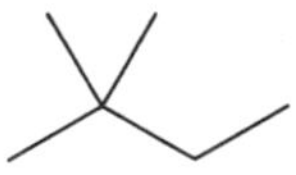	2,2-Dimethylbutane	4	12	1	2	0	0	1
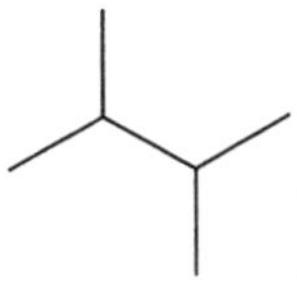	2,3-Dimethylbutane	4	12	0	0	2	2	0

EXAMPLE 5-7 All of the names below are somehow incorrect. In each case draw the skeletal structure, then propose the correct name of the compound.

(a) *tert*-butylethane (b) 4-ethylpentane (c) 3-(1-ethylpropyl)pentane

Solution

(a)  2,2-dimethylbutane (Butane is the root, not ethane)

(b) 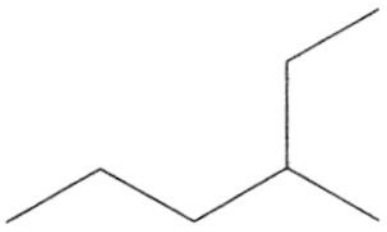 3-methylhexane (The longest chain has six carbons, not five)

(c) 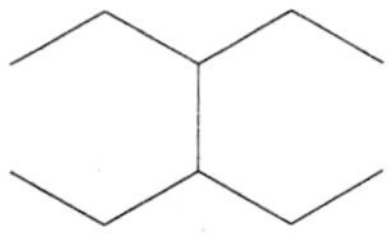 3,4-diethylhexane (The longest chain has six carbons, not five)

5-3. The Index of Hydrogen Deficiency

From Section 5-1 we know that all alkanes have the molecular formula C_nH_{2n+2}. Such a compound contains the *maximum number of hydrogens possible* for a given number of carbons, and is therefore said to be **saturated**. Still, it is possible for a (nonalkane) hydrocarbon molecule with *n* carbons to have *fewer* than $2n + 2$ hydrogens. There are two structural features in a molecule that reduce the number of hydrogens needed to satisfy the valence of each carbon: Rings (i.e., cyclic structures) and π bonds (Section 3-5).

Consider the molecular formula C_3H_6. We know this molecule cannot be an alkane because it has two hydrogens fewer than the $2n + 2$ (eight) required. How many structural isomers with this molecular formula can you write? There are just two, as shown below:

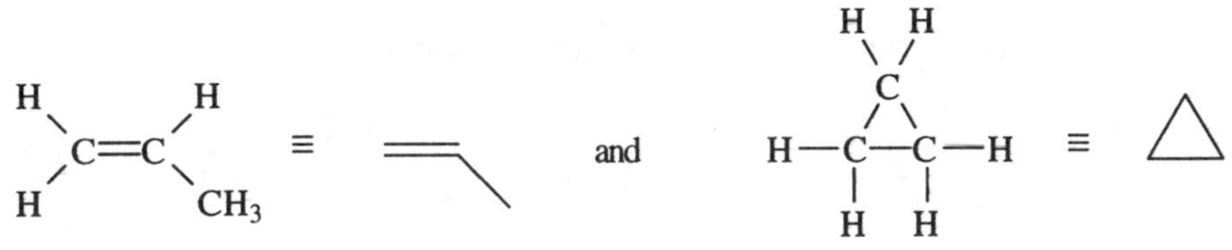

The first structure has one π bond, the second has one (three-membered) ring. (Remember from Section 3-6 that some people regard a double bond as a "two-membered" ring. But we'll adopt the usual convention that a ring must have at least three members.) A little thought should convince you that each bond or ring in a molecule reduces by *two* the number of hydrogens required to satisfy the valence of each carbon. We can express this mathematically by defining the **index of hydrogen deficiency** *I* (sometimes called the **index of unsaturation**) by the equation

INDEX OF HYDROGEN DEFICIENCY IN HYDROCARBONS

$$I = \frac{2C+2-H}{2} = \text{rings} + \pi \text{ bonds} \tag{5-1}$$

where *C* and *H* are the numbers of carbons and hydrogens, respectively, in the molecular formula. It's easy to remember this simple equation once you see where it comes from. The $2C + 2$ term is the number of hydrogens that are needed for saturation, while *H* is the number of hydrogens actually present. The difference between these is the number of "missing" hydrogens which, when divided by 2, gives the number of rings plus π bonds.

EXAMPLE 5-8 (a) Calculate the value of *I* for the molecular formula C_4H_6. (b) What combinations of rings and π bonds are possible in structures with this molecular formula? (c) Draw as many examples as you can for each of the combinations in part (b).

Solution

(a) Use Eq. (5-1), with *C* = 4 and *H* = 6:

$$I = \frac{2C+2-H}{2} = \frac{2(4)+2-6}{2} = 2$$

(b) There may be two π bonds, or two rings, or one ring and one π bond.

(c) Two π bonds require either two double bonds or one triple bond:

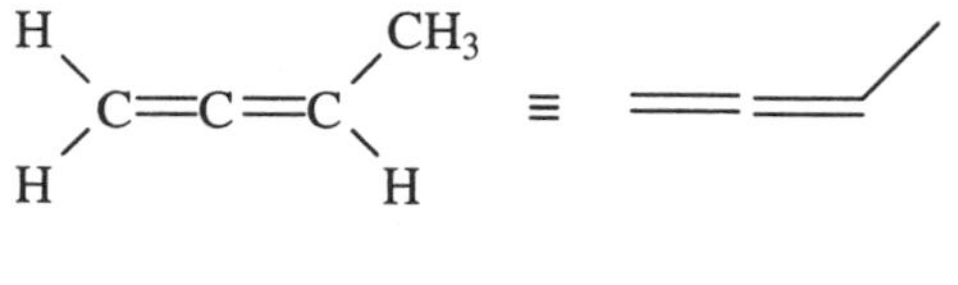

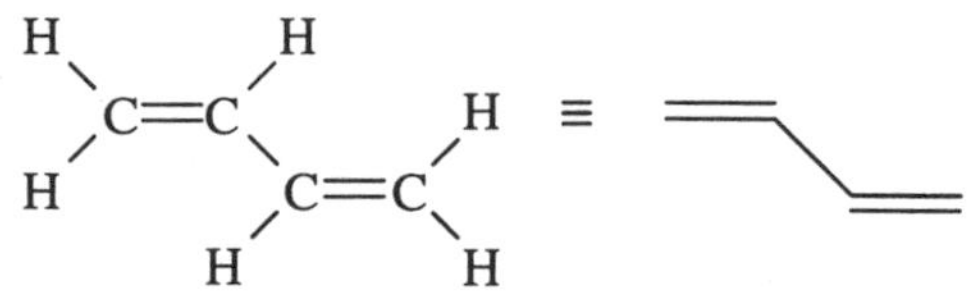

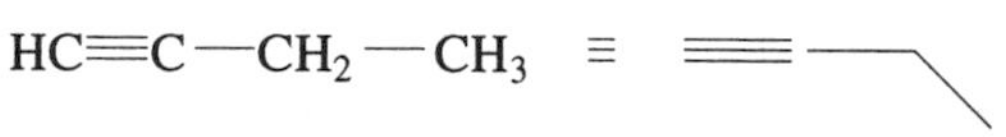

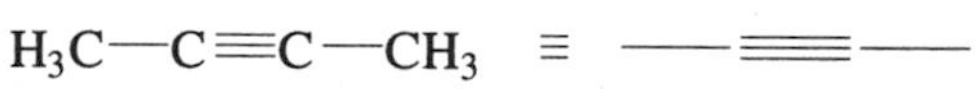

$$HC\equiv C-CH_2-CH_3 \equiv \qquad H_3C-C\equiv C-CH_3 \equiv$$

There is one possible structure that has two rings:

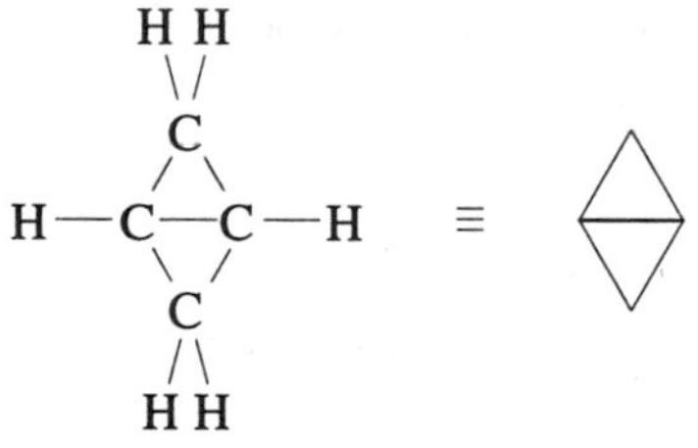

And there are four structures with one ring and one π bond:

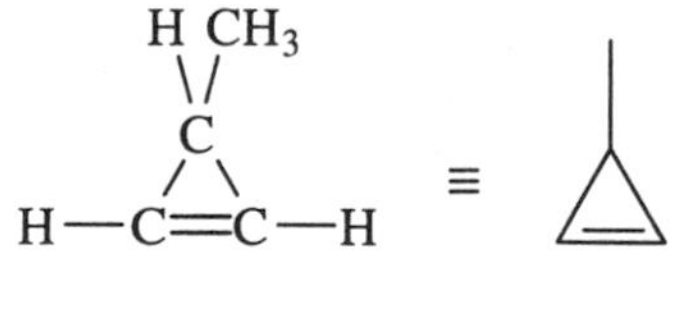

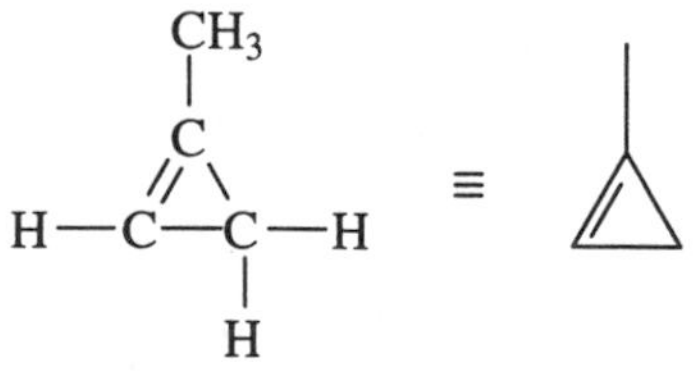

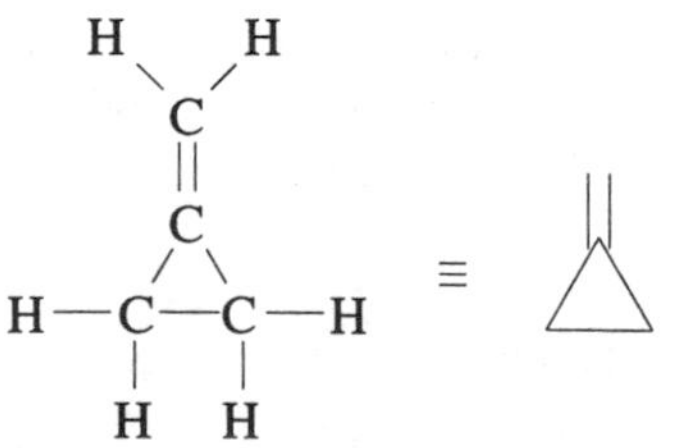

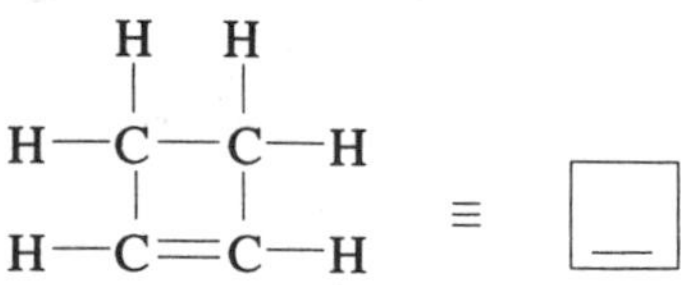

5-4. Cycloalkanes

Hydrocarbon molecules with a ring but no π bonds are called **cycloalkanes** and have the molecular formula C_nH_{2n}. Molecules that have *no* rings are said to be **acyclic**. The IUPAC name for an cycloalkane is simply the name of the *alkane* with the same number of carbons (Table 5-1), with the prefix "cyclo." Thus the simplest cycloalkane is cyclopropane (C_3H_6), the structure of which appears at the beginning of Section 5-3. Cyclobutane (C_4H_8) has a four-membered ring, cyclopentane has a five-membered ring, and so on.

In skeletal notation, cycloalkanes are usually drawn as regular polygons with the appropriate number of corners (one for each carbon).

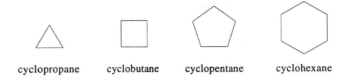

cyclopropane   cyclobutane   cyclopentane   cyclohexane

But it is important to remember that, except for cyclopropane, these rings of carbon atoms are *not* completely planar. For example, the cyclohexane molecule actually prefers a lounge-like structure called the **chair form** (Section 7-5), which looks like this:

A. Naming substituted cycloalkanes

To name a cycloalkane with only one substituent, simply add the substituent name as a prefix to the cycloalkane root, as shown below:

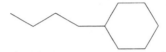

butylcyclohexane

It's not necessary to number the location of a single substituent in a cycloalkane because all ring positions are equivalent.

When there are two or more substituents, the one with the highest alphabetical priority is considered to be connected to the number-1 carbon of the ring. Counting around the ring is continued by the path that leads to the lowest sum of locator numbers. For example, in the structure below, the butyl group (which starts with *b*) defines the number-1 carbon, C-1. We continue counting toward the next nearest substituent, the methyl group at C-2, followed by the ethyl group at C-4. However, in the name, ethyl appears before methyl (*e* before *m*). Thus we have 1-butyl-4-ethyl-2-methylcyclohexane:

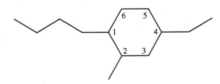

1-butyl-4-ethyl-2-methylcyclohexane

You might be wondering how to name a structure that has a ring and a chain with the same number of carbons. The rule is this:

- The ring determines the root name unless the chain has more carbons. In the latter case, the molecule is named as a derivative of the alkane, with a **cycloalkyl** substituent.

For example, the structure below is 1-cyclobutylpentane, not pentylcyclobutane:

So, Rule 1 (Section 5-2) should be amended to read: "Identify the longest carbon chain or the largest ring...."

EXAMPLE 5-9 Provide the complete IUPAC name for each structure below. Give the molecular formula for each.

(a) 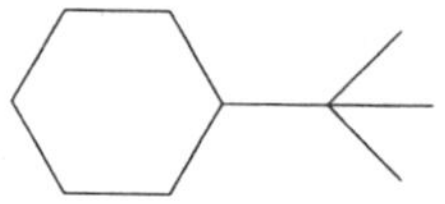   (b) 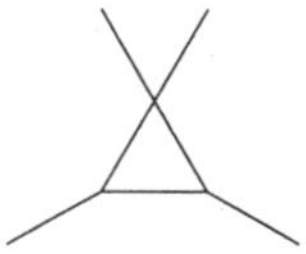   (c)

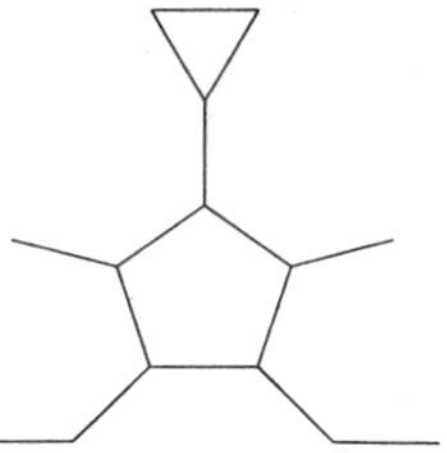

Solution

(a) *tert*-Butylcyclohexane, or (1,1-dimethylethyl)cyclohexane, $C_{10}H_{20}$
(b) 1,1,2,3-Tetramethylcyclopropane, C_7H_{14}
(c) 1-Cyclopropyl-3,4-diethyl-2,5-dimethylcyclopentane, $C_{14}H_{26}$

B. Polycyclic molecules

A molecule that has two rings sharing a single atom is said to be **spirocyclic**. One example of such a structure, spiropentane, is shown below:

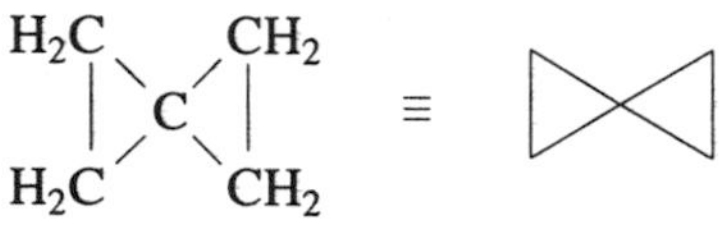

$$H_2C \diagdown \quad \diagup CH_2$$
$$\diagup C \diagdown$$
$$H_2C \diagup \quad \diagdown CH_2 \quad \equiv$$

spiropentane, C_5H_8

A molecule with two rings that share at least *two* atoms is called **bicyclic**. Structure **6** is an example of a bicyclic molecule:

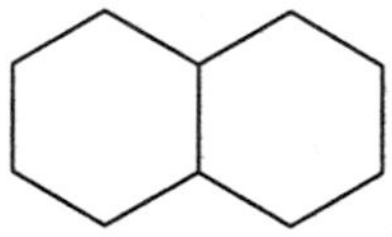

decalin, $C_{10}H_{18}$

6

Decalin (**6**) has only ten carbons, yet it has two "fused" six-membered rings that share two carbons.

> *note:* A molecule whose rings do not share two or more atoms is not considered to be bicyclic. Notice, for instance, that the structure in Example 5-9c is neither spirocyclic nor bicyclic because no atoms are shared by both rings.

C. Finding the number of rings in polycyclic structures

Although you're not expected to be able to give complete IUPAC names for spirocyclic and bicyclic structures, it is useful to be able to decide how many rings there are in certain polycylic molecules. One test you can apply is to find the *minimum* number of bond cleavages necessary to render the molecule *acyclic* (though not necessarily unbranched). Referring back to structure **6**, you can see how two cleavages—the cleavage of the central bond, along with the cleavage of any other ring bond—provide an acyclic structure:

$\longrightarrow$

6

Since it takes *two* cleavages to make an acyclic molecule of decalin, decalin must be *bicyclic*.

Now consider structures **7** and **8** and decide whether they are bicyclic, tricyclic, tetracyclic, or what.

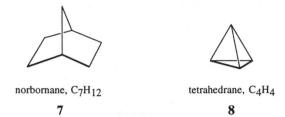

norbornane, C_7H_{12}

7

tetrahedrane, C_4H_4

8

Structure **7** can be looked on as two five-membered rings that share three atoms. That it is bicyclic can be seen from the two cleavages shown below:

Although you can no doubt find four rings in structure **8**, it takes only three cleavages to "open" it up. Thus it is actually tricyclic:

Here is another way you can quickly tell the number of rings in a complicated polycyclic structure:

- Calculate the value of I (being careful to find all the hydrogens if only a stick structure is given), then subtract the number of π bonds present, if any.

Thus for structure **8**, $I = [2(4) + 2 - 4]/2 = 3$. Since there are no π bonds, it is tricyclic.

EXAMPLE 5-10 Use either the bond-cleavage approach or the I value to determine the number of rings in each structure below:

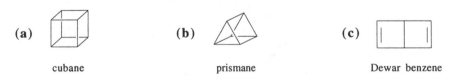

(a) cubane

(b) prismane

(c) Dewar benzene

Solution

(a)

5 cleavages: pentacyclic

$$I = \frac{2C + 2 - H}{2} = \frac{2(8) + 2 - 8}{2} = 5$$

And since there are no π bonds, cubane is pentacyclic.

(b)

4 cleavages: tetracyclic

$$I = \frac{2(6) + 2 - 6}{2} = 4$$

And since there are no π bonds, prismane is tetracyclic.

(c) 2 cleavages: bicyclic

$$I = \frac{2(6) + 2 - 6}{2} = 4$$

And since there are two π bonds, there must be 4 − 2 = 2 rings. So Dewar benzene is bicyclic.

note: The molecular formula for cubane is C_8H_8 and that for prismane is C_6H_6. Notice that these *poly*cyclic structures do not conform to the C_nH_{2n} formula for *mono*cyclic cycloalkanes; instead, each carbon in these structures is attached to three other carbons, leaving only one hydrogen per carbon. Notice also that Dewar benzene has a molecular formula of C_6H_6. In this case, however, there are two π bonds, as well as two rings, in the structure; so Dewar benzene is *not* a cycloalkane at all.

5-5. Stereoisomers

Stereochemistry describes the three-dimensional relationships of atoms in a molecule. This is a topic that causes difficulty for almost every student of organic chemistry sooner or later, but it is so important that it comes up again and again in the study of organic compounds. For now, we'll just scratch the surface. And here is a little bit of advice: You'll find it most helpful—even essential—to construct molecular models of all the structures we discuss.

A. Cis–trans isomerism

Suppose you were asked to give IUPAC names for structures **9** and **10** below:

On the basis of what we learned in the previous section, we can name both of these structures 1,2-dimethylcyclopropane. Yet, are they really the *same* structure, or are they somehow different? Careful inspection (or better yet, models) should convince you that they are indeed *different*. In structure **9** the two methyl groups lie on the *same* side of the three-membered ring plane, while in **10** they lie on *opposite* sides of the ring. Most importantly, the only way to interconvert **9** and **10** is by breaking and remaking bonds. However, even though these two structures *are* different, they are *not* structural isomers (Section 2-4) because they both have the same *sequence* of atoms and bonds. They differ only in the three-dimensional **configuration** (fixed relative arrangement) of the atoms. Therefore, we call these structures **stereoisomers**.

- Stereoisomers have the same sequence (connectivity) of atoms and bonds, but differ in the three-dimensional arrangement of the atoms.

In order to provide complete names for structures **9** and **10** we need to add a prefix that describes the spatial relationship between the methyl groups. So, the terms **cis** (identical groups on the same side of the ring) and **trans** (identical groups on opposite sides of the ring) were brought into use. Thus, the complete name for **9** is *cis*-1,2-dimethylcyclopropane, while **10** is called *trans*-1,2-dimethylcyclopropane. Notice that, in **9**, not only are the methyl groups cis to each other, but so are the methine hydrogens (the hydrogens attached to the same ring carbons as the methyls). And in **10**, the methine hydrogens are trans to each other. **Cis–trans isomers** (sometimes called **geometric isomers**) are but one type of stereoisomer.

EXAMPLE 5-11   After redrawing the structures below in side view, provide a *complete* name for each. Be sure to add stereochemical designators, as appropriate.

(a) 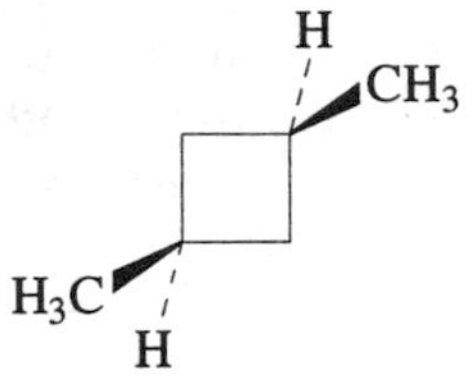   (b) 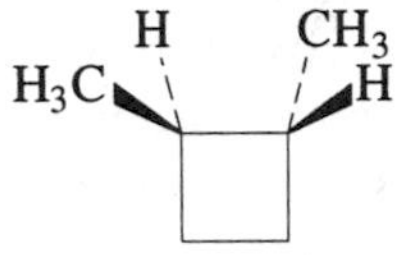   (c)

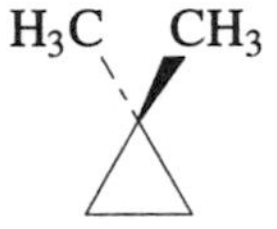

Solution   (Not all hydrogens are drawn in.)

(a) 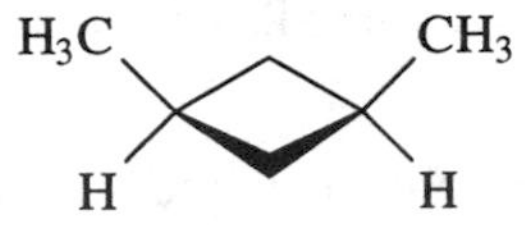   *cis*-1,3-dimethylcyclobutane

(b) 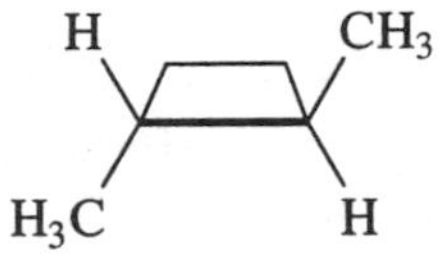   *trans*-1,2-dimethylcyclobutane

(c) 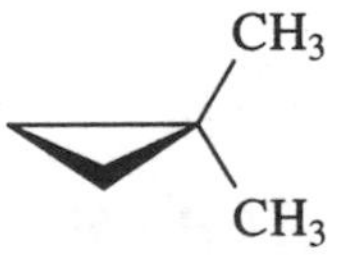   1,1-dimethylcyclopropane

note: You might have been tempted to call the last one *trans*-1,1-dimethylcyclopropane. But since both methyl groups are attached to the *same* carbon, they must be on opposite sides of the ring; i.e., there is no stereoisomer of this structure. So, the prefix trans is superfluous.

B. The E–Z system

Now consider structures **11** and **12**, which are both 1-ethyl-1-methyl-2-propylcyclobutanes:

top view

side view

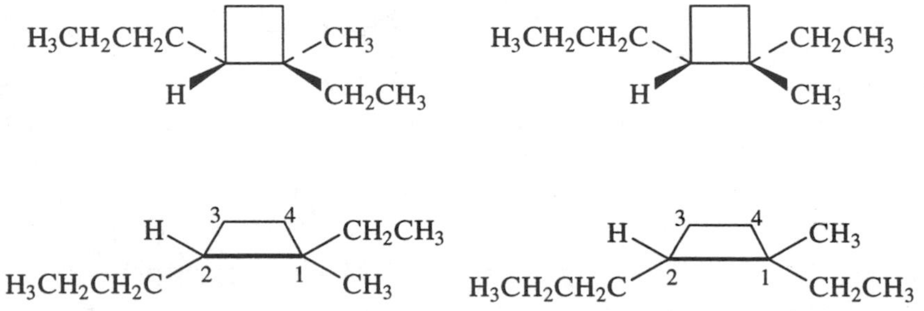

11   **12**

A problem is immediately apparent. Although **11** and **12** bear the same relationship to each other as did **9** and **10**, it's not at all clear which isomer should be called cis and which trans, because there are no identical groups whose arrangements we can compare.

To rectify this ambiguity, a system of group priorities was developed by three well-known organic chemists named **Cahn, Ingold**, and **Prelog**. This **CIP** system uses the stereochemical designators **E** and **Z**, where E stands for *entgegen* (German for "opposite," or "trans-like"), and Z stands for *zusammen* (German for" together," or "cis-like"). To assign the E or Z prefix to a given structure, it is first necessary to pick the higher priority group attached to each carbon. The CIP priority of a group is determined by the atomic number (Section 1-1) of its *first* atom (the atom directly attached to the ring). The higher the atomic number, the higher the priority. If both groups have the *same* first atom, then the determination of priority is based on the highest atomic number of the second atom(s) of the groups, and so on.

The two groups attached to carbon 2 in structures **11** and **12** are hydrogen and propyl, **H** and $CH_2CH_2CH_3$, where the boldfaced atom is the one directly attached to the ring. Clearly, the propyl group has higher priority, because carbon has a higher atomic number (6) than hydrogen (1). The groups attached to carbon 1 are CH_3 and CH_2CH_3. Since both groups have the same first atom (carbon), we must look at the second atom(s), shown in boldface below:

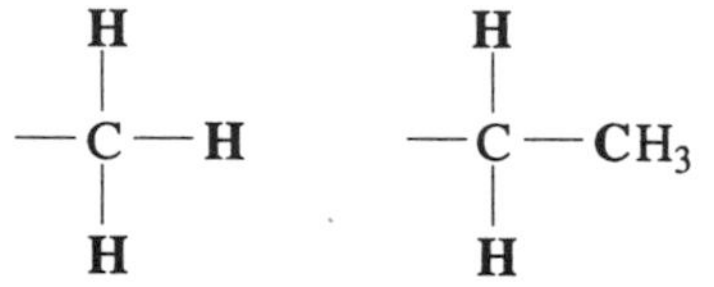

$$\begin{array}{ccc} \mathbf{H} & & \mathbf{H} \\ | & & | \\ -C-\mathbf{H} & & -C-\mathbf{CH_3} \\ | & & | \\ \mathbf{H} & & \mathbf{H} \end{array}$$

The methyl group has three hydrogens as second atoms, while the ethyl group has two hydrogens and a carbon. Therefore, the ethyl group has higher priority.

- In the CIP system, the longer *unbranched* alkyl group has priority over the shorter un-branched group, and the alkyl group with the branch nearest the point of attachment will have priority over a group with a more distant branch.

Once we have assigned priorities to the two groups attached to each ring carbon, we compare the relative positions of the higher-priority groups. If the higher-priority groups are on the same side of the ring, the structure is the Z isomer; if the higher-priority groups are on the opposite side of the ring, the structure is the E isomer. In the structures below, the higher-priority groups are shown in boldface.

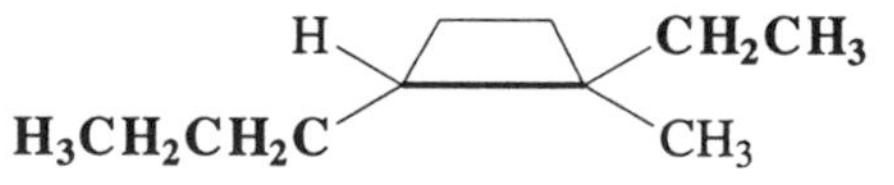

E-1-ethyl-1-methyl-2-propylcyclobutane

11

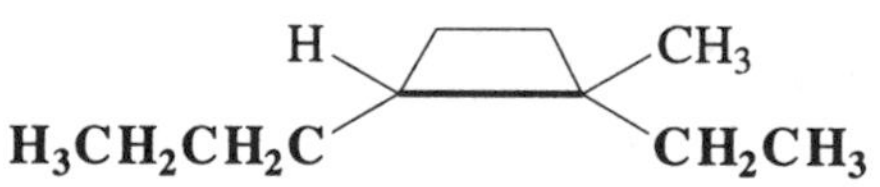

Z-1-ethyl-1-methyl-2-propylcyclobutane

12

EXAMPLE 5-12 Give a *complete* name for the structure below (not all hydrogens are drawn in):

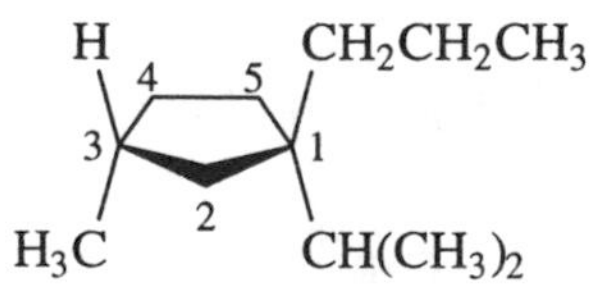

Solution It is obvious that, on carbon 3, the methyl group has a higher priority than the hydrogen. But what about the propyl vs. isopropyl groups on carbon 1? Both groups have the same first atom (carbon), so we must examine the second atom(s), shown below in boldface:

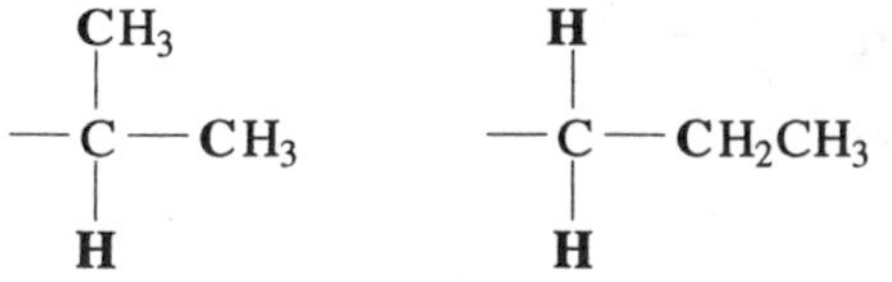

$$\begin{array}{ccc} \mathbf{CH_3} & & \mathbf{H} \\ | & & | \\ -C-\mathbf{CH_3} & & -C-\mathbf{CH_2CH_3} \\ | & & | \\ \mathbf{H} & & \mathbf{H} \end{array}$$

Because the isopropyl group has two carbons and one hydrogen as second atoms (compared to two hydrogens and one carbon for the propyl group), the isopropyl group has higher priority. Thus the complete name of this structure is Z-1-isopropyl-3-methyl-1-propylcyclopentane.

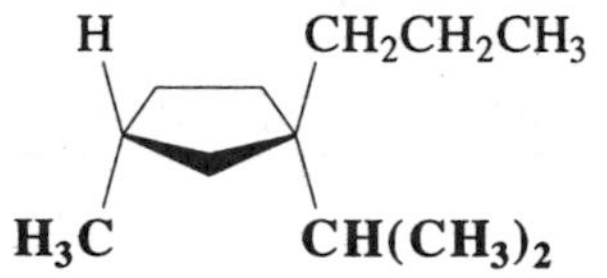

$$\begin{array}{ccc} H & & CH_2CH_2CH_3 \\ & & \\ H_3C & & CH(CH_3)_2 \end{array}$$

5-6. Alkenes

A hydrocarbon molecule with a carbon–carbon double bond belongs to the class of compounds called **alkenes**. **Alkenes** have the general molecular formula C_nH_{2n}, which is the same as that for cycloalkanes (Section 5-4). Remember this:

- The molecular formula and the associated I value do not distinguish between alkenes and cycloalkanes because the I valuer counts rings *plus* π bonds. However, only molecules with π bonds are considered to be **unsaturated.**

A. Naming alkenes

The simplest alkene is **ethene** (often called ethylene), C_2H_4, which we saw in Sections 2-5 and 3-5. To name other alkenes, we follow the same rules as we did for alkanes:

- Identify the longest chain that contains the double bond; then use the appropriate root name (Table 5-1)—except that the suffix "ene" replaces the "ane."

The position of the double bond in the alkene molecule is indicated by the number of the carbon where the double bond starts, with counting beginning at the end of the chain nearest the double bond. The examples below show how this works.

$$H_2C = CH - CH_2 - CH_3 \qquad \text{1-butene}$$
$$H_3C - CH = CH - CH_3 \qquad \text{2-butene}$$
$$H_3C - CH_2 - CH = CH_2 \qquad \text{1-butene (counting from the right-hand end)}$$
$$H_2C = CH - CH_3 \qquad \text{propene}$$

Notice that propene isn't called 1-propene. This is because the double bond must begin at one end or the other, both of which result in the same structure; so the "1" is superfluous.

B. Naming alkyl-substituted alkenes

If there are alkyl substituents attached to the chain containing the double bond, numbering still begins at the end nearest the double bond regardless of where the substituents are. That is, the double bond takes priority over the substituent. Thus, the structure below is named 3-methyl-1-butene, not 2-methyl-3-butene:

$$\overset{1}{H_2C} = \overset{2}{CH} - \overset{3}{CH} - \overset{4}{CH_3}$$
$$\qquad\qquad | $$
$$\qquad\qquad CH_3$$

EXAMPLE 5-13 What is the IUPAC name of the structure below?

$$H_3C - CH_2 - CH_2 - CH_2 - CH - CH_2 - CH_2 - CH_2 - CH_3$$
$$\qquad\qquad\qquad\qquad\qquad\quad | $$
$$\qquad\qquad\qquad\qquad\quad HC = CH_2$$

Solution Although there are nine carbons in the longest chain, there are only seven carbons in the longest chain *that contains the double bond.*

$$H_3C - \overset{6}{C}H_2 - \overset{5}{C}H_2 - \overset{4}{C}H_2 - \overset{3}{C}H - CH_2 - CH_2 - CH_2 - CH_3$$

$$\underset{2 \quad 1}{HC = CH_2}$$

Therefore, the correct name of this compound is 3-butyl-1-heptene.

C. Alkenyl substituents

Beginning in the next chapter, we'll encounter structures in which there are groups of atoms that take a higher nomenclature priority than a carbon–carbon double bond. In such structures the double bond may appear in a substituent group, rather than in the main chain. The names of such double bond-containing groups are derived in the same way as for the names of alkyl substituents, except that "en" precedes the "yl" suffix. The number of the carbon where the double bond begins (numbered from the point of the group's connection to the main chain) is added to the substituent name. Below are given several examples. Note the common names **vinyl** and **allyl**, which are generally used instead of their IUPAC names.

Group	IUPAC name	Common name
$-CH = CH_2$	ethenyl	vinyl
$-CH = CH - CH_3$	(1-propenyl)	
$-CH_2 - CH = CH_2$	(2-propenyl)	allyl
$-CH = CH - CH_2 - CH_3$	(1-butenyl)	
$-CH_2 - CH = CH - CH_3$	(2-butenyl)	
$-CH_2 - CH_2 - CH = CH_2$	(3-butenyl)	

D. Common names of alkenes

Prior to adoption of the IUPAC system, hydrocarbons with a double bond were named as derivatives of ethylene. Thus, 3-methyl-1-butene (above) would have been called isopropylethylene. Furthermore, the molecules were described as being mono-, di-, tri-, or tetrasubstituted derivatives based on the number of groups attached to the doubly bonded carbons. For example, isopropylethylene would be regarded as a monosubstituted ethylene (or as having a monosubstituted double bond), since it has only one group (the isopropyl group) attached to the doubly bonded carbons.

EXAMPLE 5-14 For each common name below, (1) draw the structural formula, (2) state whether the double bond is mono-, di-, tri-, or tetrasubstituted, and (3) give the IUPAC name of the structure.

   (a) *tert*-butylethylene   (b) tetramethylethylene   (c) 1,1-dimethylethylene

Solution

	(1)	(2)	(3)
(a)	$H_3C - \overset{\displaystyle CH_3}{\underset{\displaystyle CH_3}{C}} - CH = CH_2$	monosubstituted	3,3-dimethyl-1-butene
(b)	$\overset{H_3C}{\underset{H_3C}{}} C = C \overset{CH_3}{\underset{CH_3}{}}$	tetrasubstituted	2,3-dimethyl-2-butene
(c)	$\overset{H_3C}{\underset{H_3C}{}} C = C \overset{H}{\underset{H}{}}$	1,1-disubstituted	2-methylpropene

E. Stereochemical complication in alkenes

Alas, there is a stereochemical complication we must address. Recall from our discussion of molecular shape in the region of double bonds (Section 3-5) that the two doubly bonded carbons and the four atoms attached to them lie in a single plane. Now, either make a model of, or accurately draw, the structure of 2-butene. Notice that there are two different ways of putting the molecule together, as shown here:

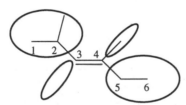

cis-2-butene
13

trans-2-butene
14

Clearly, these are isomers (same molecular formula, different structure) because there is no rotation about the double bond (why?). They can be interconverted only by breaking, then remaking, bonds. But they are *not* structural isomers because they have the same *sequence* of atoms and bonds. They differ only in the three-dimensional configuration of the atoms. So, they are *stereoisomers*. Here again, we can use the terms cis (same "side" of the double bond, as in **13**) and trans (opposite "sides" of the double bond, as in **14**) to indicate the relative positions of the identical groups, such as the methyl groups (or hydrogens) in **13** and **14**.

If there are no identical groups around the double bond (i.e., for tri- and tetrasubstituted double bonds), we use the E–Z system. For example, consider structure **15**:

15

Sometimes it helps to turn the structure so that the double bond is horizontal; next, circle the four groups around the double bond:

The four circled groups (the "empty" circle encloses a hydrogen) are each unique, so we'll use the E–Z system. The isopropyl group on the left side and the ethyl group on the right take priority; since they lie on opposite sides of the double bond (one is "up," the other "down") structure **15** has an E configuration.

Now to name it: Mentally erase the four circles you drew to highlight groups around the double bond, and apply the rules for naming alkenes. It is a 3-hexene with methyl groups on C-2 and C-4 (2,4-dimethyl-3-hexene). Because it has an E configuration, the full name is *E*-2,4-dimethyl-3-hexene.

- Cis–trans (or E–Z) designators are needed for each double bond where the two groups at one end of the double bond are different *and* the two groups at the other end are different.

EXAMPLE 5-15 Provide a complete IUPAC name for each alkene below. Also state the degree of substitution of the double bond.

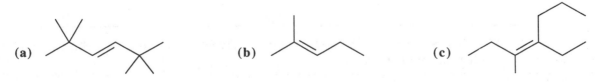

(a)   (b)   (c)

Solution

(a) 2,2,5,5-Tetramethyl-*trans*-3-hexene; disubstituted.

(b) 2-Methyl-2-pentene (no stereochemical designator required); trisubstituted.

(c) A stereochemical designator is required, so first determine which system to use: cis–trans or E–Z. Circle the groups around the double bond:

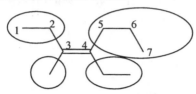

Here we can use the cis–trans system because two groups are identical (the ethyl groups on C-3 and C-4), and these are trans to each other. The name of the structure is therefore 4-ethyl-3-methyl-*trans*-3-heptene, or *trans*-4-ethyl-3-methyl-3-heptene.

5-7. Alkynes

A hydrocarbon with a carbon–carbon triple bond belongs to the class known as **alkynes**. The generic molecular formula for alkynes is C_nH_{2n-2}, but remember that molecules with two double bonds, two rings, or one ring and one double bond have the same molecular formula (see Section 5-3).

The simplest alkyne is ethyne, C_2H_2 (Sections 2-5 and 3-5), which is more commonly known by its historical name, *acetylene*. You can probably guess how IUPAC names of other alkynes are generated. Identify the longest chain that contains the triple bond, and pick the appropriate root name (Table 5-1), substituting the suffix "yne" for "ane." As with alkenes, we add a numerical prefix to specify the location of the carbon where the triple bond begins.

EXAMPLE 5-16 Provide the IUPAC name for each alkyne below:

(a) $H-C\equiv C-CH_2-CH_3$

(b) $H_3C-C\equiv C-CH_3$

(c) $(H_3C)_3C-C\equiv C-C(CH_3)_3$

(d) $H_3C-C\equiv C-H$

Solution (a) 1-butyne; (b) 2-butyne; (c) 2,2,5,5-tetramethyl-3-hexyne; (d) propyne; the "1-" prefix would be superfluous.

As was true for alkenes, most alkynes have common names that antedate their IUPAC names. The older nomenclature system regarded alkynes as alkyl derivatives of acetylene. Thus propyne (Example 5-16d) was known as methylacetylene.

Alkynes are further divided into two subcategories. An **internal alkyne** is one with carbon groups attached to both triply bonded carbons. **Terminal alkynes** are those with just one carbon group; a hydrogen is attached to the other triply bonded carbon. (Acetylene itself is regarded as a terminal alkyne.) We'll see in a later chapter that an **acetylenic hydrogen** (i.e., one attached to a triply bonded carbon) has unique chemical properties.

EXAMPLE 5-17 Give common names for the first three alkynes in Example 5-16. State whether each one is a terminal or internal alkyne.

Solution   (a) ethylacetylene, terminal; (b) dimethylacetylene, internal; (c) di-*tert*-butylacetylene, internal.

Because the two triply bonded carbons and the two atoms attached directly to them are co-linear (Section 3-5), there are no cis–trans stereochemical problems to worry about with alkynes.

5-8. Polyenes, Cycloalkenes, Polyynes, Enynes, and Cycloalkynes

There are several other classes of hydrocarbons that possess combinations of double bonds, triple bonds, and/or rings. Some of these are described briefly below.

A. Polyenes

Hydrocarbon molecules with two or more carbon–carbon double bonds are collectively referred to as **polyenes**. A **diene** has two double bonds, a **triene** three, and so on. The double bonds can be *cumulated* (sharing a carbon), *conjugated* (separated by one bond), or *isolated* (separated by two or more bonds; see Section 4-4). The IUPAC name for a polyene consists of the root name appropriate for the carbon chain containing *all* the double bonds, with the "ne" of the alkane name replaced by "diene," "triene," etc. The location of each double bond is specified by the number of the carbon where it begins. Finally, the stereochemical designators (cis–trans or E–Z) are added where appropriate. For example,

is *trans*-1,3,5-hexatriene. The trans designator refers to the configuration at the *middle* double bond. The two terminal double bonds do not require designators since there is no stereochemical ambiguity at either one (carbons 1 and 6 each have two hydrogens).

EXAMPLE 5-18   Provide a complete IUPAC name for each polyene below, and state whether the double bonds are cumulated, conjugated, or isolated.

(a) $H_2C{=}CH{-}CH{=}CH_2$

(b) $H_2C{=}C{=}C{=}CH_2$

(c)

Solution
(a) 1,3-Butadiene: The two double bonds are separated by one carbon–carbon (single) bond, so they are conjugated.
(b) Butatriene: The first and second double bonds share a carbon, so they are cumulated, as are the second and third double bonds. The first and third double bonds, however, are conjugated because they are separated by one (double) bond. The designators 1,2,3- are superfluous.
(c) *cis,cis*-2,5-Heptadiene: The two double bonds are separated by more than one bond, so they are isolated.

B. Cycloalkenes

There are many cyclic hydrocarbon molecules that have one or more double bonds between ring carbons. Such a double bond is described as **endocyclic**. The IUPAC name for such compounds is derived from the appropriate cycloalkane name (Section 5-4), replacing the "ane" suffix with "ene." Here are several examples:

cyclobutene   cyclohexene   cyclooctene

If there are substituents attached to the ring, each substituent is given the lowest possible numerical locator, with the double bond defining carbons 1 and 2. For example,

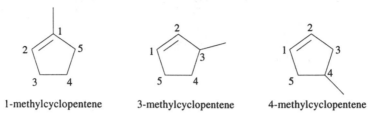

1-methylcyclopentene   3-methylcyclopentene   4-methylcyclopentene

If there are two or more endocyclic double bonds in a ring, the one closest to any substituents defines carbons 1 and 2. The number of the carbon where each double bond starts is added as a prefix to the name. For example,

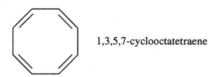

1,3,5,7-cyclooctatetraene

note: If one ring carbon is double bonded to another carbon *outside* the ring, the double bond is said to be **exocyclic** to the ring. Such a double bond may terminate, for example, with a methylene (CH_2) group, in which case the compound is a methylenecycloalkane, not a cycloalk*ene* at all. Some examples of methylenecycloalkanes are

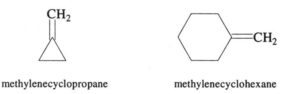

methylenecyclopropane   methylenecyclohexane

There is another subtle stereochemical feature of cycloalkenes. Consider this: Is cyclohexene the trans or cis isomer?

Clearly, the CH_2 groups attached to the doubly bonded carbons bear a cis relationship, as do the hydrogens attached to the doubly bonded carbons. Yet, we never add the designator cis to the name cyclohexene. This is because the cis configuration is understood. If you try to construct a model of (or even try to draw) *trans*-cyclohexene, you'll quickly appreciate the difficulties of having a trans double bond in a small ring.

If, however, the ring is large and flexible enough, a trans double bond *can* be tolerated. For example, *trans*-cyclooctene has been synthesized, though it's considerably less stable than its cis stereoisomer.

trans-cyclooctene   cis-cyclooctene

(We'll have more to say about the relative stabilities of alkenes and cycloalkenes in a later chapter.)

EXAMPLE 5-19 Provide IUPAC names for each structure below, and state whether the double bonds are endocyclic or exocyclic.

(a)   **(b)**   **(c)**

Solution (a) methylenecyclobutane, exocyclic; (b) 1,3-cyclopentadiene, both endocyclic (and conjugated, too!); (c) 3,6-diethyl-1,4-cyclohexadiene, endocyclic (and isolated).

C. Polyynes and enynes

Hydrocarbon molecules with two or more carbon–carbon triple bonds are known collectively as **polyynes**. A **diyne** has two triple bonds, a **triyne** three, and so on. A molecule with both a double bond *and* a triple bond is called an **enyne**. To name a polyyne, we identify the root name of the chain that contains all the triple bonds, substitute the appropriate suffix (diyne, triyne, etc.), and designate the carbon where each triple bond begins, observing the usual rule regarding the use of lowest possible numerical locators:

1,3,5-heptatriyne

To name an enyne, identify the root name similarly, but number the double and triple bonds separately. If, in numbering the chain, the same set of numbers turns out to locate the double and triple bonds, the *double* bond gets the lower number. In all other cases the usual rule of choosing the lowest locator number applies, regardless of whether the double or triple bond gets the lower one:

1-penten-4-yne   *trans*-3-penten-1-yne

D. Cycloalkynes

Because two triply bonded carbons and the two atoms attached directly to them form a linear array (Section 3-5), it's difficult to include a $CH_2C{\equiv}CCH_2$ unit within a ring—the remaining atoms would have to span a relatively great distance. Look at the structure below:

$$H_2C{-}C{\equiv}C{-}CH_2$$
$$\diagdown(CH_2)_n\diagup$$

Cycloalkynes that have seven or fewer carbons ($n \leq 3$) are extremely unstable and cannot be isolated. Nonetheless, as we found with trans cycloalkenes, it is possible to synthesize a cycloalkyne if the ring is sufficiently large and flexible. Thus, cyclooctyne (the above structure with $n = 4$) is a known compound, although it is considerably less stable than larger-ring cycloalkynes and acyclic alkynes.

5-9. Aromatic Hydrocarbons

You might be tempted to name the structure below as 1,3,5-cyclohexatriene:

But this molecule is an aromatic hydrocarbon (see Section 4-7) and is always identified by its common name, *benzene*, to remind us of the special properties of its double bonds.

If there are multiple substituents connected to the benzene ring, the location of the one with the highest alphabetical priority defines the number-1 carbon of the ring. If there are only two substituents, their locations can be labeled numerically in the usual way, or they may be designated by the terms **ortho** (1,2 substitution), **meta** (1,3), or **para** (1,4). These terms are usually abbreviated by the letters *o, m,* and *p*.

<table>
<tr><td>1,2
ortho (*o*)</td><td>1,3
meta (*m*)</td><td>1,4
para (*p*)</td></tr>
</table>

To confuse the issue, there are many aromatic hydrocarbons whose common names are much more widely used than their IUPAC names. The ones below are worth memorizing.

Structure	IUPAC name	Common name
⬡—CH₃	methylbenzene	toluene
⬡ with CH₃, CH₃	1,2-dimethylbenzene	*o*-xylene
⬡ with H₃C, CH₃	1,3-dimethylbenzene	*m*-xylene
H₃C—⬡—CH₃	1,4-dimethylbenzene	*p*-xylene

When the benzene ring is a substituent on another larger ring or chain, it is called a **phenyl group**:

4-phenyloctane

EXAMPLE 5-20 Provide IUPAC names for the hydrocarbons below.

(a)

(b)

(c)

Solution (a) isopropylbenzene; (b) 1,3,5-trimethylbenzene; (c) 3-phenylcyclooctene.

There are many other aromatic hydrocarbon ring systems that involve two or more benzene rings connected together. We'll mention only three examples—biphenyl, naphthalene, and anthracene. The structure of each of these is shown below, along with the numbering system used to locate substituents.

biphenyl   naphthalene   anthracene

SUMMARY

1. Hydrocarbons, molecules composed of only carbon and hydrogen, are divided into three basic groups:

   (a) alkane molecules (C_nH_{2n+2}) contain no multiple bonds, and are described as saturated.
   (b) alkene molecules (C_nH_{2n}) contain a carbon–carbon double bond and are therefore said to be unsaturated.
   (c) alkyne molecules (C_nH_{2n-2}) contain a carbon–carbon triple bond and are also unsaturated.

   Cyclic analogs of each group are called cycloalkanes, cycloalkenes, and cycloalkynes, respectively. Polyenes are hydrocarbons with two (dienes), three (trienes), or more double bonds. Aromatic hydrocarbons are usually derivatives of benzene.

2. The IUPAC name for a (cyclo)alkane is determined by applying the following rules:

   (a) Identify the longest carbon chain or largest ring, then select the appropriate root name, adding the prefix "cyclo" if appropriate.
   (b) Number the carbons in this chain or ring, starting at the end nearest the first substituent in the case of a chain, or at the point of attachment of the group with highest alphabetical preference in the case of a ring.
   (c) Identify the location and type of each substituent. Add this information as a prefix to the name, with substituents listed in alphabetical order.

To name (cyclo)alkenes and (cyclo)alkynes, there are three additional considerations:

(d) The root name must be derived from the longest chain or largest ring containing the multiple bond, and numbering of a chain must start at the end nearest the multiple bond.

(e) The number of the carbon where the multiple bond starts is added as a prefix to the root name.

(f) The alkane suffix "ane" is replaced by "ene" for alkenes and "yne" for alkynes.

3. Stereoisomers have the same sequence of atoms and bonds but differ in the three-dimensional configuration (fixed relative arrangement) of the atoms.

4. Geometric (cis–trans) isomers are one type of stereoisomer, differing in the relative positions (same or opposite side of the ring or double bond) of substituent groups.

5. The E (opposite) and Z (together) stereochemical designators are also used to compare the relative positions of dissimilar substituent groups. Specification of E or Z configurations requires that priorities be assigned to substituent groups. The priority is based on the atomic number of the first atom (the point of connection) of the substituent. If two substituents have the same first atom, priority is based on the second atom with the highest atomic number, and so on. Multiply bonded atoms are counted as an equivalent number of singly bonded atoms (see Problem 5-5).

6. A ring must have at least eight carbons to be flexible enough to contain a trans double bond or a triple bond.

7. Because of the special properties of their double bonds, aromatic hydrocarbons are not named as cyclic polyenes, but rather as derivatives of benzene or the appropriate aromatic ring system.

RAISE YOUR GRADES

Can you define...?

- ☑ hydrocarbon
- ☑ alkane (cycloalkane)
- ☑ alkene (cycloalkene)
- ☑ alkyne (cycloalkyne)
- ☑ internal/terminal alkyne
- ☑ aromatic hydrocarbon
- ☑ polyene (diene, triene, etc.)
- ☑ polyyne, enyne
- ☑ ortho, meta, para
- ☑ methyl, methylene, methine groups
- ☑ 1°, 2°, 3°, and 4° carbons/hydrogens
- ☑ saturated/unsaturated molecule
- ☑ acetylenic hydrogen
- ☑ substituent
- ☑ alkyl group
- ☑ branched chain
- ☑ configuration
- ☑ stereoisomer
- ☑ cis–trans isomers
- ☑ E–Z isomers
- ☑ *I* (index of hydrogen deficiency)
- ☑ spirocyclic molecule
- ☑ bicyclic molecule
- ☑ endocyclic/exocyclic double bond

Can you explain...?

- ☑ how to generate IUPAC names for all hydrocarbons and substituent groups
- ☑ how to draw skeletal structures for molecules
- ☑ the significance of *I* (the index of hydrogen deficiency) and how to calculate it
- ☑ how to identify cis–trans isomers
- ☑ how to assign CIP substituent priorities for E–Z stereochemical specification
- ☑ why a ring must have at least a certain minimum size in order to tolerate a trans double bond or a triple bond

Can you draw the structure of...?

- ☑ each alkane from methane to decane (and name each one)
- ☑ benzene
- ☑ naphthalene
- ☑ anthracene
- ☑ biphenyl
- ☑ a phenyl group

SOLVED PROBLEMS

PROBLEM 5-1 For each structure below provide the IUPAC name, the molecular formula, and the number of methyl, methylene, and methine groups, and quaternary carbons.

(a)

(b)

(c) $(H_3C)_3C$ — H, H — CH_3

Solution

Name	Molecular formula	Methyl (CH_3)	Methylene (CH_2)	Methine (CH)	Quaternary (C)
(a) 1-cyclopropyl-4-isopropylnonane	$C_{15}H_{30}$	3	9	3	0
(b) 3-ethyl-1,1-dimethylcycloheptane	$C_{11}H_{22}$	3	6	1	1
(c) *trans*-1-*tert*-butyl-4-methylcyclohexane	$C_{11}H_{22}$	4	4	2	1

PROBLEM 5-2 Provide a complete IUPAC name for each structure below. State whether each double bond is mono-, di-, tri-, or tetrasubstituted and whether each is cumulated, conjugated, isolated, or part of an aromatic system.

(a)

(b)

(c)

Solution
(a) *E*-3-Methyl-2-hexene (or *trans*-3-methyl-2-hexene); trisubstituted, isolated.

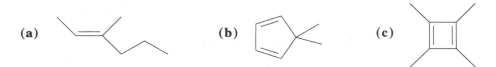

(b) 5,5-Dimethyl-1,3-cyclopentadiene; both double bonds are disubstituted and conjugated with each other.

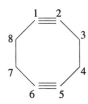

(c) Tetramethyl-1,3-cyclobutadiene (the 1,2,3,4- locators for the methyl groups are superfluous, since there is only one way for this structure to accommodate four methyls); both double bonds are tetrasubstituted and conjugated with each other.

PROBLEM 5-3 Provide the IUPAC name for each of the structures below. State whether the alkyne is terminal or internal.

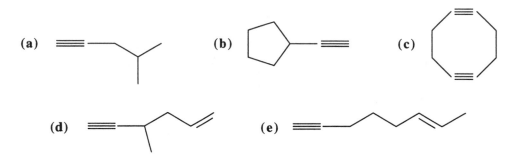

Solution
(a) 4-Methyl-1-pentyne: Terminal.

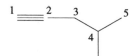

(b) This is a tougher one. The root name will be based on the cyclopentane ring. But how do you name an HC≡C− substituent? Since the H₂C=CH− group is called ethenyl (or vinyl), an HC≡C− group is called ethynyl. So, the structure is ethynylcyclopentane. Its common name would be cyclopentylacetylene, and it is a terminal alkyne.

(c) 1,5-Cyclooctadiyne: Both triple bonds are internal and endocyclic.

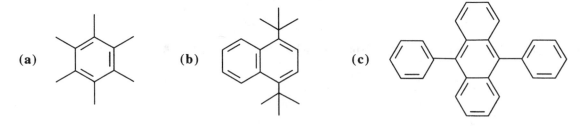

(d) 4-Methyl-1-hexen-5-yne: Because the numerical locators for the double and triple bonds are the same when numbering starts at either end of the chain, the double bond gets the lower number.

(e) *trans*-6-Octen-1-yne: The lower locator numbers require that the triple bond defines carbons 1 and 2.

PROBLEM 5-4 Provide the IUPAC name for each structure below.

Solution

(a) Hexamethylbenzene: The locators 1,2,3,4,5,6- are superfluous, since there is only one way to accommodate the six methyl groups.

(b) 1,4-Di-*tert*-butylnaphthalene.

(c) 9,10-Diphenylanthracene.

PROBLEM 5-5 For each structure below provide the appropriate E or Z stereochemical designator.

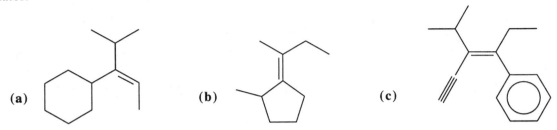

(a) (b) (c)

Solution

(a) It is clear that the methyl group attached to carbon 2 is higher priority than the hydrogen. But what about the isopropyl group on carbon 3 compared to the cyclohexyl group? The first atom (C) and the second atoms (C, C, and H) are the same for both groups. But there is a difference in the third atoms: six hydrogens for the isopropyl group, two carbons and four hydrogens for the cyclohexyl. So the cyclohexyl group has higher priority, and the configuration is therefore Z. Higher priority groups are circled below.

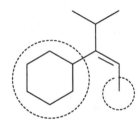

(b) The ethyl group on the upper doubly bonded carbon has higher priority than the methyl. For the five-membered ring, the methyl-bearing carbon on the left has higher priority (second atoms C, C, H) than the CH_2 group on the right (C, H, H). Therefore, this is the E stereoisomer.

(c) This is a tough one until we know how to assign priority to multiply bonded atoms. Here's the rule: When a carbon is doubly bonded to another carbon, the latter carbon is viewed as having *two* singly bonded carbons:

$$
\begin{array}{ccc}
\overset{\displaystyle H}{\underset{}{|}} & & \overset{\displaystyle H}{\underset{}{|}} \\
-C=CH_2 & \equiv & -C-CH_2 \\
& & \underset{\displaystyle CH_2}{|}
\end{array}
$$

Similarly, a triple bond is viewed as equivalent to three single bonds:

$$
\begin{array}{ccc}
& & \overset{\displaystyle CH}{\underset{}{|}} \\
-C\equiv C-H & \equiv & -C-CH \\
& & \underset{\displaystyle CH}{|}
\end{array}
$$

Using this rule, some of the commonly encountered substituents are ranked below from highest priority to lowest:

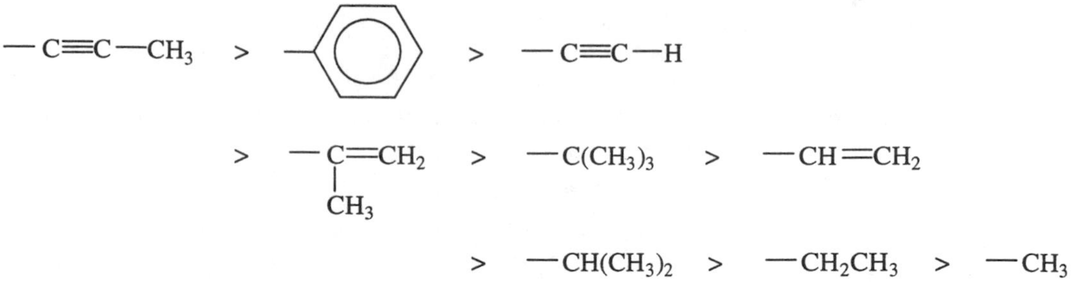

$$-C\equiv C-CH_3 \quad > \quad \bigcirc \quad > \quad -C\equiv C-H$$

$$> \quad -\underset{\underset{CH_3}{|}}{C}=CH_2 \quad > \quad -C(CH_3)_3 \quad > \quad -CH=CH_2$$

$$> \quad -CH(CH_3)_2 \quad > \quad -CH_2CH_3 \quad > \quad -CH_3$$

Thus, the molecule in question has the two higher priority groups (H-C≡C and phenyl) in a Z configuration.

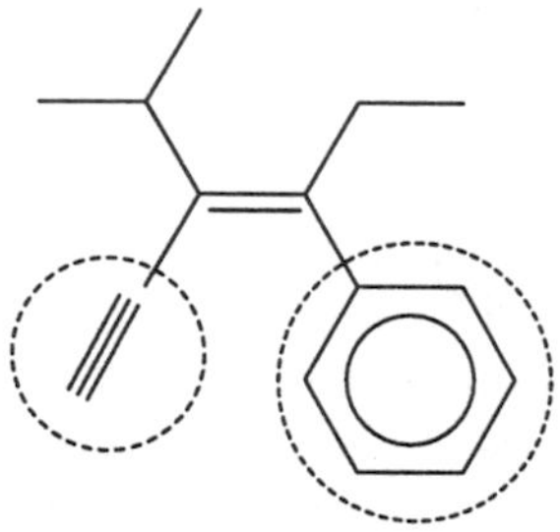

PROBLEM 5-6   Draw the structure of 1,2-cyclohexadiene, and comment on its expected stability.

Solution   The structure is misleadingly easy to draw:

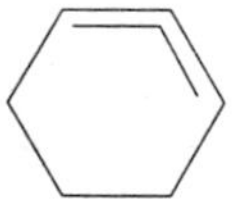

But let's look a little more closely at the cumulated double bonds. Remember from Problem 3-4 that the central carbon of two cumulated double bonds is *sp*-hybridized, and the geometry of the three doubly bonded atoms is linear. So, a more accurate representation would be

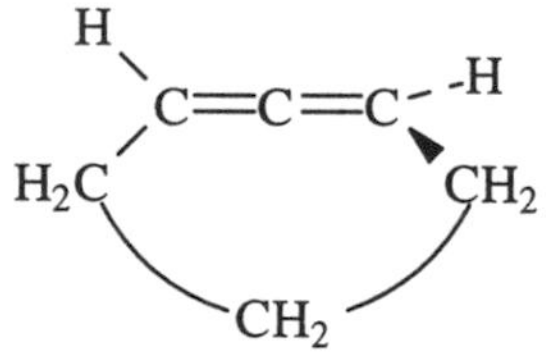

It is clear from this drawing (or from trying to make a molecular model of the molecule) that the small size of the ring precludes such a linear group of atoms. It turns out that a nine-membered ring *is* sufficiently flexible to tolerate two cumulated double bonds.

PROBLEM 5-7   (a) For the molecular formula C_8H_{12} calculate the value of I, then describe the combination(s) of π bonds and rings that must be present in any isomer with this molecular formula. (b) By inspection, give the value of I for each structure below, *without* generating the molecular formulas.

(1) 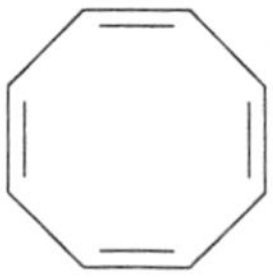   (2) 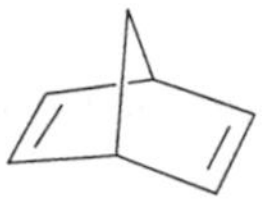   (3)

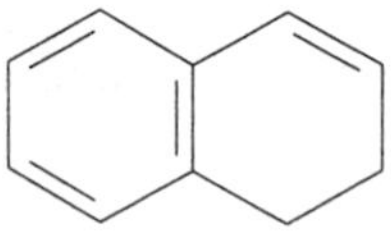

Solution
(a) Use Eq. (5-1):

$$I = \frac{2C + 2 - H}{2} = \frac{2(8) + 2 - 12}{2} = 3$$

Therefore, any molecule with this molecular formula must have one of the following combinations: three rings, three π bonds, two rings + one π bond, or one ring + two π bonds.

(b) (1) 5, (2) 4, (3) 6.

6 FUNCTIONALIZED ORGANIC MOLECULES

6-1. What Is a Functional Group?

The field of organic chemistry would be considerably less interesting than it is if it were limited only to the study of hydrocarbons such as those we discussed in Chapter 5. The most notable thing about the reactivity of alkanes and cycloalkanes is their *lack* of it! Carbon–carbon and carbon–hydrogen σ bonds are essentially inert to the vast majority of chemical reagents, though modern research is searching for ways to activate these bonds.

As we intimated in Section 3-5, π bonds are generally more reactive than σ bonds, so that unsaturated hydrocarbons (alkenes, alkynes, their cyclic analogs, and aromatic rings) have a much richer and more varied chemistry than do saturated hydrocarbons. However, there are distinct patterns to the reactivity of alkenes, alkynes, and aromatic compounds. That is, virtually any alkene, regardless of chain length or ring size, will undergo certain predictable reactions. The same thing is true of alkynes. In fact, it is the recognition of these patterns in reactivity that serves as the foundation for understanding most organic reactions.

The site (atom or group of atoms) in a molecule where a reaction takes place is called the **functional group** of that molecule. The functional group of an alkene is the double bond; for an alkyne it is the triple bond or the acetylenic hydrogen (see Section 5-7). To focus attention on these groups, the molecular structures are often written in "generic" form:

$$R{-}CH{=}CH_2$$
monosubstituted alkene

$$R_2C{=}CR_2$$
tetrasubstituted alkene

$$R{-}C{\equiv}C{-}H$$
terminal alkyne

$$R{-}C{\equiv}C{-}R$$
internal alkyne

In such structures, the **R** represents the "rest" of the molecule, that is, the carbon-containing group(s) not directly involved in the reaction. The only requirement for an R group is that its first atom (the point of connection) be a carbon.

Alkanes and cycloalkanes (except the unusually reactive cyclopropane, Section 5-4) are usually regarded as lacking functional groups. The most important thing to remember is that

● The functional group is usually the major determinant of a molecule's reactivity.

The length of the carbon chain of the molecule or the size of its ring and the nature or location of its alkyl substituents all have relatively less influence on reactivity patterns.

Except for carbon–carbon double and triple bonds, all functional groups involve **hetero-atoms,** i.e., atoms other than carbon and hydrogen (Section 4-7). The most common functional groups are those that contain halogen (fluorine, chlorine, bromine, and iodine), oxygen, and/or nitrogen. Other, less common functional groups involve sulfur, phosphorus, boron, and certain metallic elements. Our goal in this chapter is to introduce these functional groups so that you can recognize them at a glance. We'll also discuss how to provide IUPAC names for molecules that contain the common functional groups.

6-2. The Common Functional Groups in Organic Compounds

Before we begin introducing these functional groups, a few ground rules are in order. We'll be using generic structures, with **R** representing the nonfunctional residue(s) of the molecule attached to the functional group. When you see an "(H)" next to an R [i.e., R(H)], it means that the residue can be *either* a carbon-containing group *or* just a hydrogen. In some cases you'll see both an R and R′ (or even R″). These indicate that the various carbon-containing residues can be different; they need not necessarily have the same structure. Usually, an R means that the group can be either an alkyl group or an aromatic ring. The symbol **Ar** (for aryl) implies that the residue *must* be an aromatic ring (e.g., a phenyl group) at the point of attachment.

Finally, here's a bit of advice. There is no substitute for memorizing the identity of each functional group and the class of compounds to which it belongs. One good way to do this is to make "flash cards" (3 × 5 in. index cards), one for each functional group. Draw the generic structure of a compound with the group appearing on one side of the card, and write the functional group name and class on the other. Table 6-1 at the end of this section will help in this labor.

A. Organic halides

Alkane molecules in which one or more of the hydrogens have been replaced by halogens (fluorine, chlorine, bromine, iodine) belong to the class called **alkyl halides.** The functional group is the halogen atom (and the carbon to which it is attached). The generic structure for alkyl halides is RX, which includes RF (alkyl fluorides), RCl (alkyl chlorides), RBr (alkyl bromides), and RI (alkyl iodides). Alkyl halides are further subdivided into three groups according to the degree of substitution of the halogen-bearing carbon (X = halogen), i.e., the number of carbons attached to the carbon bearing the halogen:

primary (1°) halide   secondary (2°) halide   tertiary (3°) halide

A molecule with two halogen atoms bonded to the *same* carbon is a **geminal dihalide**, while a molecule with two halogen atoms attached to neighboring carbons is a **vicinal dihalide**.

geminal dihalide   vicinal dihalide

If the halogen atom is attached to a doubly bonded carbon (e.g., from an alkene), the molecule is an example of a **vinyl halide.**

vinyl halide

Finally, if the halogen atom is connected directly to an aromatic ring (ArX), the resulting molecule is an **aryl halide**.

B. Singly bonded oxygen

If an oxygen atom is inserted into a carbon–hydrogen bond of an alkane, the resulting molecule becomes ROH, which is the generic formula for a class of compounds known as **alcohols**. When an organic molecule contains an –OH group, OH is *not* a hydroxide ion—rather, it is a **hydroxy** (or **hydroxyl**) **group**, which is covalently bonded to the carbon. Alcohols, then, can be viewed as organic derivatives (or analogs) of water (HOH).

Alcohols, like alkyl halides, can be classified according to the number of carbons attached to the hydroxy-bearing carbon:

$$
\begin{array}{ccc}
\quad\text{H} & \quad\text{R}' & \quad\text{R}' \\
\quad| & \quad| & \quad| \\
\text{R}-\text{C}-\text{OH} & \text{R}-\text{C}-\text{OH} & \text{R}-\text{C}-\text{OH} \\
\quad| & \quad| & \quad| \\
\quad\text{H} & \quad\text{H} & \quad\text{R}''
\end{array}
$$

primary (1°) alcohol   secondary (2°) alcohol   tertiary (3°) alcohol

When a hydroxy group is directly attached to a carbon in an aromatic ring instead of an alkane, the resulting molecule (ArOH) is classed as a **phenol** (from phenyl alcohol), not as an alcohol. The properties of phenols are somewhat different from those of alcohols.

Insertion of an oxygen atom into a carbon–carbon bond of an alkane leads to the COC linkage characteristic of **ethers** (ROR'). The –OR group found in ethers is called an **alkoxy group.** If one of the R groups in an ether is changed to an aromatic residue, the resulting molecule ArOR is called an **aryl ether.** If both R groups are replaced, the resulting ArOAr is called a **diaryl ether.**

C. Singly bonded nitrogen

Organic derivatives of ammonia (NH_3) are called **amines**, which can be subdivided according to the number of carbons attached to the *nitrogen* (as opposed to the carbon bearing the functional group, as in alcohols and alkyl halides):

$$
\begin{array}{ccc}
\quad\text{H} & \quad\text{R}' & \quad\text{R}' \\
\quad| & \quad| & \quad| \\
\text{R}-\text{N:} & \text{R}-\text{N:} & \text{R}-\text{N:} \\
\quad| & \quad| & \quad| \\
\quad\text{H} & \quad\text{H} & \quad\text{R}''
\end{array}
$$

primary (1°) amine   secondary (2°) amine   tertiary (3°) amine

If the nitrogen becomes tetracoordinated through attachment of a fourth atom or group, the resulting cation is an **aminium ion**, or substituted **ammonium ion.**

$$
\begin{array}{c}
\text{R}'(\text{H}) \\
| \\
\text{R}-\overset{+}{\text{N}}-\text{R}''(\text{H}) \\
| \\
\text{R}'''(\text{H})
\end{array}
$$

aminium (ammonium) ion

If all four groups attached to the nitrogen are alkyl or aryl groups, the aminium (ammonium) ion is described as *quaternary*.

D. Simple carbonyl derivatives

The C=O functional group is called a **carbonyl** or **oxo** group. If the carbonyl carbon is attached to two hydrogens, or to one hydrogen and one R group (i.e., $\frac{\text{H}}{\text{H}}{>}\text{C}{=}\text{O}$ or $\frac{\text{R}}{\text{H}}{>}\text{C}{=}\text{O}$), the molecule is an **aldehyde**. Aldehydes are often written RCHO. If *both* are carbon residues ($\frac{\text{R}}{\text{R}}{>}\text{C}{=}\text{O}$), the molecule is a **ketone**. An R–C=O group is called an **acyl** group, while an H–C=O group is called a **formyl** group.

E. Carboxylic acids and their derivatives

A hydroxy group directly attached to a carbonyl group constitutes a **carboxyl group**, and a molecule possessing this group is called a **carboxylic acid**, $RC\lessgtr^O_{OH}$ (often written RCO_2H). The OH hydrogen of a carboxylic acid is relatively acidic. When it is removed (as H^+), the remaining anion (RCO_2^-) is called a **carboxylate ion**.

There is a family of other carboxylic acid derivatives, in which the OH group is altered or replaced by other groups. If, for example, the acidic hydrogen is replaced by an alkyl or aryl group, the resulting structure is a **carboxylate ester**, $(H)RC\lessgtr^O_{OR'}$ or $(H)RCO_2R'$.

note: There are several other types of esters (see Section 6-5), so it's best to include the qualifier "carboxylate." Nonetheless, an organic chemist's use of the term "ester" usually implies carboxylate ester.

The $-CO_2R'$ functional group is called a **carboalkoxy** group. Notice that the R' *must* be an alkyl or aryl group: if it were a hydrogen, we would have a carboxylic acid!

When the –OH group of a carboxylic acid is replaced by a halogen [$(H)RC\lessgtr^O_X$], the compound is an **acid halide**, with a **haloformyl** ($-C\lessgtr^O_X$) functional group. Similarly, replacement of the –OH group by an $-NR_2$ group gives rise to an **amide**. The $-C\lessgtr^O_{N[R(H)]_2}$ functional group is called a **carbamoyl** (or **carbamyl**) group. Amides are classified as either 1°, 2°, or 3°, depending on the number of carbons attached to the nitrogen.

$$
\begin{array}{cc}
\overset{\displaystyle O}{\overset{\displaystyle \|}{}}\quad \overset{\displaystyle H}{\overset{\displaystyle |}{}} & \overset{\displaystyle O}{\overset{\displaystyle \|}{}}\quad \overset{\displaystyle R'}{\overset{\displaystyle |}{}} \\
(H)R-C-N\text{:} & (H)R-C-N\text{:} \\
\overset{\displaystyle |}{\overset{\displaystyle H}{}} & \overset{\displaystyle |}{\overset{\displaystyle H}{}} \\
\text{primary (1°) amide} & \text{secondary (2°) amide}
\end{array}
$$

$$
\begin{array}{c}
\overset{\displaystyle O}{\overset{\displaystyle \|}{}}\quad \overset{\displaystyle R'}{\overset{\displaystyle |}{}} \\
(H)R-C-N\text{:} \\
\overset{\displaystyle |}{\overset{\displaystyle R''}{}} \\
\text{tertiary (3°) amide}
\end{array}
$$

F. Other common nitrogen-containing functional groups

The nitrogen analog of a ketone or aldehyde has generic structure $[(H)R]_2C=NR'(H)$, and is called an **imine**. Although you might view an imine as the nitrogen analog of an alkene, it turns out to be more reasonable, from the standpoint of reactivity, to regard the C=N bond of imines as an analog of the C=O bond of aldehydes and ketones.

The $-C\equiv N$ group is called the **cyano** group, and molecules that contain this group belong to the class known as **nitriles**. And finally, an $-NO_2$ group is called a **nitro** group. Note that the point of connection is the *nitrogen*. Nitro compounds are also subdivided as 1°, 2°, and 3° according to the degree of substitution of the carbon bearing the nitro group. Furthermore, remember from our discussions of formal charge and resonance in Chapter 4 that the nitro group has a positive charge on the nitrogen and a negative charge shared by the two oxygens.

$$
R-\overset{+}{N}\overset{\displaystyle \nearrow O}{\underset{\displaystyle \searrow O^-}{}} \quad\longleftrightarrow\quad R-\overset{+}{N}\overset{\displaystyle \nearrow O^-}{\underset{\displaystyle \searrow O}{}}
$$

All the functional groups discussed so far are summarized in Table 6-1. Now it's up to you to memorize this list before proceeding any further.

TABLE 6-1.   The Common Organic Functional Groups

Generic structure	Functional group name	Functional group class
RX (X = F, Cl, Br, I)	Halo	Alkyl halide (1°, 2°, 3°)
$\overset{(H)R}{\underset{(H)R}{\diagup}}C=C\overset{}{\underset{(H)R}{\diagdown}}X$	Halo	Vinyl halide
ArX	Halo	Aryl halide
ROH	Hydroxy	Alcohol
ArOH	Hydroxy	Phenol
ROR′	Alkoxy	Ether
ArOR	Aryloxy or alkoxy	Aryl ether
ArOAr	Aryloxy	Diaryl ether
RN[R′(H)]$_2$	Amino	Amine (1°, 2°, 3°)
R$\overset{+}{N}$[R′(H)]$_3$	Amino	Aminium ion (1°, 2°, 3°, 4°) [alkyl ammonium ion (1°, 2°, 3°, 4°)]
$(H)RC\overset{O}{\underset{H}{\diagdown}}$	Formyl (oxo), acyl*	Aldehyde
$RC\overset{O}{\underset{R'}{\diagdown}}$	Keto (oxo), acyl*	Ketone·
$(H)RC\overset{O}{\underset{OH}{\diagdown}}$	Carboxyl*	Carboxylic acid
$(H)RC\overset{O}{\underset{O^-}{\diagdown}}$	Carboxylate*	Carboxylate (anion)
$(H)RC\overset{O}{\underset{O-R'}{\diagdown}}$	Carboalkoxy*	Carboxylate ester
$(H)RC\overset{O}{\underset{X}{\diagdown}}$	Haloformyl*	Acid halide
$(H)RC\overset{O}{\underset{N}{\diagdown}}\overset{R'(H)}{\underset{R''(H)}{}}$	Carbamoyl*	Amide (1°, 2°, 3°)
[(H)R]$_2$C=NR′(H)	Imino	Imine
RCN	Cyano	Nitrile
RNO$_2$	Nitro	Nitro compound

*Each of these functional groups includes a carbonyl group.

EXAMPLE 6-1   For each structure below, identify the functional group by drawing a box around it; then name the functional group class of each.

H$_3$CCH=NH   (H$_3$C)$_3$CCHO

Solution

ketone	imine	aldehyde
3° alkyl bromide	carboxylate ion	3° amide
2° alkyl iodide	carboxylic acid	ether
1° alcohol	carboxylate ester	acid chloride
nitro compound	nitrile	2° amine

6-3. Naming Monofunctionalized (Cyclo)Alkanes

Now the problems really begin. We'll find that many organic compounds go by a variety of aliases. There can be a historical name, a common name, and an IUPAC name, all for one innocent compound! To make matters worse, *Chemical Abstracts* (CA), the constantly updated encyclopedia of all known chemical compounds, uses a system that is sometimes at variance even with IUPAC. In the discussion that follows, we'll attempt to follow the slightly more systematic CA system, pointing out certain common names that are still used by IUPAC.

A. Functional groups named as prefixes

It should relieve you to learn that the basic IUPAC nomenclature rules you learned for hydrocarbons will, with only slight modification, provide systematic names for functionalized organic compounds. We begin by identifying those functional groups that are *always* regarded as substituents. These include halogens, nitro groups, and alkoxy groups from Table 6-1, as well as certain others we'll mention in Section 6-5. Such groups will always appear in alphabetical order (along with the number of the carbon where they are attached) as a prefix to the systematic name of the parent (cyclo)alkane. The appropriate prefixes for the halogens are fluoro, chloro, bromo, and iodo. The nitro is simply called nitro. Alkoxy group names are formed from the root name of the corresponding alkyl group, with a suffix "oxy." Examples are methoxy (CH_3O-), ethoxy (CH_3CH_2O-), and so on. Example 6-2 gives you some structures to practice on.

EXAMPLE 6-2   Provide a systematic name for each structure below:

(a)   (b)   (c)   —NO_2

Br   I

(d)   (e)

H
CH$_3$

H
OCH$_3$

Solution

(a) 2-bromobutane (the carbons are numbered from the end of the chain nearest the substituent)
(b) iodocyclohexane (the locator "1-" isn't needed, since the iodine defines the number-1 carbon)
(c) 2-nitropropane
(d) *cis*-1-methoxy-2-methylcyclopentane (methoxy has a higher alphabetical priority than methyl, so it defines the number-1 carbon of the ring; note that the placement of the two substituents imparts a cis configuration to the structure)
(e) 2-ethoxy-2-methylpropane (the R group with the shorter carbon chain is treated along with the oxygen as the substituent)

Although the systematic names for organic compounds are generally preferred, you'll often have to confront their common (sometimes called "trivial") names. For example, alkyl halides were originally named by first naming the alkyl group, followed by the name of the appropriate halide. So, CH_3I was called methyl iodide (systematic: iodomethane), $(CH_3)_3CCl$ was *tert*-butyl chloride (systematic: 2-chloro-2-methylpropane), and so on.

Ethers can also be named in several different ways. One way is to name both alkyl groups (as separate words), followed by the word "ether." Thus, $CH_3CH_2OCH_2CH_3$ (ethoxyethane) is usually called diethyl ether (or just ethyl ether), while 2-ethoxy-2-methylpropane (the structure in Example 6-2e) is called *tert*-butyl ethyl ether. Here's another system by which ethers are named, especially those with complicated structures. You can regard the ether as an alkane in which one of the CH_2 groups is replaced by an "oxa" group (an oxygen); i.e., you consider the oxa group as a member of the chain or ring. Thus diethyl ether becomes 3-oxapentane (chain of five members). And in like manner, the cyclic ether below would be called oxacyclohexane, where the oxygen defines the number-1 atom of the ring.

O

Cyclic ethers with three-membered rings could be called oxacyclopropanes, but they are more commonly called **epoxides,** or **oxiranes.** Notice that both of these cyclic ethers are examples

O

of *heterocyclic* compounds, that is, cyclic compounds where one (or more) of the atoms in the ring is a heteroatom.

EXAMPLE 6-3 Provide a systematic name for each of the following common names and structures:

(a) ethyl iodide; (b) ethyl methyl ether; (c) propyl bromide; (d) O⬦$-CH_3$

Solution It often helps to write out the structural formula first:
(a) CH_3CH_2I = iodoethane
(b) $CH_3OCH_2CH_3$ = methoxyethane (or 2-oxabutane)
(c) $H_3CCH_2CH_2Br$ = 1-bromopropane
(d) 3-methyloxacyclobutane

B. Functional groups that are named as suffixes

Compounds containing one of the remaining functional groups in Table 6-1 (that is, those other than halide, nitro, and alkoxy) are named by replacing the "e" from the name of the corresponding (cyclo)alkane with the appropriate suffix from Table 6-2. Note that the carbon of the functional groups marked with an asterisk in Table 6-2 defines the number-1 carbon of the alkane chain. Furthermore, be sure to pick the longest chain of which the functional group is a part for the root name. Several examples will help clarify these points.

TABLE 6-2. Functional Group Suffixes

Generic structure	Functional group suffix	Sample compounds	
		Formula	Name
$[R(H)]_4\overset{+}{N}$	-aminium ion (-ammonium ion)	$CH_3NH_3^+$	Methaminium ion (methylammonium ion)
$RC\!\!-\!\!O^-$ $\overset{\|}{O}$	-oate (ion)*	$CH_3CH_2CH_2CO_2^-$	Butanoate (ion)
$RC\!\!-\!\!OR'$ $\overset{\|}{O}$	-oate* (with R' name as separate preceding word)	$CH_3CH_2CH_2CO_2CH_3$	Methyl butanoate
$RC\!\!-\!\!OH$ $\overset{\|}{O}$	-oic acid*	$CH_3CH_2CH_2CO_2H$	Butanoic acid
$RC\!\!-\!\!X$ $\overset{\|}{O}$ (X = halide)	-oyl halide*	$CH_3CH_2CH_2COCl$	Butanoyl chloride
$RC\!\!-\!\!NR_2'$ $\overset{\|}{O}$	-amide*	$CH_3CH_2CH_2CONHCH_3$	*N*-Methylbutanamide**
$RC\!\!\equiv\!\!N$	-enitrile*	$CH_3CH_2CH_2CN$	Butanenitrile
$RC\!\!-\!\!H$ $\overset{\|}{O}$	-al*	$CH_3CH_2CH_2CHO$	Butanal
RCR' $\overset{\|}{O}$	-one	$CH_3CH_2COCH_3$	2-Butanone
ROH	-ol	$CH_3CH_2CHCH_3$ $\mid$ OH	2-Butanol
RNH_2	-amine	$CH_3CH_2CH_2CH_2NH_2$	Butanamine (butylamine)

*Carbon atom of the functional group defines the number-1 carbon (C-1) of the alkane chain.
**The locant "N" indicates the methyl substituent is attached to the nitrogen.

EXAMPLE 6-4 Provide the systematic name for each of the following structures:

(a) $(CH_3CH_2CH_2CH_2)_4N^+Br^-$

(b) $CH_3CH_2\underset{\underset{O}{\|}}{C}OCH_2CH_3$

(c) 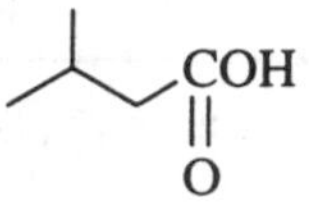 [structure: branched chain with COH and =O]

(d) 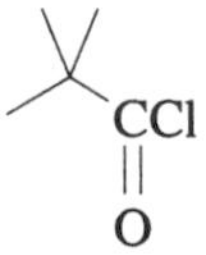 [structure: CCl with =O]

(e)  [structure: chain with CNH₂ and =O]

(f) 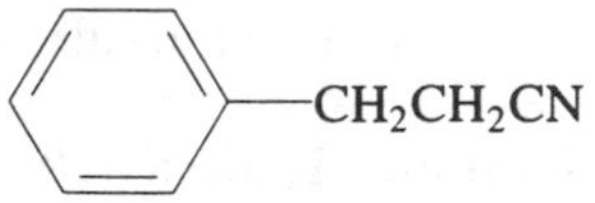 [benzene ring]—CH_2CH_2CN

(g) 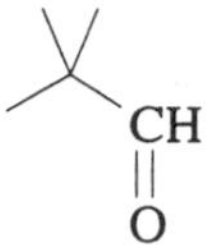 [structure: CH with =O]

(h) 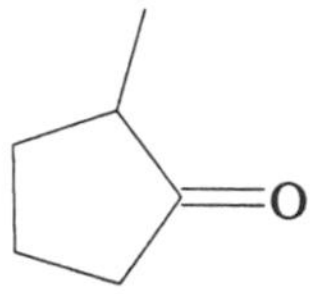 [methylcyclopentanone structure =O]

(i) 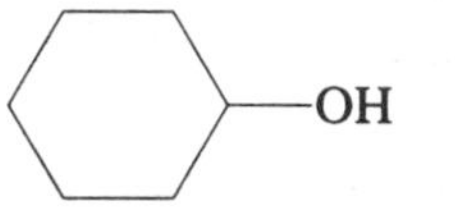 [cyclohexane ring]—OH

(j)  [chain]—$N(CH_3)_2$

(k) 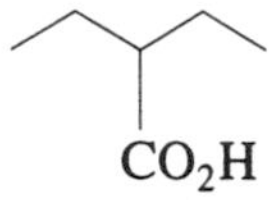 [branched chain with CO_2H]

(l) 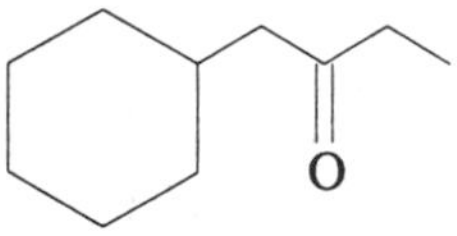 [cyclohexane ring with ketone chain, =O]

Solution
(a) *N,N,N*-tributylbutanaminium bromide (tetrabutylammonium bromide)
(b) ethyl propanoate   (c) 3-methylbutanoic acid
(d) 2,2-dimethylpropanoyl chloride (e) hexanamide   (f) 3-phenylpropanenitrile
(g) 2,2-dimethylpropanal   (h) 2-methylcyclopentanone (i) cyclohexanol
(j) *N,N*-dimethylbutanamine   (k) 2-ethylbutanoic acid   (l) 1-cyclohexyl-2-butanone

Before we practice naming more monofunctionalized organic compounds, let's discuss some of the common names that are sufficiently important to justify memorizing them. First, the one- and two-carbon molecules that possess the functional groups in Table 6-2 are usually known by their common names, shown in Table 6-3.

TABLE 6-3. Some Common Names of One- and Two-Carbon
Organic Compounds

Structure	Systematic name	Common name
HCO_2^- *	Methanoate	Formate
HCO_2H	Methanoic acid	Formic acid
$HCOCl$	Methanoyl chloride	Formyl chloride
$HCONH_2$	Methanamide	Formamide
H_2CO	Methanal	Formaldehyde
$CH_3CO_2^-$ *	Ethanoate	Acetate (acetoxy)
CH_3CO_2H	Ethanoic acid	Acetic acid
CH_3COCl	Ethanoyl chloride	Acetyl chloride
CH_3CONH_2	Ethanamide	Acetamide
CH_3CN	Ethanenitrile	Acetonitrile
CH_3CHO	Ethanal	Acetaldehyde
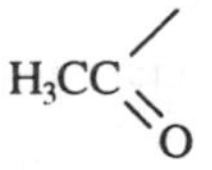 $H_3CC\underset{\diagdown O}{\diagup}$	1-Oxoethyl (group)	Acetyl (group)

*Groups such as formate, acetate, and other homologs may also exist as uncharged groups covalently bound to organic residues, as in esters.

For aldehydes, carboxylic acid derivatives, and nitriles, where the functional group is directly attached to a ring, the name of the structure takes the form cycloalkanecarboxaldehyde, cycloalkanecarboxylic acid, or cycloalkanecarbonitrile, respectively. Thus cyclopropanecarboxylic acid has the structure

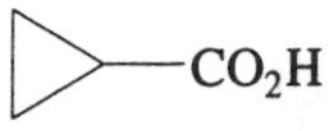

$\triangleright\!\!-\!\!CO_2H$

The common names of ketones are constructed in a manner similar to the common names of ethers: name both alkyl groups (separately, in alphabetical order), followed by the word ketone. So, 2-butanone (Table 6-2) could be named ethyl methyl ketone (or, less satisfactorily, methyl ethyl ketone). The common name of 2-propanone is **acetone** (dimethyl ketone). Armed with this information, let's try to name some other functionalized molecules.

EXAMPLE 6-5 Drawing on the list of common names in Table 6-3, provide a name for each structure below:

$$\text{(a)} \quad CH_3\overset{\displaystyle O}{\overset{\displaystyle \|}{C}}OCH_2CH_3 \qquad \text{(b)} \quad H\overset{\displaystyle O}{\overset{\displaystyle \|}{C}}N(CH_3)_2$$

Solution (a) ethyl acetate (systematic: ethyl ethanoate); (b) *N,N*-dimethylformamide (systematic: *N,N*-dimethyl methanamide)

6-4. Naming Polyfunctionalized Organic Molecules

A. Functionalized (cyclo)alkenes and alkynes

The method for naming functionalized (cyclo)alkenes and alkynes follows logically from the rules we developed in Section 6-3 for functionalized (cyclo)alkanes. However, we must remember that the multiple bond of the alkene or alkyne is itself another functional group.

When a (cyclo)alkene or alkyne contains one or more of the functional groups that always appear as prefixes (halo, nitro, alkoxy), the functional groups are simply added alphabetically as prefixed substituents (with appropriate locators) to the alkene or alkyne name. If the alkene or alkyne is *acyclic*, numbering of the carbons begins at the chain end nearest the multiple bond, just as we saw for nonfunctionalized alkenes and alkynes in Sections 5-6 and 5-7. For cycloalkenes and cycloalkynes, the multiple bond defines carbons 1 and 2 of the ring (Section 5-8). Here are a few examples:

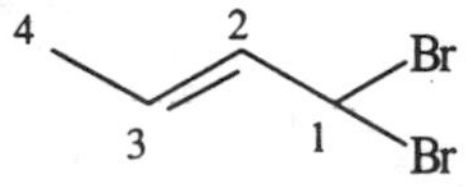

trans-1,1-dibromo-2-butene

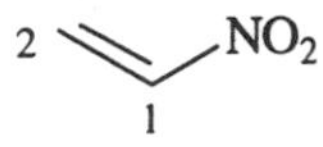

nitroethene (nitroethylene); the locator "1-" is superfluous

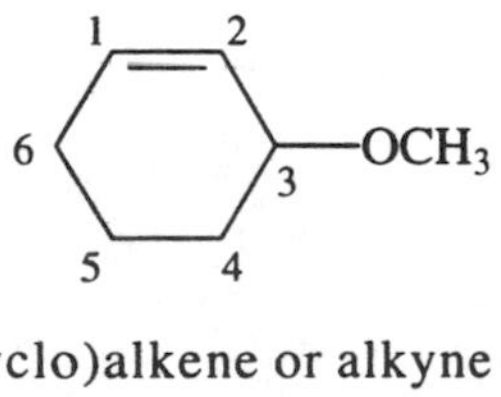

3-methoxycyclohexene (in a cycloalkene the doubly bound carbons are numbered 1 and 2)

When a (cyclo)alkene or alkyne contains one of the suffixed functional groups from Table 6-2, we use the same procedures as we did for functionalized alkanes (Section 6-3): delete the final "e" from the hydrocarbon name and add the appropriate suffix from Table 6-2. Once again, though, we must be careful to pick the longest chain (or largest ring) that contains the functional group. Also, this functional group takes precedence over the multiple bond when numbering carbons. Finally, remember that those groups marked with an asterisk in Table 6-2 define carbon 1 of the chain. Here are some examples.

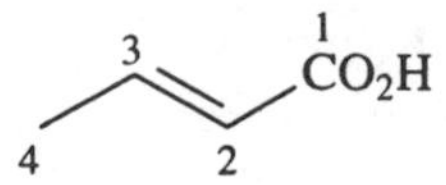

trans-2-butenoic acid (the "2-" indicates where the double bond begins)

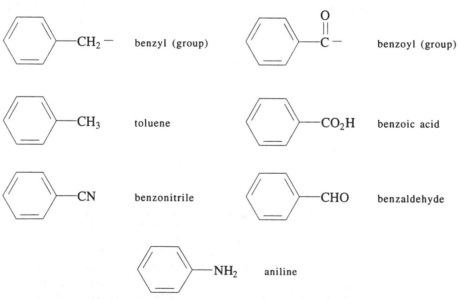

3-butynal

methyl 1-cyclohexenecarboxylate

In the case of the other functional group suffixes in Table 6-2 (-one, -ol, -amine), it is necessary to specify the location of the functional group by inserting its locator directly before the suffix. Numbering of the carbons begins at the end nearest the functional group.

3-butyn-2-ol (or but-3-yn-2-ol)

trans-3-penten-2-one (or *trans*-pent-3-en-2-one)

2-cyclohexenone; the "1-" locator for the "-one" suffix is superfluous

Finally, the structures below show how some of the common functionalized aromatic compounds are named.

benzyl (group)

benzoyl (group)

toluene

benzoic acid

benzonitrile

benzaldehyde

aniline

EXAMPLE 6-6   Provide a systematic name for each structure below:

(a)

(b) O_2N—⬡—CH_3

(c) 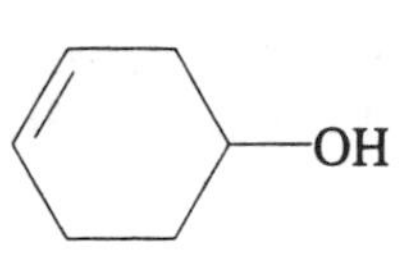 [structure: cyclohexene with —OH] **(d)** 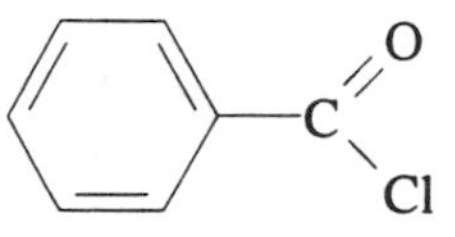 [structure: benzene with —C(=O)Cl]

(e) 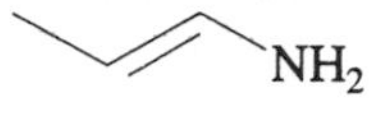 [structure: CH=CH—NH₂] **(f)** 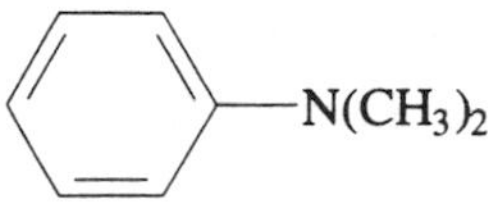 [structure: benzene with —N(CH₃)₂]

(g) 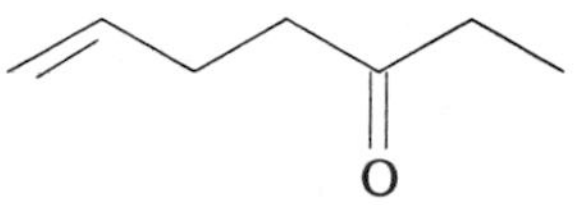 [structure: alkene chain with ketone O] **(h)** 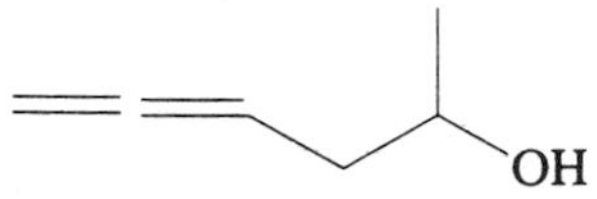 [structure: diene chain with OH]

Solution (**a**) 3-butenoyl chloride, (**b**) 4 (or para)-nitrotoluene, (**c**) 3-cyclohexenol, (**d**) benzoyl chloride, (**e**) *trans*-1-propenamine, (**f**) *N,N*-dimethylaniline, (**g**) 6-hepten-3-one, (**h**) 4,5-hexadien-2-ol

B. Compounds containing one prefixed and one suffixed functional group

We can used the methodology we've already described to handle structures with both a suffixed functional group *and* one or more prefixed functional groups. First, identify the longest chain that contains the suffixed group, as we did in Sections 6-3 and 6-4A, and add the appropriate suffix to the root name. Then add all prefixes (with locators) in alphabetical order. Here are two examples.

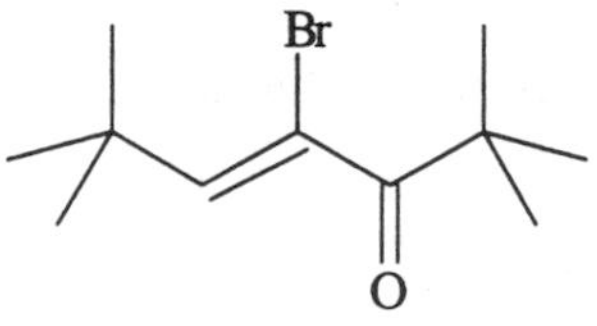

[structure] Z-4-bromo-2,2,6,6-tetramethyl-4-hepten-3-one

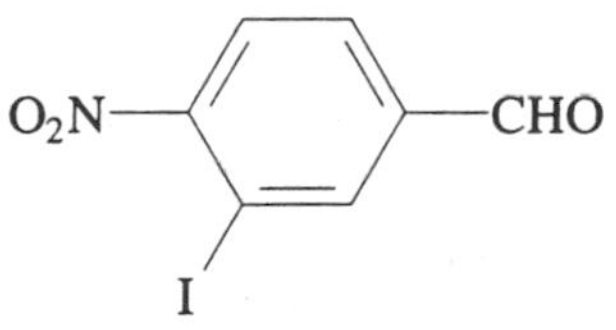

[structure] 3-iodo-4-nitrobenzaldehyde

Now, you try some.

EXAMPLE 6-7 Provide a systematic name for each structure below.

(a) 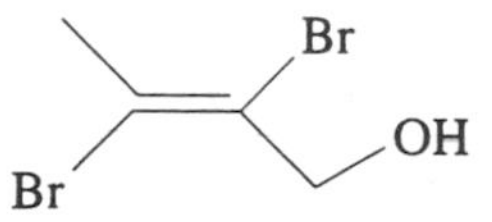 [structure] **(b)** 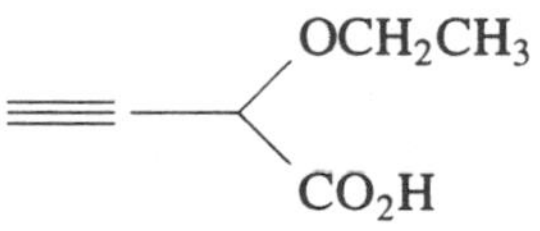 [structure]

(c) 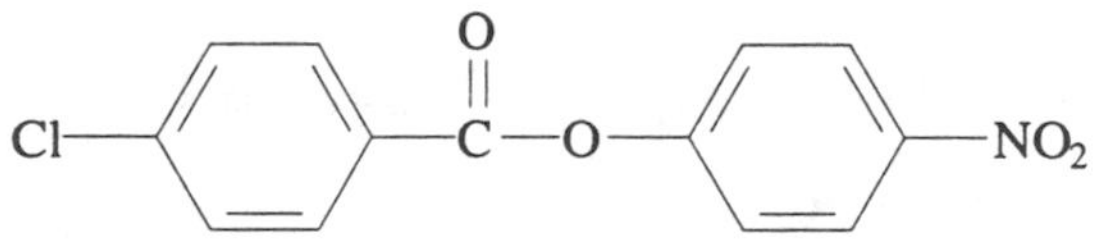 [structure: Cl—benzene—C(=O)—O—benzene—NO₂]

Solution (**a**) *trans*-2,3-dibromo-2-buten-1-ol, (**b**) 2-ethoxy-3-butynoic acid, (**c**) 4 (or para)-nitrophenyl 4 (or para)-chlorobenzoate

C. Compounds with two or more suffixed functional groups

Now we have a problem. What do we do when there are two (or more) of the suffixed functional groups from Table 6-2 in the same molecule? Clearly, one of the groups will have to be assigned highest priority, and the chosen suffix will correspond to that group. But how do we assign relative priority to the functional groups, and how do we handle the remaining group(s)? Not to worry! The functional groups in Table 6-2 are listed from highest priority $[R(H)]_4N^+$ to lowest (RNH_2).

EXAMPLE 6-8 Circle the functional group with highest priority in each structure below:

(a) 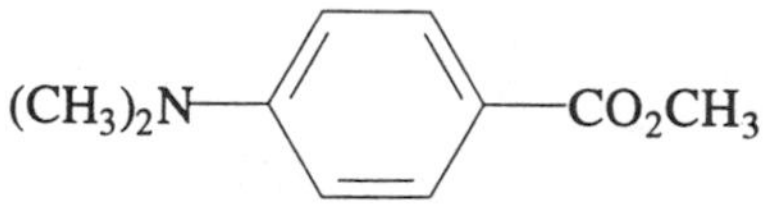 $(CH_3)_2N$—[benzene ring]—CO_2CH_3 **(b)** 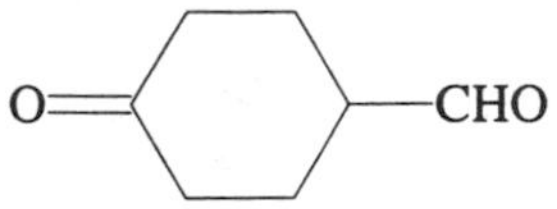 O=[cyclohexane ring]—CHO **(c)** 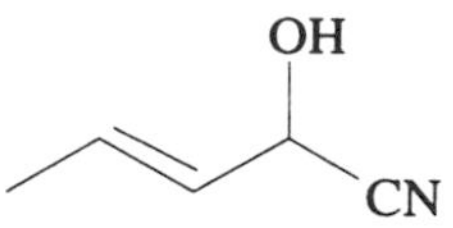 [structure with OH and CN]

Solution

(a) 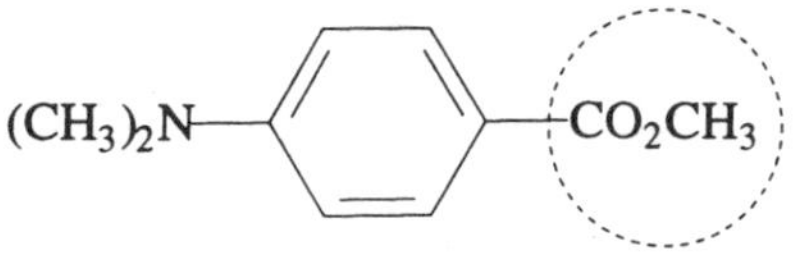 $(CH_3)_2N$—[benzene ring]—(CO_2CH_3) **(b)** 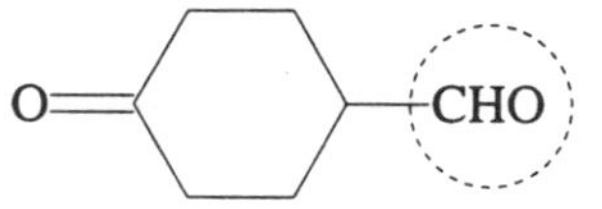 O=[cyclohexane ring]—(CHO)

(c) 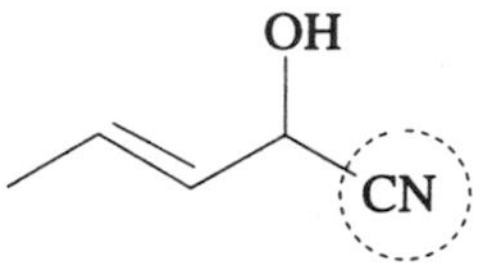 [structure with OH and (CN)]

All remaining functional groups will be named as prefixes, using the functional group names from Table 6-1. Here are a few examples.

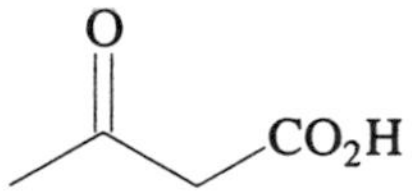

[structure with O and CO_2H]   3-ketobutanoic acid (or 3-oxobutanoic acid)

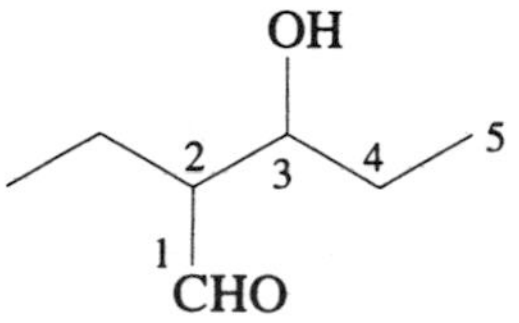

[structure with OH and CHO, numbered 1-5]   2-ethyl-3-hydroxypentanal (pick the chain that contains the suffixed functional group)

Now it's your turn.

EXAMPLE 6-9 Provide a systematic name for each structure below.

(a) 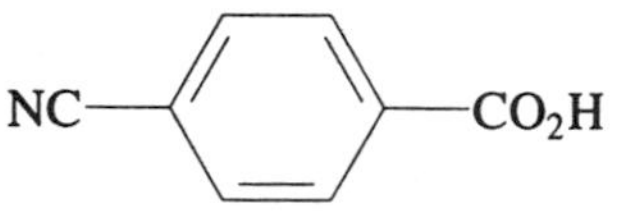 NC—[benzene ring]—CO_2H **(b)** 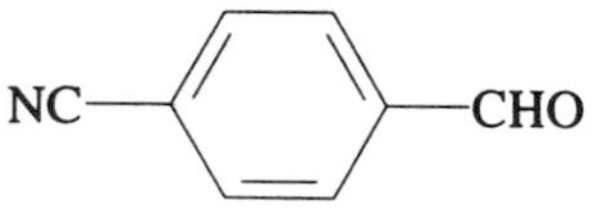 NC—[benzene ring]—CHO

(c) 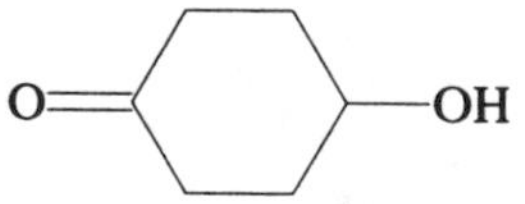 O=[cyclohexane ring]—OH **(d)**  H_2N—[chain]—OH

Solution **(a)** 4 (or para)-cyanobenzoic acid, **(b)** 4 (or para)-formylbenzonitrile, **(c)** 4-hydroxycyclohexanone, **(d)** 2-aminoethanol (the common name is ethanolamine)

Dicarboxylic acids such as those in Table 6-4 have systematic names of the form alkanedicarboxylic acid, where both carboxy carbons are considered part of the chain. However, these compounds are better known by their common names shown in Table 6-4. There is a mnemonic device to help you remember the first letter of each of these common names: **Oh, my,** such **g**ood **a**pple **p**ie, **s**weet **a**s **s**ugar.

TABLE 6-4. The First Nine Dicarboxylic Acids

Structure	Systematic name	Common name
HO_2CCO_2H	Ethanedicarboxylic acid	Oxalic acid
$HO_2CCH_2CO_2H$	Propanedicarboxylic acid	Malonic acid
$HO_2C(CH_2)_2CO_2H$	Butanedicarboxylic acid	Succinic acid
$HO_2C(CH_2)_3CO_2H$	Pentanedicarboxylic acid	Glutaric acid
$HO_2C(CH_2)_4CO_2H$	Hexanedicarboxylic acid	Adipic acid
$HO_2C(CH_2)_5CO_2H$	Heptanedicarboxylic acid	Pimelic acid
$HO_2C(CH_2)_6CO_2H$	Octanedicarboxylic acid	Suberic acid
$HO_2C(CH_2)_7CO_2H$	Nonanedicarboxylic acid	Azelaic acid
$HO_2C(CH_2)_8CO_2H$	Decanedicarboxylic acid	Sebacic acid

The common names of the three isomeric benzenedicarboxylic acids are also worth knowing:

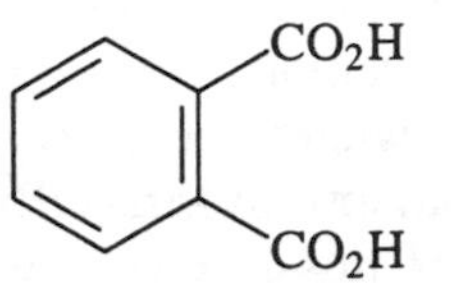

phthalic acid 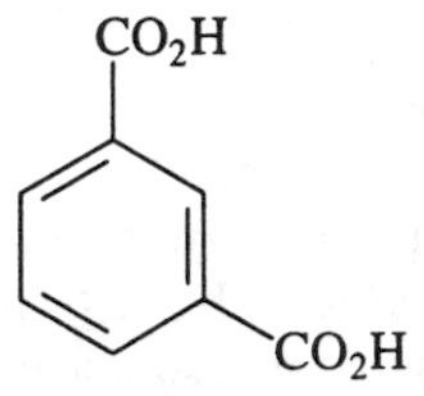   isophthalic acid 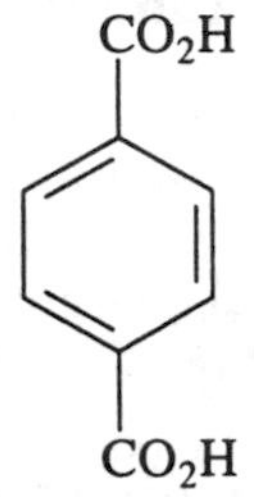   terephthalic acid

Esters of dicarboxylic acids can be either monoesters (with one carboxylic acid group remaining) or diesters. Monoesters have just one R′ group name, while diesters have two.

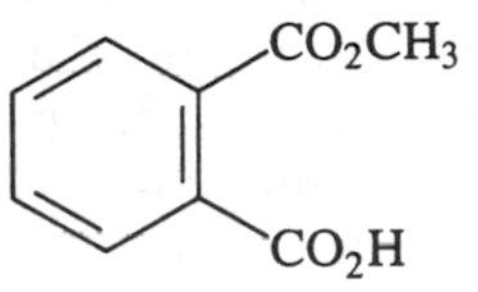

methyl phthalate 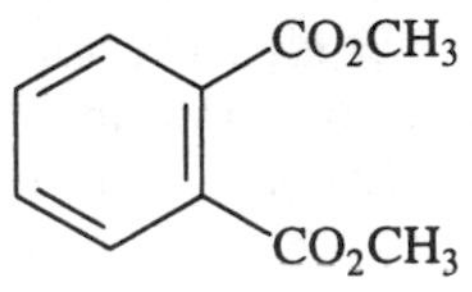   dimethyl phthalate 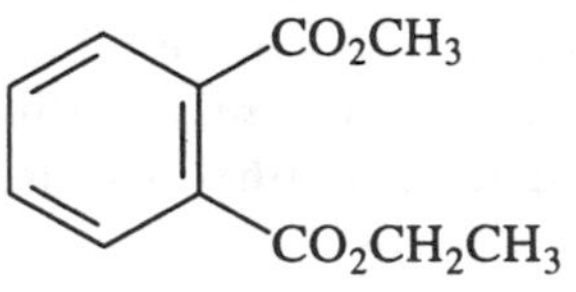   ethyl methyl phthalate

EXAMPLE 6-10 Provide a systematic name for each ester below:

(a) 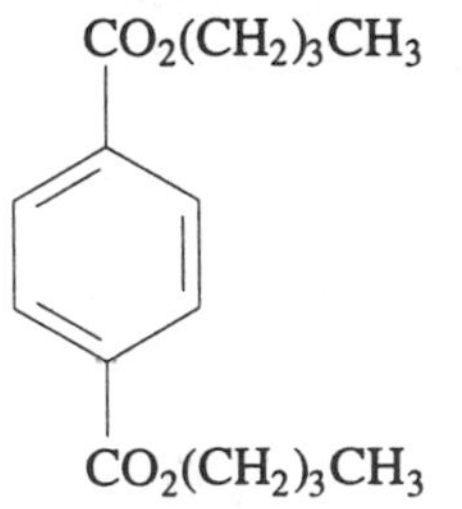 (b) 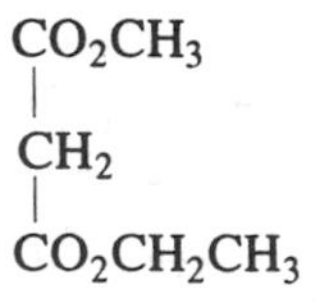 (c)

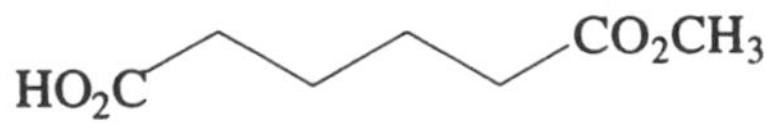

Solution (a) dibutyl terephthalate, (b) ethyl methyl malonate, (c) methyl succinate

As you review these nomenclature sections, don't despair. There is a certain amount of tedium to the task of nomenclature, but a working knowledge of it is necessary if you want to become conversant in organic chemistry. Just think how embarrassing it would be to complete a spectacular multistep organic synthesis, only to be unable to name the product!

6-5. Other, Less Common Functional Groups

The vast majority of organic reactions involve the functional groups described in Section 6-2 (Table 6-1). However, there are many other functional groups that we'll encounter from time to time. While it's not our purpose to provide an exhaustive (and exhausting!) listing of *all* organic functional groups, it is worthwhile to have at least a nodding acquaintance with some of these less common functional groups. But take heart—in this book you'll not be expected to provide systematic names for compounds containing these functional groups. Just be able to identify them when they occur in a molecule.

A. Some other functional groups containing oxygen

The anion that results when a hydrogen ion (H^+) is removed from an alcohol molecule is called an **alkoxide ion**, RO^-. Two oxygen atoms connected by a single bond constitute a **peroxide linkage**. A molecule with structure ROOR is called a **peroxide**, while one with structure ROOH is a **hydroperoxide**.

Two acyl groups connected by an oxygen atom constitute a **carboxylic acid anhydride** (RCO_2COR'), or just *anhydride* for short. And finally, molecules with a C=C=O linkage are known as **ketenes.**

B. Some other functional groups containing nitrogen

A **nitroso** group (–NO) resembles a nitro group ($–NO_2$), but it doesn't have the second oxygen. An **azo compound** has two doubly bonded nitrogen atoms connecting two organic residues, as in RN=NR′. There are several related functional groups where one or more of the nitrogens possess formal charge: **diazo compounds** ($R_2C=\overset{+}{N}=\overset{..}{N}$), **diazonium ions** ($R\overset{+}{N}\equiv N$), and **azides** ($RN=\overset{+}{N}=\overset{..}{N}$). For each of these, several resonance forms are possible (see, e.g., Problem 4-3).

C. Sulfur-containing functional groups

The sulfur analog of an alcohol, RSH, is called a **thiol** (or **mercaptan**). Similarly, the sulfur analog of an ether, RSR′, is a **sulfide**, and the analog of a peroxide, RSSR′, is a **disulfide**. These functional groups are especially important in certain biochemical reactions.

Sulfur, being a third-period element, has a set of empty $3d$ orbitals (Section 1-6). Because of this, sulfur can form bonds that, at first glance, seem to violate the octet rule. For example, organic sulfides react with oxygen to form **sulfoxides**. The S–O bond in such compounds can be written as either a zwitterionic single bond, or as a double bond with no formal charges.

In the doubly bonded resonance form, the sulfur has ten valence electrons around it. But this is allowed because the double bond actually results from the overlap of a filled p orbital on oxygen with an empty d orbital on sulfur, giving what's known as a p–d π bond.

A second oxygen can be added to the sulfoxide sulfur to form a **sulfone**, $R\overset{\overset{O}{\|}}{\underset{\underset{O}{\|}}{S}}R'$.

D. Functional groups containing phosphorus

Phosphorus analogs of amines are **phosphines**, R_3P. Like sulfur, phosphorus is a third-period element capable of expanding its valence beyond eight electrons. Compounds of the type $R_3P=O$ are known as **phosphine oxides**.

E. Esters of inorganic acids

You are undoubtedly familiar with mineral acids such as HCl. There are many inorganic oxyacids (in which the acidic hydrogen is attached to oxygen) that bear a similarity to carboxylic acids. Here are a few of them:

$$\text{sulfonic acid:} \quad R\overset{\displaystyle O}{\underset{\displaystyle O}{\overset{\|}{\underset{\|}{S}}}}OH$$

$$\text{sulfuric acid:} \quad HO\overset{\displaystyle O}{\underset{\displaystyle O}{\overset{\|}{\underset{\|}{S}}}}OH$$

$$\text{phosphonic acid:} \quad R\overset{\displaystyle O}{\underset{\displaystyle OH}{\overset{\|}{\underset{|}{P}}}}OH$$

$$\text{phosphoric acid:} \quad HO\overset{\displaystyle O}{\underset{\displaystyle OH}{\overset{\|}{\underset{|}{P}}}}OH$$

$$\text{nitrous acid:} \quad O{=}NOH$$

$$\text{nitric acid:} \quad O{=}\overset{\displaystyle O^-}{\underset{+}{\overset{|}{N}}}OH$$

Each of these can form esters by substituting one or more of the acidic hydrogens by an R′ group. For the most part, these "inorganic" esters are more reactive than typical carboxylate esters.

$$\text{alkyl sulfonate:} \quad R\overset{\displaystyle O}{\underset{\displaystyle O}{\overset{\|}{\underset{\|}{S}}}}OR'$$

$$\text{dialkyl sulfate:} \quad RO\overset{\displaystyle O}{\underset{\displaystyle O}{\overset{\|}{\underset{\|}{S}}}}OR'$$

$$\text{dialkyl phosphonate:} \quad R\overset{\displaystyle O}{\underset{\displaystyle OR''}{\overset{\|}{\underset{|}{P}}}}OR'$$

$$\text{trialkyl phosphate:} \quad RO\overset{\displaystyle O}{\underset{\displaystyle OR''}{\overset{\|}{\underset{|}{P}}}}OR'$$

$$\text{alkyl nitrite:} \quad O{=}NOR$$

$$\text{alkyl nitrate:} \quad O{=}\overset{\displaystyle O^-}{\underset{+}{\overset{|}{N}}}OR$$

F. Organometallic compounds

One of the most important areas of modern chemistry involves a marriage between organic and inorganic chemistry. **Organometallic compounds** are those containing at least one carbon–metal bond.

One of the most common types of organometallic compound has the structure RM, where M represents a group IA metal such as Li, Na, or K. Examples include methyllithium (CH_3Li), butyllithium ($CH_3CH_2CH_2CH_2Li$), and phenyllithium (C_6H_5Li). Another important group is that of Grignard reagents RMgX (X = Cl, Br, I), named after the Nobel prize-winning French chemist who discovered them. Both of these types of organometallic compounds involve a highly polar carbon–metal σ bond, with carbon bearing the δ– charge, and the metal bearing the δ+ charge (see Section 2-6).

Notice that ionic compounds of the type M^+ ^-OR (metal alkoxides) are *not* considered organometallic compounds because the metal is attached to oxygen, not carbon.

Admittedly, this has been only a brief introduction to some of these more exotic functional groups. But at least you won't be caught completely off guard if you should encounter one some day. All the groups mentioned in this section are summarized in Table 6-5.

6-6. The Index of Hydrogen Deficiency Revisited

In Section 5-3 we introduced the index of hydrogen deficiency [*I*, Eq. (5-1)], which tells us the number of rings plus π bonds for any valid structure with the given hydrocarbon molecular formula. This equation can be modified so it's valid for any molecular formula containing carbon, hydrogen, oxygen, sulfur, nitrogen, phosphorus, and halogen.

INDEX OF HYDROGEN DEFICIENCY IN ORGANIC HETEROCOMPOUNDS

$$I = \frac{2C + 2 + N + P - H - hal}{2} \qquad (6\text{-}1)$$

Notice that neither oxygen nor sulfur appears in this equation, even though the equation is still valid for compounds containing these elements. This is because both of these elements are divalent. Furthermore, zwitterionic bonds such as P=O (P$^+$–O$^-$) and S=O (S$^+$–O$^-$) are not counted as having π bonds in this treatment.

For example, the molecular formula C_3H_6O has $I = [2(3) + 2 - 6]/2 = 1$. So, any valid structure with this molecular formula will have one ring or one π bond. There are nine possible isomeric structures that meet this criterion:

Note that if you ever come up with a noninteger (or worse, a negative) value for I, either the molecular formula is not valid for a neutral molecule, or you've made some sort of mistake in the calculation. A corollary to this is the so-called **Nitrogen Rule**: If the number of nitrogen atoms in a molecule is *even* (0, 2, 4,...), the number of hydrogens plus halogens must also be even.

EXAMPLE 6-11 Calculate the I value for the molecular formula $C_3H_4NO_2Br$, and indicate the number of rings plus bonds present in any valid structure with this molecular formula. Which of the functional group classes described in this chapter can be accommodated in structures with this molecular formula?

Solution Using Eq. (6-1), we calculate the value of I:

$$I = \frac{2C + 2 + N + P - H - hal}{2} = \frac{2(3) + 2 + 1 + 0 - 4 - 1}{2} = 2$$

Thus any valid structure with this molecular formula must have a total of two rings plus π bonds: two rings, one π bond and one ring, or two π bonds. Remember that the π bonds include not only C–C π bonds, but also C–O, C–N, N–O, and N–N π bonds. Reviewing the functional group classes in Tables 6-1 and 6-5 shows that the following could be accommodated by the molecular formula:

(1) C-C double and triple bonds (8) carboxylic acid   (14) peroxide and hydroperoxide
(2) halides (except aryl)   (9) ester   (15) ketene
(3) alcohol   (10) acid halide   (16) nitroso
(4) ether   (11) imine   (17) nitrite
(5) amine   (12) nitrile   (18) nitrate
(6) aldehyde   (13) nitro   (19) ammonium carboxylate
(7) ketone

Admittedly, some will have multiple functional groups, and some may have rather bizarre structures.

Here is a list of the functional group classes that cannot be accommodated by the given molecular formula:

(1)	aromatic ring	(4)	diazonium ion	(7)	all sulfur functional groups
(2)	anhydride	(5)	azo compound	(8)	all phosphorus functional groups
(3)	diazo compound	(6)	azide	(9)	organometallic compounds

TABLE 6-5. Less Common Functional Group Classes

Element	Structure	Class	Element	Structure	Class
Oxygen	RO^-	Alkoxide ion	*Sulfur (contd.)*		
	$ROOR'$	Peroxide		$\underset{\underset{O}{\overset{\overset{O}{\|\|}}{\|\|}}}{RSR'}$	Sulfone
	$ROOH$	Hydroperoxide			
	$\underset{\overset{\|\|}{O}}{(H)RC}\!-\!O\!-\!\underset{\overset{\|\|}{O}}{CR'(H)}$	Anhydride		$\underset{\underset{O}{\overset{\overset{O}{\|\|}}{\|\|}}}{ROSOR'}$	Sulfate (ester)
	$R_2C\!=\!C\!=\!O$	Ketene			
Nitrogen	$RN\!=\!O$	Nitroso compound		$\underset{\overset{\|\|}{O}}{RSOR'}$	Sulfonate (ester)
	$RN\!=\!NR'$	Azo compound			
	$R_2C\!=\!\overset{+}{N}\!=\!\bar{N}$	Diazo compound	Phosphorus	$R_3P{:}$	Phosphine
	$R\overset{+}{N}\!\equiv\!N$	Diazonium ion		$R_3P\!=\!O$	Phosphine oxide
	$RN\!=\!\overset{+}{N}\!=\!\bar{N}$	Azide		$\underset{\underset{OR''}{\|}}{\overset{\overset{O}{\|\|}}{RPOR'}}$	Phosphonate (ester)
	$RON\!=\!O$	Nitrite (ester)			
	$RONO_2$	Nitrate (ester)		$\underset{\underset{OR''}{\|}}{\overset{\overset{O}{\|\|}}{ROPOR'}}$	Phosphate (ester)
Metal	RM	Organometallic compound			
	$RMgX$	Grignard (organometallic) reagent			
Sulfur	RSH	Thiol (mercaptan)			
	RSR'	Sulfide			
	$RSSR'$	Disulfide			
	$\underset{\overset{\|\|}{O}}{RSR'}$	Sulfoxide			

SUMMARY

1. The functional group of a molecule is the atom or group of atoms at which reactions take place. Therefore, the type of functional group(s) present in a molecule, more than any other structural feature, determines the chemical behavior of a molecule. Other than carbon–carbon multiple bonds, all other functional groups involve one or more heteroatoms (atoms other than carbon or hydrogen).

2. The common organic functional groups are listed in Table 6-1, which shows the structure, functional group name, and functional group class of each. Some of the less common but nonetheless important functional groups are listed in Table 6-5. Before proceeding beyond this chapter, you should be able to recognize and identify these functional groups from memory.

3. In order to generate systematic names of functionalized organic molecules, we use a modified version of the nomenclature rules developed in Chapter 5 for hydrocarbons. Some functional groups (e.g., halides, alkoxy groups, and nitro groups) always appear in alphabetical order as prefixes (with numerical locators) to the appropriate hydrocarbon name. The remaining functional groups in Table 6-1 are named as suffixes (see Table 6-2) to the hydrocarbon name. When there are two or more suffixed functional groups, the one with the highest priority (the highest one in Table 6-2) appears as the suffix, while any others are named as prefixes using the functional group names (Table 6-1).

4. Most smaller organic molecules have common names as well as systematic names. You should

be familiar with the common names in Tables 6-3 and 6-4 and the names of the common functionalized aromatic compounds (Section 6-4).

5. The index of hydrogen deficiency (I) can be calculated, using Eq. (6-1), for molecular formulas with heteroatoms:

$$I = \frac{2C + 2 + N + P - H - hal}{2} \qquad (6\text{-}1)$$

The resulting I value gives the number of rings plus π bonds in any valid structure with that molecular formula. Note that the π bonds include not only carbon–carbon multiple bonds, but also carbon–heteroatom and heteroatom–heteroatom multiple bonds, but *not* zwitterionic bonds (e.g., P^+-O^-).

RAISE YOUR GRADES

Can you define...?

☑ functional group
☑ Ar (group)
☑ geminal
☑ heterocyclic
☑ organometallic compound
☑ primary (1°), secondary (2°), and tertiary (3°) as applied to halides, alcohols, amines, and amides

☑ R (group)
☑ heteroatom
☑ vicinal
☑ oxirane (epoxide)

Can you explain...?

☑ how to generate systematic names for molecules with one or more of the functional groups in Table 6-1
☑ the significance of I (the index of hydrogen deficiency) and how to calculate it for molecular formulas with heteroatoms

Can you draw structures for....?

☑ all the functional groups listed in Tables 6-1 and 6-5
☑ all the common names in Tables 6-3 and 6-4, and the functionalized aromatic compounds in Section 6-4
☑ the following common names:

diethyl ether	acetone
a typical organometallic compound	a Grignard reagent

SOLVED PROBLEMS

PROBLEM 6-1 Each structure below is monofunctionalized. Without referring to Table 6-1, circle each functional group and identify the functional group class. Be sure to add the designators 1°, 2°, and 3° as appropriate.

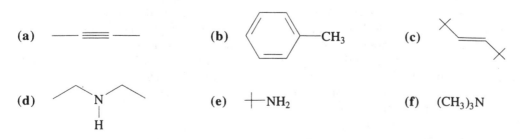

(g)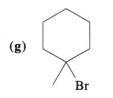

(h) CH₃CH₂I

(i)

(j)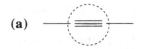

(k) $\leftthreetimes$—OH

(l) ▷—OH

(m)

(n) CH₃CHO

(o) HCO₂H

(p) H₃CCOC(CH₃)₃

(q) 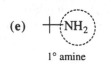 CO₂⁻

(r) H₃CCOCl

(s)

(t)

(u)

(v) (CH₃)₄N⁺

(w)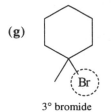

(x)

(y) ◻—CN

Solution

(a) alkyne

(b) aromatic ring

(c) alkene

(d) 2° amine

(e) 1° amine

(f) (CH₃)₃—N 3° amine

(g) Br 3° bromide

(h) CH₃CH₂—I 1° iodide

(i) Cl 2° chloride

(j) OH 1° alcohol

(k) OH 3° alcohol

(l) ▷—OH 2° alcohol

(m) [cyclohexanone structure] ketone

(n) CH_3—CHO aldehyde

(o) H—CO_2H carboxylic acid

(p) H_3C—C(=O)—O—$C(CH_3)_3$ carboxylate ester

(q) [pentyl chain]—CO_2^- carboxylate (anion)

(r) H_3C—C(=O)—Cl acid chloride

(s) H—C(=O)—NH_2 1° amide

(t) H_3CC(=O)—NHCH₃ 2° amide

(u) [structure] C(=O)N with ethyl groups 3° amide

(v) $(CH_3)_4N^+$ quaternary aminium (ammonium) ion

(w) [ether structure with O] ether

(x) [structure]—NO_2 2° nitro compound

(y) [cyclobutane]—CN nitrile

PROBLEM 6-2 Provide a name for each of the structures in Problem 6-1.

Solution **(a)** 2-butyne; **(b)** toluene; **(c)** *trans*-2,2,5,5-tetramethyl-3-hexene; **(d)** *N*-ethylethanamine (diethylamine); **(e)** 1,1-dimethylethanamine (*tert*-butylamine); **(f)** *N,N*-dimethylmethanamine (trimethylamine); **(g)** 1-bromo-1-methylcyclohexane; **(h)** iodoethane; **(i)** 2-chlorobutane; **(j)** 1-pentanol; **(k)** 2- methyl-2-propanol (*tert*-butyl alcohol); **(l)** cyclopropanol; **(m)** 4-methylcyclohexanone; **(n)** acetaldehyde (ethanal); **(o)** formic (methanoic) acid; **(p)** *tert*-butyl acetate (ethanoate); **(q)** hexanoate (anion); **(r)** acetyl (ethanoyl) chloride; **(s)** formamide (methanamide); **(t)** *N*-methylacetamide (*N*-methylethanamide); **(u)** *N,N*-diethylbutanamide; **(v)** *N,N,N*-trimethylmethanaminium ion (tetramethylammonium ion); **(w)** 3-ethoxypentane; **(x)** 2-nitropropane; **(y)** cyclobutanecarbonitrile

PROBLEM 6-3 Each structure below is monofunctionalized. Without referring to Table 6-5, circle each functional group and identify the functional group class.

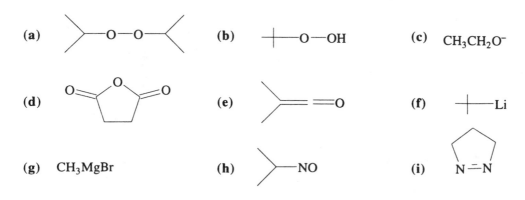

(a) [structure] O—O

(b) [structure] O—OH

(c) $CH_3CH_2O^-$

(d) [cyclic anhydride structure]

(e) [structure] =O

(f) [structure] Li

(g) CH_3MgBr

(h) [structure]—NO

(i) [five-membered ring] N—N

(j) 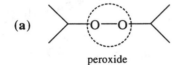 **(k)** $H_3C\overset{+}{N}\equiv\overset{-}{N}$ **(l)** $CH_3CH_2N_3$

(m) ◇—ONO_2 **(n)** 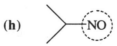 ONO **(o)** ⟍⟍⟋SH

(p) (ring with S) **(q)** ⟍S—S⟋ **(r)** H_3CSCH_3, with $\overset{O}{\|}$

(s) (ring with $S\overset{O}{\underset{O}{\lessgtr}}$) **(t)** $H_3COSOCH_3$ with O above and O below **(u)** $H_3CS—OCH_2CH_3$ with O above and O below

(v) $(CH_3)_3P$ **(w)** $(CH_3)_3P{=}O$ **(x)** $H_3CP(OCH_2CH_3)_2$ with $\overset{O}{\|}$

Solution

(a) ⟩—(O—O)—⟨

peroxide

(b) ╪—(O—OH)

hydroperoxide

(c) $CH_3CH_2O^-$

alkoxide (ion)

(d) (O=⟨ring⟩=O with ring O)

(cyclic) anhydride

(e) ⟩=C=O

ketene

(f) ╪—(Li)

organometallic compound

(g) CH_3MgBr

Grignard reagent

(h) ⟩—(NO)

nitroso compound

(i) (ring with N=N)

(cyclic) azo compound

(j)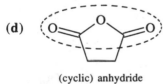

diazo compound

(k) $H_3C—(\overset{+}{N}\equiv\overset{-}{N})$

diazonium ion

(l) $CH_3CH_2N_3$

azide

(m) ◇—(ONO_2)

nitrate ester

(n)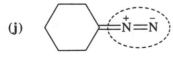
ONO

nitrite ester

(o) ⟍⟍⟋(SH)

thiol (mercaptan)

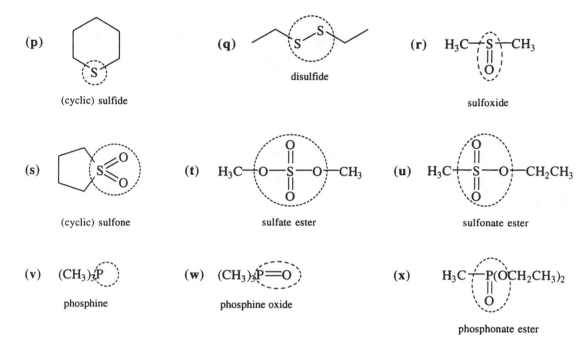

(p) (cyclic) sulfide

(q) disulfide

(r) sulfoxide

(s) (cyclic) sulfone

(t) sulfate ester

(u) sulfonate ester

(v) $(CH_3)_3P$ phosphine

(w) $(CH_3)_3P=O$ phosphine oxide

(x) $H_3C-P(OCH_2CH_3)_2$ phosphonate ester

PROBLEM 6-4 Provide a systematic name for each of the polyfunctional structures below:

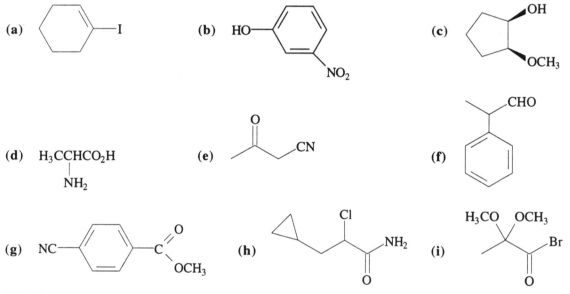

(a)

(b)

(c)

(d) H_3CCHCO_2H with NH_2

(e)

(f)

(g) $NC-$

(h)

(i)

Solution (a) 1-iodocyclohexene; (b) 3 (or meta)-nitrophenol; (c) *cis*-2-methoxycyclopentanol; (d) 2-aminopropanoic acid; (e) 3-ketobutanenitrile (3-oxobutanenitrile); (f) 2-phenylpropanal; (g) methyl 4-cyanobenzoate; (h) 2-chloro-3-cyclopropylpropanamide; (i) 3,3-dimethoxypropanoyl bromide

PROBLEM 6-5 Calculate the I value for each molecular formula below. What can you deduce from each value of I?

$$\text{(a) } C_5H_4BrNO_2 \qquad \text{(b) } C_3H_7ClNO_3 \qquad \text{(c) } C_6H_{14}I_2$$

Solution In each case use Eq. (6-1) to calculate the I value:

(a) $I = \dfrac{2(5)+2+1-4-1}{2} = 4$ Thus, any valid structure with this molecular formula will have a total of four rings plus π bonds.

(b) $I = \dfrac{2(3)+2+1-7-1}{2} = \dfrac{1}{2}$ Thus, this molecular formula cannot be valid for any neutral structure with all valences filled.

(c) $I = \dfrac{2(6)+2-14-2}{2} = -1$ This is an impossible molecular formula; there are too many atoms

for the available bonds.

PROBLEM 6-6 Take ten minutes to draw as many structural isomers of C_3H_5NO as you can. Identify the functional group classes represented by each of your structures. (*Hint*: Begin by calculating the I value.)

Solution The I value is $[2(3) + 2 + 1 - 5]/2 = 2$, so there must be a total of two rings plus π bonds in each valid structure. The list below is not exhaustive, and some of the structures would have questionable stability.

Functional group class	Structural isomers

Functional group classes listed (top to bottom): Nitroso; Amine + ether; Amine + alcohol; Nitrile + alcohol; Nitrile + ether; Amide; Imine + ether; Imine + alcohol; Ketene + amine.

Epoxide + amine		
Amine + aldehyde		
Imine + aldehyde		
Ketone + imine		
Ketone + amine		

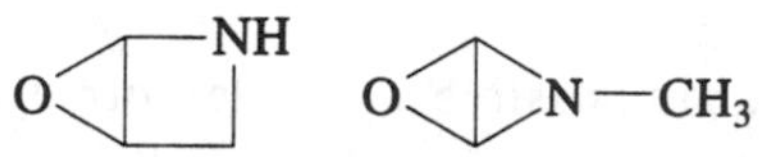

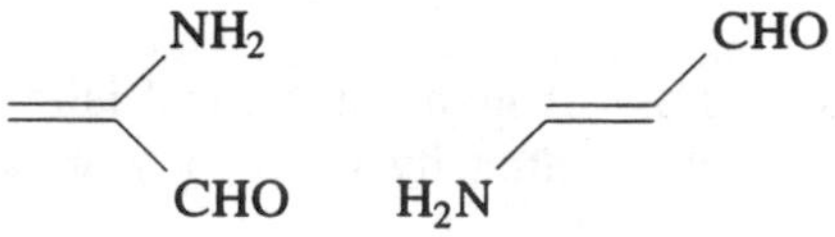

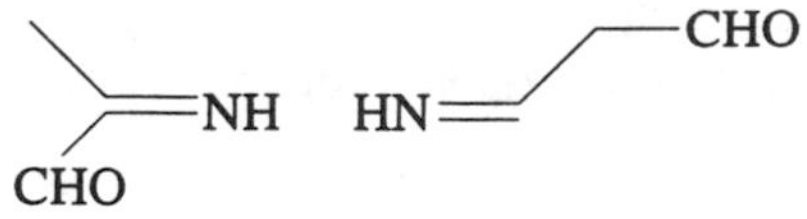

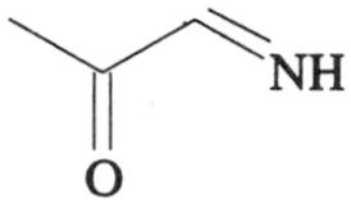

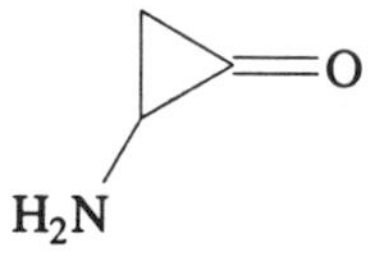

Notice how even a relatively simple molecular formula can have a wide variety of structural isomers.

PROBLEM 6-7 Review Problem: In each structure below draw in all nonbonding pairs. Then, using the concepts described in Chapter 3, indicate the nominal (VSEPR) hybridization of each internal atom.

(a)  R—N—R
        |
        R

(b) 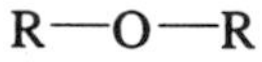 R—O—R

(c) 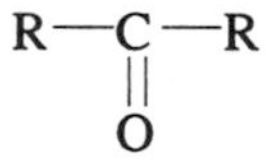 R—C—R
          ‖
          O

(d) 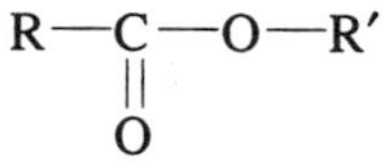 R—C—O—R′
          ‖
          O

(e) 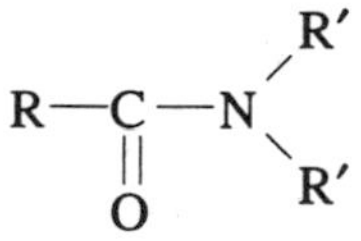 R—C—N
          ‖   R′
          O   R′

(f) 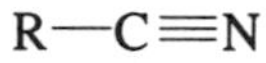 R—C≡N

(g) 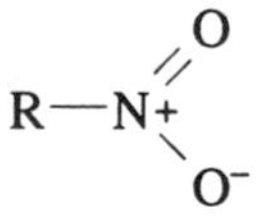 R—N+
              O
              O⁻

(h) R   R
      C=N
    R

(i) R—C—O—C—R′
          ‖   ‖
          O   O

(j) R
      C=C=O
    R

(k) R
      N=N
          R

(l) R
      C=N=N
    R

(m) R—N=N=N

(n) R—O—N=O

(o) R—O—N
                  O
                  O⁻

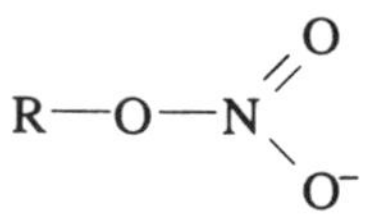

(p) 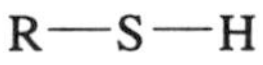 R—S—H

(q) 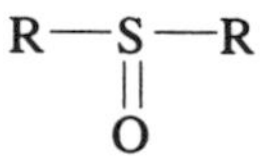 R—S—R
          ‖
          O

(r) R—S—O—R′
          ‖
          O
          O

(s) R—P—R
          |
          R

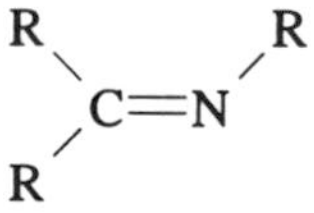

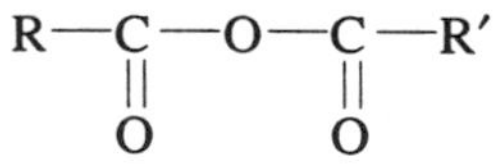

(t) 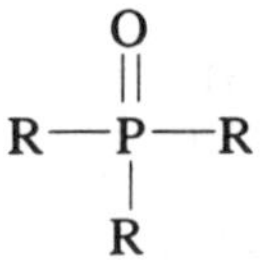 R—P—R
          ‖
          R
          O

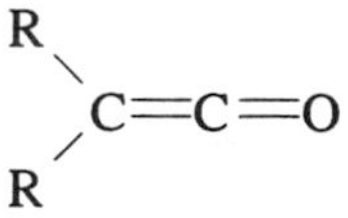

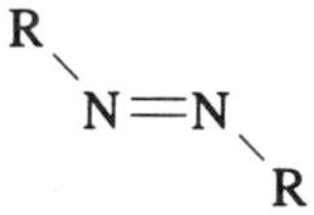

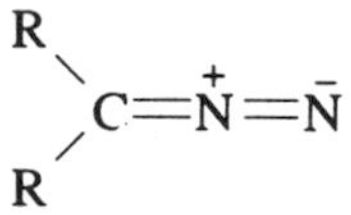

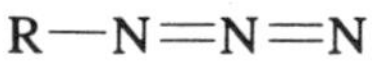

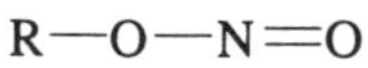

(u) 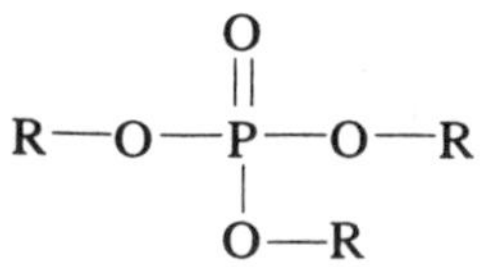 R—O—P—O—R
                  |
                  O—R
          O

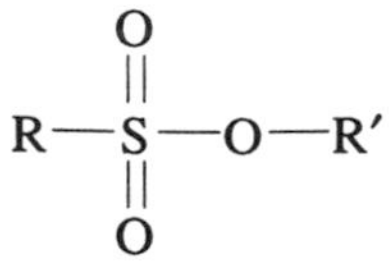

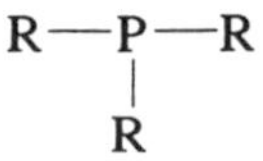

Solution

(a) $\overset{\displaystyle sp^3}{R-\underset{\displaystyle |}{\overset{\displaystyle ..}{N}}-R}$ R

(b) $R-\overset{sp^3}{\underset{..}{\overset{..}{O}}}-R$

(c) $R-\overset{sp^2}{\underset{\displaystyle :O:}{C}}-R$

(d) $\overset{sp^2}{R-\overset{sp^3}{\underset{\displaystyle \underset{:O:}{||}}{C}}-O-R'}$

(e) $R-\overset{sp^2}{\underset{\underset{:O:}{||}}{C}}-\overset{sp^3}{\underset{R'}{\overset{R'}{N}}}$

(f) $R-\overset{sp}{C}\equiv N:$

(g) $R-\overset{..}{\underset{sp^2}{N}}\overset{:\overset{..}{O}\cdot}{+}\;:\overset{..}{O}:^-$

(h) $\overset{R}{\underset{R}{C}}=\overset{sp^2}{\underset{..}{N}:}$

(i) $R-\overset{sp^2}{\underset{\underset{:O:}{||}}{C}}-\overset{sp^3}{\underset{..}{\overset{..}{O}}}-\overset{sp^2}{\underset{\underset{:O:}{||}}{C}}-R'$

(j) $\overset{R}{\underset{R}{\overset{sp^2}{C}}}=C=\overset{sp}{\underset{..}{\overset{..}{O}}:}$

(k) $\overset{R}{\underset{sp^2}{N}}=\overset{sp^2}{\underset{R}{\overset{..}{N}}}$

(l) $\overset{R}{\underset{R}{\overset{sp^2}{C}}}=\overset{sp}{\underset{+}{N}}=\overset{..}{\underset{..}{N}}:$

(m) $R-\overset{..}{\underset{sp^2}{N}}=\overset{+}{\underset{sp}{N}}=\overset{..}{\underset{..}{\overline{N}}}:$

(n) $R-\overset{sp^3}{\underset{..}{\overset{..}{O}}}-\overset{sp^2}{N}=\overset{..}{\underset{..}{O}}:$

(o) $R-\overset{sp^3}{\underset{sp^2}{\overset{..}{\underset{..}{O}}}}-\overset{:\overset{..}{O}\cdot}{N}+\;:\overset{..}{O}:^-$

(p) $R-\overset{..}{\underset{sp^3}{\overset{..}{S}}}-H$

(q) $R-\overset{sp^3}{\underset{\underset{O}{||}}{S}}-R$

(r) $R-\overset{\overset{\cdot\cdot\overset{..}{O}}{||}}{\underset{\underset{\cdot\cdot\overset{..}{O}\cdot}{||}}{\overset{sp^3}{S}}}-\overset{sp^3}{\overset{..}{\underset{..}{O}}}-R'$

(s) $\overset{sp^3}{R-\underset{\underset{R}{|}}{P}-R}$

(t) $R-\overset{\overset{\cdot\overset{..}{O}\cdot}{||}}{\underset{\underset{R}{|}}{\overset{sp^3}{P}}}-R$

(u) $R-\overset{sp^3}{\underset{..}{\overset{..}{O}}}-\overset{\overset{\cdot\overset{..}{O}\cdot}{||}}{\underset{\underset{sp^3}{\underset{:\overset{..}{O}-R}{|}}}{P}}-\overset{sp^3}{\underset{..}{\overset{..}{O}}}-R$

PROBLEM 6-8 Below are the structures of four important organic compounds. Which functional group classes are represented in each?

(a)

$$HSCH_2\underset{\underset{+NH_3}{|}}{CH}CO_2^-$$

cysteine (an amino acid)

(b)

acetylsalicylic acid (aspirin)

cortisone

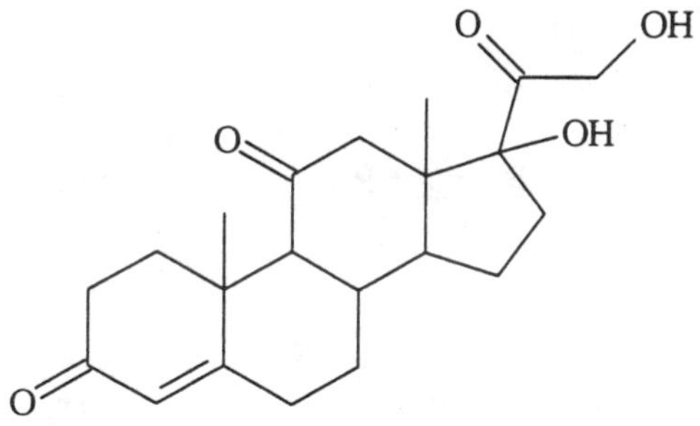

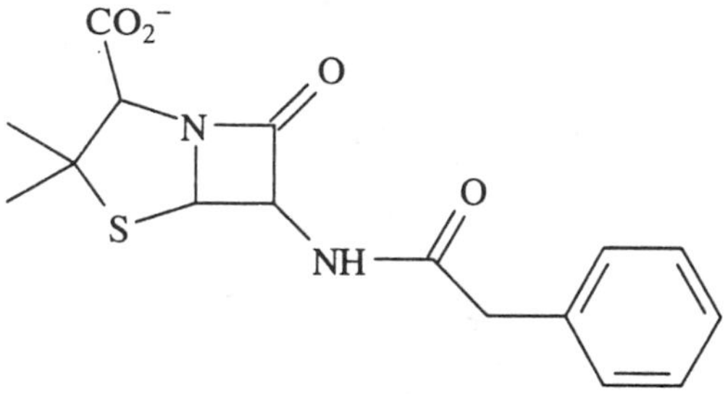

penicillin G

Solution   **(a)** aminium (ammonium) ion, carboxylate ion, thiol; **(b)** aromatic ring, carboxylic acid, ester; **(c)** C=C, ketone (×3), 1° alcohol, 3° alcohol; **(d)** cyclic sulfide, cyclic 3° amide, 2° amide, carboxylate ion, aromatic ring.

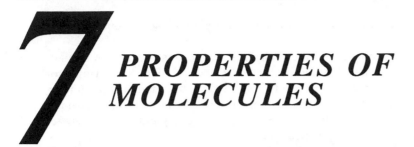

7 PROPERTIES OF MOLECULES

THIS CHAPTER IS ABOUT

☑ **Physical Properties and Elemental Composition**
☑ **Spectroscopic Properties**
☑ **Intramolecular Forces**
☑ **Conformations of Alkanes**
☑ **Conformations of Cycloalkanes**
☑ **Conformational Dynamics in Cyclohexane Derivatives**
☑ **Intermolecular Forces and Trends in
Physical Properties**

7-1. Physical Properties and Elemental Composition

Suppose you were given a small vial containing one gram of an unknown organic compound. How would you go about identifying the compound? This question is one that confronts organic chemists on a daily basis. But where do we begin?

A. Determining purity

Before we can begin to identify the compound, we must be sure it is pure, and not a mixture of two or more compounds. Techniques for purifying or establishing the purity of compounds include *fractional distillation* (for liquids), *recrystallization* (for crystalline solids), as well as chromatographic methods such as *thin-layer chromatography* (TLC), *gas–liquid partition chromatography* (GLPC or GC), and *high-pressure liquid chromatography* (HPLC or LC). Because these techniques are usually described in detail in an organic laboratory course, they are only mentioned here.

B. Determining physical properties

Once the compound is demonstrated to be pure, we'll note its physical state (solid, liquid, or even gaseous), any detectable color, and, if crystalline, the microscopic nature of its crystals. Before the hazards of doing so were fully appreciated, chemists used to note the odor (and even taste!) of the compound, but this practice is now highly discouraged. Next, we'll measure the standard physical properties of the compound: *melting point* (for solids), *boiling point* and *index of refraction* (for liquids), *density*, and *dielectric constant*. The normal melting point (mp) of a solid is the temperature range at which the solid and liquid can coexist (at a pressure of one atmosphere). A pure crystalline compound usually has a sharp melting point, with a range of less than 0.5°C. The normal boiling point (bp) of a liquid is the temperature at which its vapor pressure equals one atmosphere. Compounds with relatively low boiling points (less than, say, 100°C) are said to be **volatile**. The index of refraction (η) of a liquid is an indirect measure of the velocity of light through the substance and is related to the polarizability of atoms and bonds in the molecules. Density (ρ) is the ratio of mass to volume (g/cm³). Most organic compounds (except those that contain "heavy" atoms from the third period and beyond) have densities slightly less than that of water (1.0 g/cm³). The dielectric constant is a bulk property of a compound that measures its capacity to prevent the discharge of an electric field (voltage) gradient. The magnitude of the dielectric constant generally increases with a compound's dipole moment.

Each pure compound has its own set of these physical properties, which are tabulated in such references as the CRC *Handbook of Chemistry and Physics*. Of course, trying to look through a list of, say, 10,000 organic compounds to find the one that matches the unknown's properties would not be a pleasant task. And since there are nearly 10 million known organic compounds, even a relatively fast computer-assisted search might yield a list of dozens of compounds with nearly the same set of physical properties.

C. The molecular formula

If we have no idea of the chemical background of our unknown, the molecular formula is the next logical step. Historically, molecular formulas were deduced from two pieces of information: molecular weight (*MW*, Section 2-3) and elemental composition. The determination of molecular weight was accomplished by one or more of a family of techniques involving measurement of *colligative properties*. One of these, known as *vapor phase osmometry*, involved measuring the vapor pressure of a dilute solution of the unknown compound in a known solvent. The magnitude of the vapor pressure *decrease* was directly related to the molecular weight of the unknown compound. Two other related colligative properties, *melting point depression* and *boiling point elevation*, could also be used. These techniques are usually discussed in first-year college chemistry courses, so they are only mentioned here.

The elemental composition of a compound, expressed as percentage by weight of each element present, is determined by *combustion analysis* (for carbon and hydrogen) and *gravimetric analysis* of the treated residue for such elements as halogen and some others. (Because oxygen is used in the combustion, it is impossible to measure percent oxygen by this method. Nonetheless, it is usually a reasonable assumption that any missing mass can be ascribed to oxygen.) These combustion measurements can be done with very high precision, using just a few milligrams of the unknown compound.

From the elemental composition we can generate an **empirical formula** by dividing each weight-percent value by the corresponding average atomic mass (Section 1-1), then finding the simplest integer ratio of the atoms involved. Here is an example for you to try.

EXAMPLE 7-1   A compound whose structure is unknown is found be 24.2% carbon, 4.04% hydrogen, and 71.7% chlorine (all by weight). What is the empirical formula of the compound?

Solution   First, we note that 24.2% + 4.0% + 71.7% = 99.9%, so that there is no significant amount of material unaccounted for. The atomic ratios are determined as follows:

	carbon	hydrogen	chlorine
wt%	24.2	4.04	71.7
atomic mass	12.0	1.00	35.5
decimal ratio	2.02	4.04	2.02
simplest integer ratio	1	2	1

The decimal ratio is obtained by dividing the weight data by the atomic mass. The simplest integer ratio is obtained by dividing all the decimal ratios by the smallest one (2.02 in this case). These integer ratios indicate an empirical formula of $C_1H_2Cl_1$, or simply CH_2Cl.

It's important to remember that the empirical formula is not necessarily the same as the molecular formula, because the empirical formula expresses the *simplest* ratio of atoms in the molecule. Thus, C_2H_4, C_3H_6, and C_4H_8 would all have the same empirical formula, namely CH_2. So, we can express the *molecular* formula as some integer multiple of the empirical formula, e.g., $(CH_2)_n$. To find the value of *n*, all we need to know is the molecular weight because of the relationship

$$MW = n(EFW) \tag{7-1}$$

or

$$n = \frac{MW}{EFW} \qquad (7\text{-}1')$$

where *EFW* is the **empirical formula weight**, the sum of the atomic weights in the empirical formula. Suppose, for example, that the actual molecular weight of the unknown in Example 7-1 was found to be 100 ± 1; what is the *molecular formula* of the unknown? The *EFW* of CH_2Cl is $12.0 + 2(1.0) + 35.5 = 49.5$. From Eq. (7-1')

$$n = \frac{MW}{EFW} = \frac{100}{49.5} \approx 2$$

So, the molecular formula of our unknown is $(CH_2Cl)_2 = C_2H_4Cl_2$. Now you try one.

EXAMPLE 7-2 Another unknown compound is found to be 62.1% carbon and 10.3% hydrogen (by weight), and its molecular weight is 59 ± 1. What is its molecular formula?

Solution First, we note that 62.1% + 10.3% does not equal 100%. We'll assume that the missing 27.3% is probably due to oxygen. Next we generate the empirical formula:

	carbon	hydrogen	oxygen
wt%	62.1	10.3	27.6
atomic mass	12.0	1.00	16.0
decimal ratio	5.18	10.3	1.73
simplest integer ratio	3	6	1

The empirical formula is therefore C_3H_6O, and the *EFW* is $3(12.0) + 6(1.0) + 16.0 = 58.0$. So, from Eq. (7-1'), $n = MW/EFW = 59/58$ 1. In this case the molecular formula *is* the same as the empirical formula, namely C_3H_6O.

By the way, do you see one quick test to determine whether a given empirical formula is possibly a molecular formula? Calculate its *I* value (Section 6-6)! For CH_2Cl (Example 7-1) $I = [2(1) + 2 - 2 - 1]/2 = 1/2$, so this empirical formula cannot be a valid *molecular* formula for a neutral, valence-filled structure.

D. Identification of an unknown compound

With both the molecular formula *and* physical properties of an unknown in hand, it is usually fairly easy to identify the compound from tables of physical properties, provided the unknown compound has been previously identified and its properties tabulated. In such a case, the identification can be further confirmed by chromatographic comparison with a sample of the "authentic" material.

But what if the unknown compound is truly unknown, never before reported in the chemical literature? Prior to the 1950s such an identification task could become quite a challenge. Chemists relied on a battery of so-called "wet chemical" tests for individual functional groups, degradative reactions that chopped the big molecule into smaller, identifiable molecular fragments, preparation of one or more derivatives, as well as some of the most brilliant deductive reasoning this side of a Sherlock Holmes story. But these days, there is a new family of sophisticated analytical techniques that make the task much more straightforward. These are the topic of the next section.

EXAMPLE 7-3 Suppose the unknown compound in Example 7-2 gave a positive test for the ketone functional group. Propose a structure for the compound.

Solution There is only one ketone with molecular formula C_3H_6O, and that is acetone, H_3CCOCH_3.

7-2. Spectroscopic Properties

In a modern chemical laboratory there are new kinds of sophisticated instruments that can provide detailed structural information about molecules in a small fraction of the time once necessary with the classical techniques described in the previous section. We've already mentioned one of these techniques in Section 3-1, X-ray crystallography. Using just one microscopic crystal of the compound, the computer-controlled X-ray diffractometer can (in principle) locate the exact position of every atom in the molecule, thereby providing not only the complete structure, but also all bond lengths and bond angles. This process is somewhat time-consuming (a couple of days to several weeks) compared to the spectroscopic techniques described below and requires a highly trained crystallographer. It also helps to know the compound's molecular formula, density, and at least something about its structure before beginning a crystallographic analysis. Nonetheless, this technique gives the most unambiguous picture possible of the molecule's actual structure in the crystalline state. Even a liquid can be studied by performing the X-ray analysis of a crystal below its melting point.

The other modern techniques for structure elucidation are collectively referred to as **spectroscopic methods**, and (except for mass spectrometry) they involve absorption of electromagnetic radiation by the sample molecules. A detailed discussion of these techniques is beyond the scope of this book, but a brief description of the four most common spectroscopic techniques is presented below.

A. Mass spectrometry

The task of generating a molecular formula from molecular weight and elemental analysis information through colligative properties and combustion analysis was discussed in Section 7-1. These days it is possible to establish the exact molecular weight and molecular formula of an unknown compound in just minutes, using **mass spectrometry** (MS). A minuscule amount, about one microgram, of the compound is ionized in the gas phase. The resulting **molecular ions** (molecules that have lost one or more electrons), together with any **fragment ions** from the decomposition of the molecular ions, are magnetically separated on the basis of their mass-to-charge (m/e) ratio, and the **exact mass** (to 0.0001 amu) of the ions is measured. From its exact mass, the molecular formula of each ion can be determined, because each combination of atoms has a different exact mass. Thus it is possible to distinguish between two ions of the same nominal mass, such as CH_2O^+ and CH_4N^+ (both of which have a nominal mass of 30 amu), because of significant differences in their exact masses (CH_2O, 30.0106; CH_4N, 30.03445). A graph of mass (actually m/e) vs. abundance of all the ions is called the **mass spectrum** of the compound. To someone familiar with mass spectral fragmentation patterns, it is even possible to deduce structural information about the unknown molecule.

EXAMPLE 7-4 The mass spectrum of a certain compound indicates a molecular formula of C_2H_6O. **(a)** Draw two possible structures for this compound. **(b)** The mass spectrum also exhibits fragments with molecular formulas $C_2H_5O^+$, CH_3O^+, and CH_3^+, but no $C_2H_5^+$. Suggest a structure for the compound.

Solution **(a)** The only possible structures for C_2H_6O are CH_3CH_2OH (ethanol) and CH_3OCH_3 (dimethyl ether). **(b)** Although you haven't had any real training in mass spectral fragmentation patterns, see if you can draw likely structures for the fragments. The formula $C_2H_5O^+$ could have several possible structures:

$$H_3C\overset{+}{\underset{\cdot\cdot}{O}}=CH_2 \qquad H_3CCH_2\overset{+}{\underset{\cdot\cdot}{O}}\colon \qquad H_3CCH=\overset{+}{\underset{\cdot\cdot}{O}}H$$

The formula CH_3O^+ could have either of two different structures:

$$H_3C\overset{+}{\underset{\cdot\cdot}{O}}\colon \qquad H_2C=\overset{+}{\underset{\cdot\cdot}{O}}H$$

So, any of these fragments, including CH_3^+, could have been formed from either the alcohol or the ether. Perhaps the most significant fact is the *absence* of $C_2H_5^+$ ($CH_3CH_2^+$), which would have been expected from the alcohol but not the ether. So, our tentative identification for the compound would be dimethyl ether, which we could confirm by looking up the published mass spectra of both compounds.

B. Ultraviolet (electronic) absorption spectroscopy

Ultraviolet (UV) or **electronic absorption spectroscopy** involves the absorption of electromagnetic radiation in the UV/visible region (200–700 nm) by the sample molecules. It is generally true that, in organic molecules, only delocalized electrons (Chapter 4) absorb in this region. Thus, while conjugated multiple bonds (Section 4-4) and aromatic molecules (Section 4-7) give characteristic **UV spectra** (a graph of wavelength absorbed vs. intensity of absorption), most isolated multiple bonds and single bonds do not absorb. Therefore the greatest value of UV spectroscopy is in determining the presence and type of conjugation, if any, in the sample molecules.

note: Isolated multiple bonds do absorb in the so-called *far* UV region (<200 nm), but this absorption is not detected in commonly used UV spectrophotometers.

EXAMPLE 7-5 Which of the isomeric structures below absorbs ultraviolet radiation? Why?

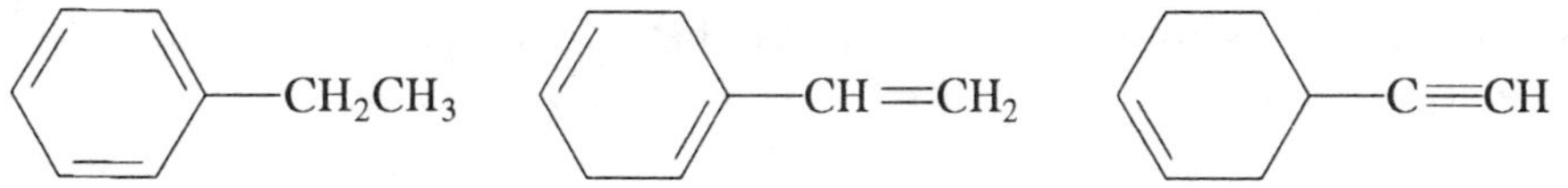

Solution The first and second structures have aromatic or conjugated multiple bonds (circled in the structures below), and will therefore absorb UV radiation. The third structure has only isolated multiple bonds and will be essentially transparent in the UV.

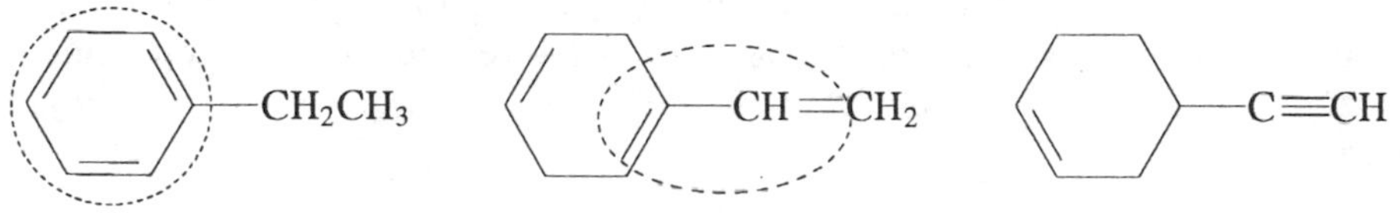

C. Infrared absorption spectroscopy

Each bond in an organic molecule (indeed, in *any* molecule) vibrates continuously at a characteristic frequency that depends on the strength of the bond and the masses of the bonded atoms. When the molecule is irradiated with electromagnetic radiation from the infrared (IR) region (5000–200 cm^{-1}), it can absorb radiation at those frequencies that match the frequencies of the bond vibrations. The **IR spectrum** (again, frequencies absorbed vs. absorption intensity) thus shows the kinds of bonds present in the molecule.

Polar bonds and functional groups (such as C=O, NO$_2$, and OH) absorb very strongly in characteristic regions of the IR spectrum. In fact, by noting the exact frequency of a C=O absorption, one can determine whether it belongs to a ketone, aldehyde, acid, or acid derivative. Other multiple bonds, such as C=C, aromatic C=C, and C≡C, give weaker but nonetheless characteristic absorptions. The C–H region of the spectrum is usually complex because there are so many different C–H bonds. However, certain C–H bonds including C≡C–H, C=C–H, aromatic C–H, and cyclopropyl C–H give absorptions at characteristic positions. In summary, the IR spectrum is highly diagnostic for the types of functional groups present (or absent) in the molecule. And, as is true for mass spectra, the IR spectrum can be used as a compound's fingerprint with which to compare published IR spectra.

EXAMPLE 7-6 A certain compound has molecular formula C$_3$H$_4$O. Its IR spectrum exhibits absorptions characteristic of OH, C≡C, and ≡CH. Propose a structure for the compound.

Solution The only possibility is 2-propyn-1-ol, HC≡CCH$_2$OH.

D. Nuclear magnetic resonance spectroscopy

Of the four types of modern spectroscopic analysis, **nuclear magnetic resonance** (NMR) **spectroscopy** provides the most complete structural information about the sample molecules, in the least amount of time. In this technique, the sample is immersed in a strong magnetic field, and its nuclei are irradiated with radio-frequency electromagnetic radiation. The exact frequency of absorption by a nucleus, called its **chemical shift**, is indicative of its molecular environment. Only one element can be examined at a time, and not all elements are amenable to study by NMR. But those that *are* include the most ubiquitous ones in organic chemistry, carbon and hydrogen.

The carbon NMR spectrum of a compound exhibits one signal for each carbon (or set of symmetry-equivalent carbons). From the exact chemical shift of each signal the hybridization of the carbon can be determined, as well as what other elements (e.g., hydrogen, oxygen, and halogen) are directly attached to it. Similarly, the hydrogen (or "proton") NMR spectrum indicates the type of atom to which each hydrogen is attached, how many of each type (environment) of hydrogen there are, as well as something about the stereochemical relationships between hydrogens (e.g., cis or trans).

Armed with just a molecular formula and a set of carbon and hydrogen NMR spectra, the organic chemist can usually deduce the structure of all but the most complex molecules. Furthermore, NMR is an especially useful tool for monitoring the progress of chemical reactions and dynamic molecular processes such as those described in Sections 7-4 and 7-5.

EXAMPLE 7-7   A compound with molecular formula $C_4H_5BrO_2$ exhibits carbon NMR signals for a carbonyl carbon, two different sp^2-hybridized vinyl carbons (those attached by a carbon–carbon double bond), and an sp^3-hybridized carbon attached to oxygen. The hydrogen NMR spectrum has signals attributable to three equivalent methoxy (H_3CO) hydrogens, and two different vinyl hydrogens (those attached to vinyl carbons) bearing a trans relationship to each other. (a) Suggest a structure for the compound. (b) What features would you expect to find in the IR and UV spectra of this compound?

Solution

(a) The solution to this problem requires us to fit the pieces of the puzzle together. The molecular formula has an *I* value of 2, so there are two rings plus π bonds in the structure. The carbon NMR spectrum suggests the presence of the following fragments: C=O, C=C, and C–O. The hydrogen NMR indicates two types of hydrogens:

There are two possible structures that fit this collection of information:

Even though these structures appear quite similar, it is true that a very careful analysis of the chemical shifts would probably allow us to make a distinction between them.

(b) The UV spectrum of either compound would show an absorption for the conjugated carbonyl (C=C–C=O). Furthermore, conjugation of the nonbonding electrons on Br and on the methoxy oxygen with the C=C and C=O π bonds would also give rise to UV absorptions. The IR spectrum could differentiate an ester carbonyl from an acid bromide carbonyl.

7-3. Intramolecular Forces

In Chapters 2 and 4 we described the most important type of **intramolecular interactions** (i.e., interactions *within* a molecule) between atoms, namely, localized and delocalized chemical bonds. In this section we'll discuss certain weaker forces between the atoms in a molecule. These forces can have a direct impact on the energy content of the molecule, either increasing it or decreasing it. Remember from Section 2-1 that *increasing* the energy of a molecule *decreases* its relative stability and *increases* its potential reactivity.

A. Angle strain

We saw in our discussion of molecular shape (Chapter 3) that electron pairs around an atom prefer to occupy certain geometric shapes that minimize repulsions between the electrons (VSEPR, Section 3-2). However, in some molecules these preferred shapes and bond angles are precluded by other geometric requirements of the structure. For example, while carbon atoms surrounded by four groups prefer a tetrahedral geometry and bond angles of 109.5°, small cyclic molecules such as cyclopropane and cyclobutane require internuclear angles much smaller than this ideal value (see Problem 3-9). This departure from ideal bond angles gives rise to what is called **angle strain**, making such molecules less stable and more reactive than their strain-free counterparts.

EXAMPLE 7-8   Which molecule would you expect to suffer more angle strain, cyclopropane or cyclobutane? Why? (You may assume the cyclobutane ring is planar.)

Solution   From the structures below, it is clear that each carbon in cyclopropane has

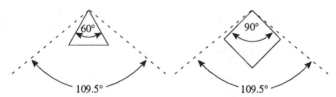

109.5° − 60° = 49.5° of angle strain, while each carbon in cyclobutane has 109.5° − 90° = 19.5° of angle strain. But remember that there are only three carbons in cyclopropane, while there are four in cyclobutane. So, the total angle strain in cyclopropane is 3(49.5°) = 148.5°, while in cyclobutane it is 4(19.5°) = 118°. We can conclude that cyclopropane suffers somewhat more angle strain than cyclobutane. But see Example 7-11.

B. Steric strain

Another very important type of strain results when two or more atoms in a molecule are forced to be so close that their respective electron clouds begin to repel one another. The resulting crowding, called **steric strain** or **nonbonded repulsion**, occurs when atoms are closer than their atomic (van der Waals) radii. Steric strain increases the energy of a structure, making it less stable. It is often difficult to assess the degree of steric interactions in a molecule by simply looking at a line–bond drawing of its structure. A better way is to look at a space-filling model of the structure, which more accurately depicts any unfavorable crowding in the molecule.

A classic example of the operation of intramolecular steric effects is found in the cis and trans isomers of di-*tert*-butylethylene (2,2,5,5-tetramethyl-3-hexene). The trans isomer is a reasonably normal alkene, but the cis isomer is much less stable and difficult to synthesize because of the unfavorable nonbonded repulsions between the methyl groups. This is most readily seen with space-filling models, as seen in Figure 7-1. Other examples of steric effects can be seen in Problem 4-13, and Problem 5 in the Final Exam.

EXAMPLE 7-9 Which of the two following structures do you expect to suffer the greater steric strain? Why? (You may assume the rings remain planar.)

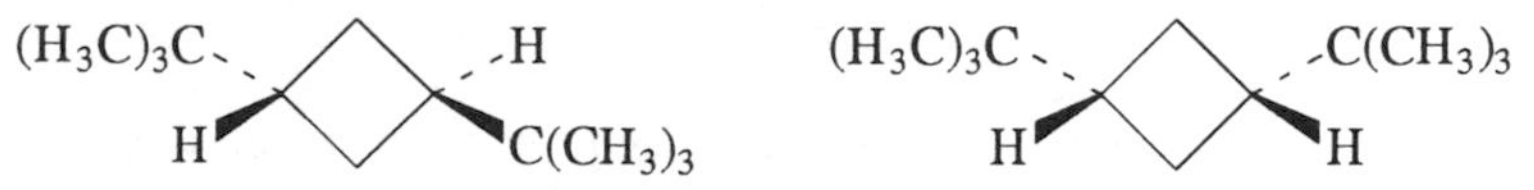

Solution By drawing these structures slightly differently we can see that with planar rings the cis isomer will suffer the same type of unavoidable interaction between the two bulky *tert*-butyl groups that was seen in *cis*-di-*tert*-butylethylene. (See also Section 7-5D.)

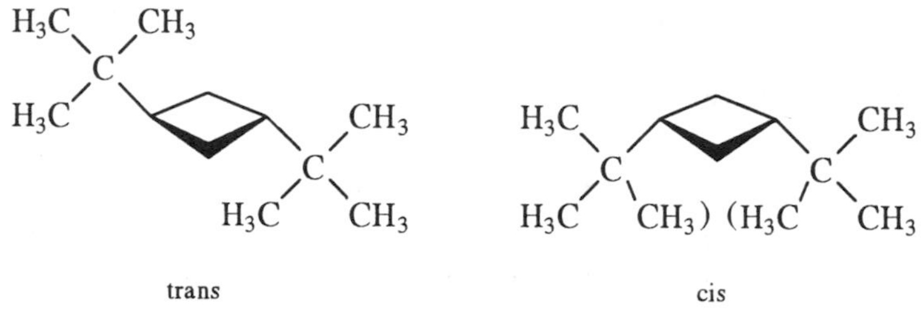

trans   cis

Steric effects can also operate **intermolecularly** (i.e., *between* molecules).

C. Electrostatic interactions and hydrogen bonding

In contrast to strain, there can be other types of intramolecular interactions that *stabilize* a molecule. One of these is related to the subtle way that the bond dipoles of polar covalent bonds (Section 2-6) can interact. Where possible, a molecule will arrange its individual bond dipoles in such a way that the positive ($\delta+$) end of each dipole is near the negative ($\delta-$) end of another. This *attractive* interaction between the ends tends to stabilize the structure and often decreases the net dipole moment (Section 3-1) of the molecule. Conversely, when two dipoles are forced to be aligned more or less parallel to each other, the resulting *repulsive* interaction is destabilizing. The ways a molecule can contort itself in order to find the most stable arrangement of dipoles will become clear in the next two sections.

A very special subcategory of dipole–dipole interactions involves the *attraction* of a positively polarized hydrogen (one attached to an electronegative atom such as nitrogen or oxygen) toward the nonbonding pairs of another electronegative atom in the molecule. In a very real sense, the hydrogen is shared between the two electronegative atoms, and this is depicted with a dotted bond between the hydrogen and the second electronegative atom. Clearly, an intramolecular **hydrogen bond** requires the molecule to coil up into a "ring." It turns out that a "ring" size of five or six members (including the hydrogen) is usually the most stable.

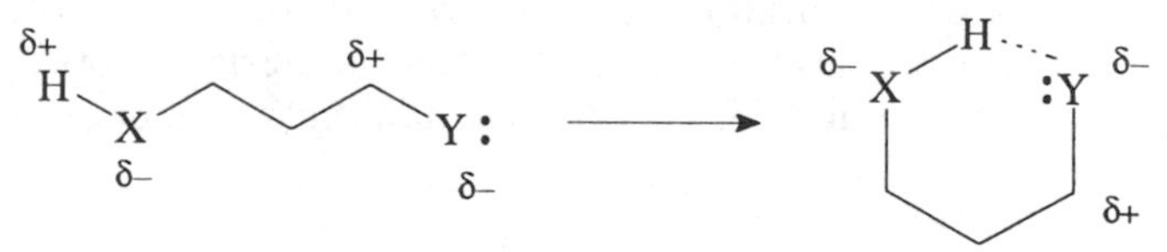

Hydrogen bonding is especially important in biomolecules such as proteins, DNA, and RNA.

EXAMPLE 7-10 The molecule $H_3CCOCH=C(OH)CH_3$ forms a very strong intramolecular hydrogen bond. Draw the hydrogen-bonded structure and account for its stability.

Figure 7-1. Space-filling CPK molecular models of **(a)** *trans*-2,2,5,5-tetramethyl-3-hexene and **(b)** its cis isomer.

Solution   The molecule is ideally constituted to form a six-membered hydrogen-bonded ring:

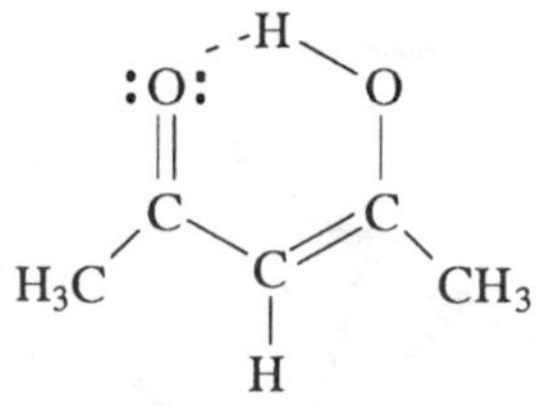

$$
\begin{array}{c}
\text{-H} \\
:\ddot{O}: \qquad O \\
\| \qquad\qquad | \\
C \qquad\qquad C \\
H_3C \qquad\ \ C \qquad CH_3 \\
\| \\
H
\end{array}
$$

We shall see in Section 7-6 that hydrogen bonding and other dipole–dipole interactions can also operate *between* molecules.

7-4. Conformations of Alkanes

A. Torsional strain

Now, let's talk about another type of intramolecular force. But before you begin this section, it will be most helpful if you make a molecular model of ethane, H_3C–CH_3. Below are two drawings ("sawhorse" formulas) of ethane.

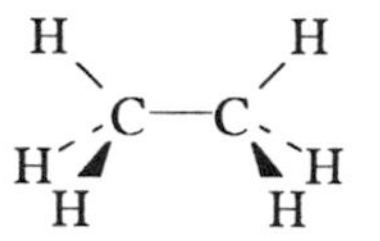

A 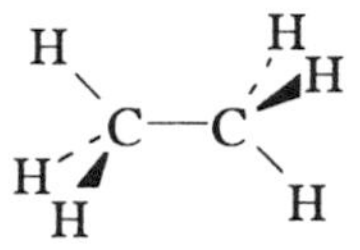   **B**

What is the relationship between the structures? Are they identical? Are they stereoisomers, or what? To help you answer this question, let's redraw the structures as they would appear if viewed directly along the C–C bond.

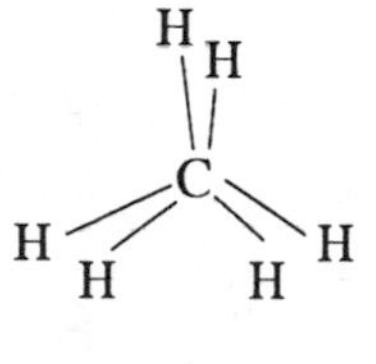

A 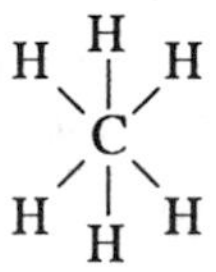   **B**

These latter drawings are a bit cumbersome because we can't tell the hydrogens in front from those in back, so a chemist named Melvin Newman developed a way to depict such structures more clearly. In these drawings (called **Newman projections**) the carbons are not explicitly shown.

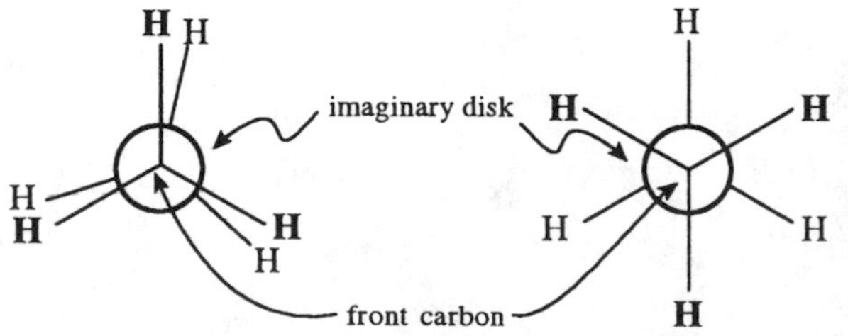

imaginary disk

front carbon

A   **B**

Newman projections (front hydrogens are shown in boldface)

Instead, there is an imaginary disk between the two carbons (and perpendicular to the C–C bond) that partially blocks the view of the bonds from the "back" carbon. This perspective allows us to tell which bonds are connected to the front carbon, and which are connected to the back one.

Getting back to our original question: Are **A** and **B** the same? The answer is clearly *no*. In **A**, the three C–H bonds in front are exactly aligned with the three in back. Such a structure is said to have the C–H bonds **eclipsed**. In **B**, the front C–H bonds are spaced midway between the back ones, leading to a **staggered** arrangement of C–H bonds. It turns out that the staggered arrangement is the more stable one, because it allows the C–H bonding electrons to be further apart (and hence less repulsive) than in the eclipsed structure. In fact, each set of eclipsed C–H bonds "costs" 4 kJ/mol, so eclipsed ethane (with three sets of eclipsed bonds) is 3(4) = 12 kJ/mol *less* stable than the staggered form. The energetic cost of eclipsing bonds is called **torsional strain** (or eclipsing strain) because eclipsing is encountered as a consequence of torsion (twisting, or rotation) around the C–C bond.

EXAMPLE 7-11   How many eclipsing interactions can you find in cyclopropane? Cyclobutane? (You may assume cyclobutane is planar.)

Solution   Each C–H bond in cyclopropane is eclipsed to *two* other C–H bonds. Thus, there is a total of *six* such interactions—three on each side of the ring. There is also subtle eclipsing of the C–C bonds, which contributes to the overall ring strain.

In cyclobutane, there are eight C–H eclipsing interactions, as well as eclipsing interactions of the C–C bonds.

Referring back to Example 7-8, we see that cyclopropane has more angle strain but less eclipsing strain than cyclobutane. On balance, the two molecules have nearly the same total strain.

D. Dihedral angle

To describe structures **A** and **B** fully, we need to define a new term, **dihedral** (or **torsional**) **angle**, the angle between two planes. If you open a book and lay it on your desk, the angle between the facing pages is 180°; if closed, the angle between facing pages is 0° (see Figure 7-2).

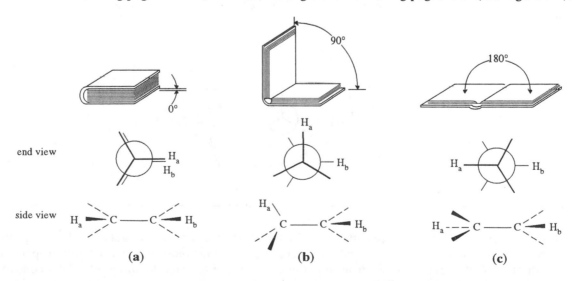

Figure 7-2.   Dihedral angles between vicinal C–H bonds: **(a)** 0°, **(b)** 90°, **(c)** 180°. (Reprinted with permission from R. S. Macomber, *NMR Spectroscopy*, Harcourt Brace Jovanovich, San Diego, California, 1988.)

In the same way, the front and back C–H bonds in ethane describe a dihedral angle. The planes of interest are those defined by the bonds H_a–C–C and C–C–H_b (Figure 7-2). Thus in the eclipsed structure **A** (below), the dihedral angle between eclipsed C–H bonds is 0°. Also notice that each front hydrogen makes a dihedral angle of 120° with the other two back hydrogens.

(A)

In the staggered structure **B** (below), each C–H bond in front is separated from the two nearest back hydrogens by a dihedral angle of 60°. Two bonds (or groups) separated by a dihedral angle of 60° are said to be **gauche** to each other. Also in **B**, the two hydrogens separated by a dihedral angle of 180° are said to be **s-trans** (or **anti**) to each other.

(B)

Of course, there is an infinite number of other structures for ethane, each with a different dihedral angle between the C–H bonds, all of which can be interconverted by rotation around the C–C bond.

EXAMPLE 7-12 Draw a Newman projection of ethane with a dihedral angle of 30°. What other dihedral angles are present in this structure?

Solution

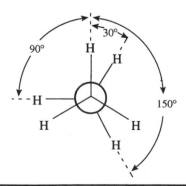

We already know that eclipsed ethane is 12 kJ/mol higher in energy (less stable) than staggered ethane. As you might have guessed, the energy of the ethane molecule is a continuous periodic function of the angle of rotation around the C–C bond (i.e., the dihedral angles between front and back bonds). Figure 7-3 is a graph of this relationship.

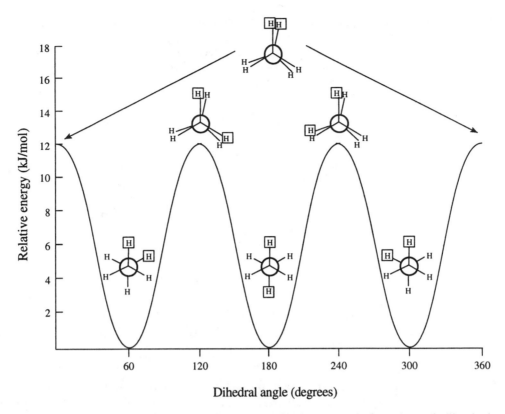

Figure 7-3. The conformational energy of ethane as a function of dihedral angle between the "boxed" hydrogens.

Notice how the structures with dihedral angles of 0°, 120°, 240°, and 360° are all identically eclipsed in the case of ethane, as are the staggered structures with dihedral angles of 60°, 180°, and 300°. From this graph, we can see that rotation around a C–C single bond is not completely free. Instead, there is a 12-kJ/mol energy barrier to rotation, encountered every 120°. This barrier is not large enough to suppress rotation completely, but it does mean that rotation around the C–C bond resembles a ratchet rather than a free-spinning top. Nonetheless, the rate of ratcheting rotation around the C–C bond at room temperature is millions of times per second.

EXAMPLE 7-13 What is the relative energy of the ethane structure described in Example 7-12?

Solution   In Figure 7-3, locate the dihedral angle of 30° (or 90°, or 150°), and find the corresponding energy, about 6 kJ/mol.

C. Conformations vs. stereoisomers

So far, we know that structures **A** and **B**, as well as the intermediate structure in Example 7-12, *are* different, but they differ only in the angle of rotation around the C–C bond. Do you think it would be appropriate to call these structures stereoisomers, since each one has the same sequence of atoms and bonds but a different three-dimensional arrangement of the atoms? This questions brings up a very important point:

● If two structures that would otherwise be considered stereoisomers undergo rapid interconversion by means of rotations around bonds (precluding their separation), the structures are called **conformations** (or **conformational isomers,** or just **conformers**).

Thus the distinction between stereoisomers and conformations *depends on the rate of interconversion*. Take, for example, the case of *cis*- and *trans*-2-butene. At any reasonable temperature, the rate of interconversion between them is immeasurably slow, because rotation around a *double* bond requires breakage of the π bond at an energetic cost of about 270 kJ/mol. Therefore, these structures *are* stereoisomers. However, the two analogous structures of butane

are interconverted at millions of times per second at room temperature, so they represent two conformations of butane (see Problem 7-10).

There is one more very important fact to remember when dealing with various conformations of a molecule. Any sample of ethane, for example, is actually a dynamic equilibrium mixture of all possible conformations, each undergoing rapid interconversion with all the others. But at any instant, there is a greater fraction of the more stable conformation(s) compared to the less stable conformations. The ratio of the amounts of any two conformations in an equilibrium mixture is a function of their relative energy and the absolute temperature (degrees Kelvin), according to the equation:

$$\frac{\text{less stable}}{\text{more stable}} = e^{-\Delta E/RT} \tag{7-2}$$

where the quantities in the numerator and denominator represent the relative amounts (moles, %, etc.), ΔE equals the relative energy of the less stable conformations minus the relative energy of the more stable conformation (in kJ/mol), and $R = 0.0083$ kJ/K-mol.

EXAMPLE 7-14 (a) What is the ratio of eclipsed conformation to staggered conformation of ethane at 25°C (298 K)? (b) What is the ratio at −173°C (100 K)?

Solution

(a) From Figure 7-3 we know that the relative energies of the eclipsed and staggered conformations of ethane are 12 and 0 kJ/mol, respectively. Thus $\Delta E = 12 - 0 = 12$ kJ/mol. Substituting these values into Eq. (7-2)

$$\frac{\text{eclipsed}}{\text{staggered}} = e^{(-12 \text{ kJ/mol})/(0.0083 \text{ kJ/K-mol})(298 \text{ K})}$$

$$= 0.0078$$

Thus for every 1000 molecules with the staggered conformation there are about 8 with the eclipsed conformation.

(b) Repeating the calculation with a temperature of 100 K gives

$$\frac{\text{eclipsed}}{\text{staggered}} = e^{(-12 \text{ kJ/mol})/(0.0083 \text{ kJ/K-mol})(100 \text{ K})}$$

$$= 5.3 \times 10^{-7}$$

So, at this lower temperature, for every 10^7 staggered conformations there will be only about 5 eclipsed ones. Remember: The lower the temperature, the lower will be the fraction of the less stable conformation.

7-5. Conformations of Cycloalkanes

A. Are cyclic molecules planar?

You might recall from your last geometry course that the internal angle at any corner of a regular polygon with *n* sides is given by

$$\angle = 180° - \left(\frac{360°}{n}\right) \tag{7-3}$$

Applying this equation to cycloalkanes, we would predict the following internuclear C–C–C bond angles: cyclopropane ($n = 3$), 60°; cyclobutane ($n = 4$), 90°; cyclopentane ($n = 5$), 108°; cyclohexane ($n = 6$), 120°.

EXAMPLE 7-15   Predict the internuclear C–C–C angles in cycloheptane and cyclooctane.

Solution   Use Eq. (7-3), with $n = 7$ and 8, respectively: For cycloheptane, $\angle = 180° - (360°/7) = 129°$; for cyclooctane, $\angle = 180° - (360°/8) = 135°$.

But this treatment of cycloalkane bond angles assumes that in each case the carbon rings are *planar* (flat). While this fact is of necessity true for cyclopropane (because three points determine a plane), *all* other cycloalkanes have carbon rings that are, to some extent, nonplanar (see Section 5-4). This is *not* to say that there are no planar carbon rings with more than three carbons. Benzene (Section 4-7) is just one of many compounds with planar rings—a direct consequence of the arrangement of multiple bonds and the attendant sp^2 hybridization at each ring atom. But cycloalkanes beyond cyclopropane are nonplanar. Why?

Example 7-11 showed us that *planar* rings suffer both angle strain *and* eclipsing strain. Focusing our attention on cyclohexane (and now is as good a time as any to build a molecular model of it), here is how it would look if it were planar:

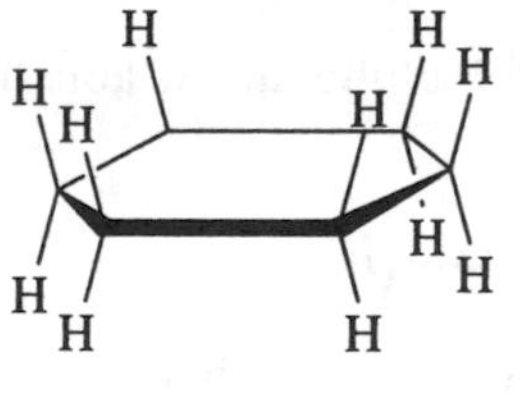

In addition to six C–C–C bond angles, each suffering 120°– 109.5° = 10.5° of angle strain, there are 12 C–H eclipsing interactions. This planar structure represents a highly strained conformation of cyclohexane. But, as you may have discovered when you made your model (you *did* make it, didn't you?) cyclohexane has a beautiful way to avoid *both* the angle strain and the eclipsing strain. The molecule relaxes into a chaise lounge-shaped conformation called a **chair form:**

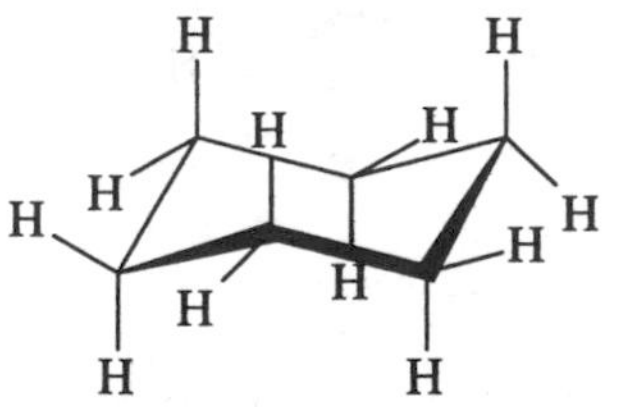

Not only does this chair form permit each C–C–C angle to be nearly 109.5°, but all eclipsing interactions have become staggered, as is most easily seen from your model or a "double"

Newman projection along two parallel C–C bonds.

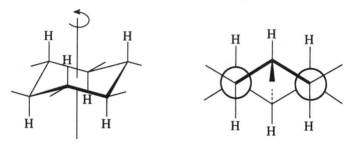

B. Axial and equatorial positions in cyclohexane

As you examine the chair form of cyclohexane, you'll notice that six of the C–H bonds are parallel, three pointing up and three down. Because these bonds are also parallel to an imaginary (symmetry) axis of the molecule (shown as the long vertical line, below), these six hydrogens are said to occupy **axial** positions. Only the axial hydrogens are shown in the drawings below:

The other six hydrogens, which are situated around the "equator" of the ring, are said to occupy **equatorial** positions. Only the equatorial hydrogens are shown in the following structures:

Because the chair conformation of cyclohexane is able to relieve all angle and eclipsing strain, it is strain free.

C. Other cyclohexane ring conformations

There is, of course, an infinite number of other, less stable conformations of cyclohexane, differing from each other in the dihedral angles (angles of twist) around the various C–C bonds. We'll mention just one of these, the so-called **boat form**, drawn in three ways:

boat conformation

It's immediately apparent from this structure that the boat conformation, though essentially free of angle strain, suffers from both C–H (four) and C–C (two) eclipsing interactions. In addition, there is a subtle steric repulsion between the two **flagpole** hydrogens (well, what would *you* call these positions on a boat?).

For these reasons, the boat form of cyclohexane is about 29 kJ/mol less stable than the chair form.

EXAMPLE 7-16   What is the ratio of boat form to chair form in cyclohexane at 298 K?

Solution   Use Eq. (7-2) with ΔE = 29 kJ/mol:

$$\frac{\text{boat}}{\text{chair}} = e^{(-29 \text{ kJ/mol})/(0.0083 \text{ kJ/K-mol})(298 \text{ K})}$$

$$= 8.1 \times 10^{-6}$$

Thus for every 10^6 molecules in the chair conformation, there are 8 in the boat conformation.

It is important to realize that the boat form is readily accessible from the chair form by simply "folding up" (or down) one of the CH_2 groups relative to the diagonally opposite CH_2 group. Try this with your model.

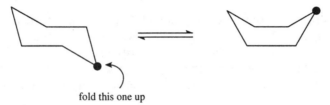

fold this one up

D. Conformations of other cycloalkanes

Other cycloalkanes twist and bend to find the best compromise between angle strain and eclipsing strain. In the case of cyclobutane, the most stable conformation is one in which the molecule is slightly folded along the diagonal (the so-called "butterfly" conformation), as shown in the drawing below.

The most stable conformation of cyclopentane is often described as envelope-shaped, with four carbons in one plane, and the fifth one bent up to form the flap of the envelope.

Again, it must be emphasized that these conformations are the most stable (and hence most prevalent), but that each cycloalkane molecule (except cyclopropane) exists as a dynamic mixture of all possible conformations, and that interconversion of these conformations is extremely rapid at room temperature.

7-6. Conformational Dynamics in Cyclohexane Derivatives

So far we've learned that cyclohexane itself exists as a mixture of rapidly interconverting conformations, with the chair form being the most stable. But there is another very important type of conformational interconversion seen most clearly with substituted cyclohexane derivatives.

A. Monosubstituted cyclohexane

Take, for example, methylcyclohexane. Should the methyl group occupy an axial position or an equatorial one? The answer is...both! Methylcyclohexane exists predominantly as a mixture of two different chair conformations, one with the methyl axial and the other with the methyl equatorial. Furthermore, these two conformations undergo rapid interconversion (at room temperature) by a process called **ring flipping** (one carbon is labeled with a dot to help you keep track of it):

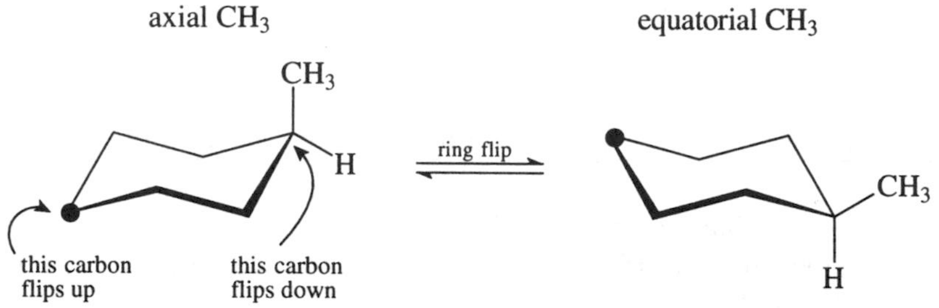

axial CH_3 equatorial CH_3

this carbon flips up   this carbon flips down

Try this ring flipping with your molecular model. If you examine the process carefully, you'll see that ring flipping actually involves a series of rotations (called **connected rotations**) around several of the C–C bonds and that the methyl substituent remains on the same side of the ring. But even more importantly, notice that the ring-flip process causes an exchange of all positions. That is, those positions that were equatorial are now axial, and vice versa, as shown by the drawings below:

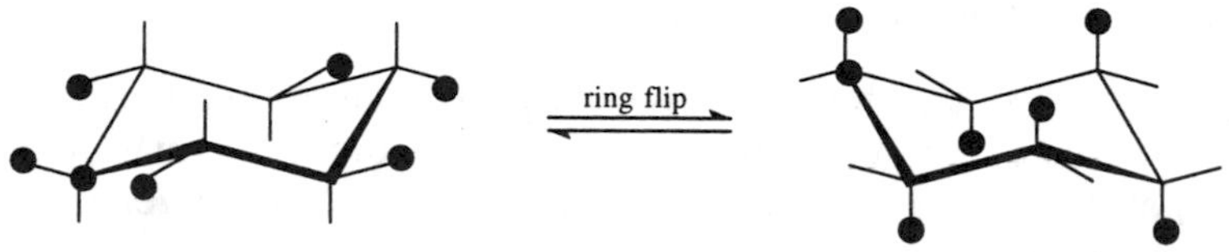

Of course, this same type of ring flipping goes on in cyclohexane itself, but because there are no substituents to serve as "flags," the two conformations are equivalent.

Now here is the next question: Which of the two chair conformations of methylcyclohexane is more stable and why? Another way to ask the same question is: Does a substituent attached to a cyclohexane ring prefer to be axial or equatorial? The answer is that, given a choice, most groups prefer the *equatorial* position, and the larger the group the greater the preference. Whenever the size of a group is the dominant factor we're probably looking at the manifestations of a steric effect at work, and that's exactly the case here. When a group such as a methyl group is forced to occupy an axial position, it encounters unfavorable steric repulsions from the two nearby axial hydrogens. This interaction is shown by parentheses here:

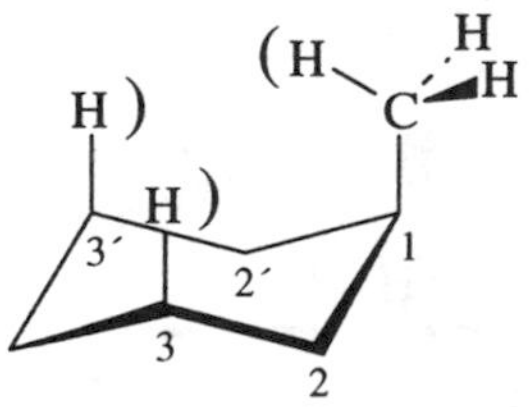

This type of unfavorable steric interaction is called a **1,3-diaxial repulsion** because the axial group (on carbon 1) is repelled by the axial hydrogens on carbons 3 and 3'. This nonbonded repulsion is completely relieved when a ring flip places the group back in an equatorial position. The energetic cost of an axial methyl group (compared to an equatorial one) is 7.6 kJ/mol. For the much larger *tert*-butyl group, the cost is 23 kJ/mol.

EXAMPLE 7-17 (a) Calculate the ratio of axial methyl chair conformations to equatorial methyl chair conformations of methylcyclohexane at 298 K. (b) Perform the same calculation for *tert*-butylcyclohexane. (c) In the case of methylcyclohexane, which will be more prevalent: Chair conformations with the methyl axial, or boat conformations? Why?

Solutions

(a) We'll use Eq. (7-2) with $\Delta E = 7.6$ kJ/mol:

$$\frac{axial}{equatorial} = e^{(-7.6 \text{ kJ/mol})/(0.0083 \text{ kJ/K-mol})(298 \text{ K})}$$

$$= 0.046$$

Thus, for every 100 equatorial methyls, there will be about 5 axials.

(b) Again, we'll use Eq. (7-2) with $\Delta E = 23$ kJ/mol:

$$\frac{axial}{equatorial} = e^{(-23 \text{ kJ/mol})/(0.0083 \text{ kJ/K-mol})(298 \text{ K})}$$

$$= 9.2 \times 10^{-5}$$

In the case of the *tert*-butyl group, there are only 9 axials per 100,000 equatorials! Clearly, for all intents and purposes, the *tert*-butyl group is essentially "locked" into an equatorial position, and ring flipping is essentially completely suppressed.

(c) Since the boat form is 29 kJ/mol less stable than the chair form (boat/chair = 8.1×10^{-6}, Example 7-16), while the axial methyl is only 7.6 kJ/mol less stable than the equatorial, there will be approximately $0.046/8.1 \times 10^{-6} = 5.7 \times 10^{3}$ times as many axial chair forms as boat forms.

EXAMPLE 7-18 (a) Draw the two chair conformations of cyclohexanecarbonitrile, and label each as to whether the cyano group is axial or equatorial.

(b) At 298 K the ratio of chair conformations with the cyano group axial to those with an equatorial cyano group is 0.72 (72 axial per 100 equatorial). Which is the more stable form, and by how much?

(c) What does the value calculated in part (b) indicate about the relative size of CN vs CH_3?

Solution

(a)

equatorial   axial

(b) As expected, the axial/equatorial ratio is less than one, so the equatorial is slightly more prevalent and hence somewhat more stable. We can rearrange Eq. (7-2) and solve for ΔE, provided we remember from our last math course that $\ln e^x = x$.

$$\ln[(axial)/(equatorial)] = \ln e^{-\Delta E/RT} = -\Delta E/RT \qquad (7\text{-}2')$$

or,

$$\Delta E = -RT \ln[(axial)/(equatorial)]$$

$$= -(0.0083 \text{ kJ/mol-K})(298 \text{ K}) \ln 0.72$$

$$= 0.8 \text{ kJ/mol}$$

(c) Thus, the energetic cost of putting a cyano group into the axial position is only about a tenth of putting a methyl group axial (7.6 kJ/mol). This indicates that a cyano group suffers less severe 1,3-diaxial repulsions than does a methyl group. Such a result should not surprise you, because the cyano group is linear and occupies considerably less volume than a methyl group, as shown by the outline of the van der Waals radius of each atom.

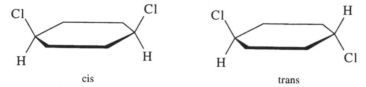

B. Disubstituted cyclohexane

Cyclohexane derivatives with two or more substituents present even more challenges. Consider, for example, the case of 1,4-dichlorocyclohexane. We know from our discussions of stereochemistry in Section 5-5 that the compound can exist in two different stereoisomeric configurations, cis or trans. With the rings drawn as flat, these two structures look like this:

cis   trans

However, we know that cyclohexane rings are *not* flat, but rather prefer to adopt chair conformations. So, what do these structures actually look like? Make models and see for yourself!

When the cis isomer "relaxes" into a chair conformation, one of the chlorines occupies an equatorial (e) position, while the other one finds itself in an axial (a) position. We'll call this an (ea) conformation.

cis (ea)

Admittedly, it's a little hard to see how the two chlorines are still cis in this structure. But if you flatten out the ring, you'll see they are still cis to each other. A ring flip of the cis isomer brings the previously equatorial chlorine into an axial position, and vice versa, leading to the equivalent (ae) conformation. However, a ring flip of the cis isomer does *not* convert it to the trans isomer, or vice versa.

cis (ea)   ring flip   cis (ae)

Now, how does this compare to the situation with the trans isomer? When it "relaxes" into a chair conformation, there are two dissimilar conformations that are possible: both chlorines equatorial (ee), or both axial (aa). Obviously, the latter one is much less stable.

trans (ee)   ring flip   trans (aa)

Notice that in both conformations the two chlorines are on opposite sides of the ring, and are therefore trans. This is somewhat more obvious in the (aa) conformation, even without flattening the ring. By comparing the most stable conformation of the trans isomer (ee) with the most stable conformation of the cis isomer (ea or ae), we deduce that the trans isomer is more stable than the cis because in the trans isomer both chlorines can simultaneously occupy equatorial positions.

Next, consider the stereoisomers of 1,3-dichlorocyclohexane. How many chair conformations of each are possible? Make a model! In this case it is the cis isomer that has a chair form in which both groups are equatorial (or both axial):

The trans-1,3 isomer, on the other hand, exists in two equivalent chair conformations, (ea) and (ae):

So, in this instance, it is the cis isomer (in its ee conformation) that is more stable than the trans isomer.

EXAMPLE 7-19   Perform the same type of analysis for 1,2-dichlorocyclohexane as we did for the 1,4 and 1,3 isomers. Which 1,2 stereoisomer is more stable?

Solution   The two stereoisomers of 1,2-dichlorocyclohexane look like this:

As you can see, this is the same result we had with the 1,4 isomer, and the trans (in its ee conformation) is more stable than the cis.

EXAMPLE 7-20   Draw the most stable chair conformation of *cis*-1-*tert*-butyl-4- methylcyclohexane.

Solution   Any cis 1,4-disubstituted cyclohexane will exist in two possible chair forms, the (ea) and the (ae). Although these two conformations are of equal energy and stability when the two groups are the same (as was the case with 1,4-dichlorocyclohexane), they are not equally stable when the two groups are different. Here is what these two conformations look like in this case:

Clearly, the conformation labeled here as (ea) is the more stable one, for it allows the larger *tert*-butyl group to occupy an equatorial position. We can even estimate that the (ea) conformation is more stable than the (ae) one by about $23 - 7.6 = 15$ kJ/mol.

7-7. Intermolecular Forces and Trends in Physical Properties

The physical properties described in Section 7-1 (especially the melting point and boiling point) are direct indicators of the strength of *inter*molecular (i.e., between molecules) interactions. For example, the higher the melting point of a molecular crystal, the stronger are the intermolecular forces that hold each molecule in place in the crystal lattice. Similarly, the higher the normal boiling point of a liquid, the more attractive are the interactions between molecules in the liquid state.

A. Dipole–dipole interactions

Most of the forces that occur *between* molecules have already been introduced in our discussion of *intra*molecular forces (Section 7-3). These include attractive interactions such as dipole–dipole forces and hydrogen bonding, as well as repulsive forces such as steric effects. (Of course, torsional strain and angle strain are purely intramolecular.) There is another weakly *attractive* force that operates between molecules that are fairly close together (but not so close as to encounter steric repulsion). This so-called **van der Waals attraction** is believed to be due to the attraction between two **instantaneous dipoles,** which arise in nonpolar molecules through temporary polarization in the distribution of their electrons. Let's describe a few specific examples of these forces, and then we'll discuss some more generalized trends in physical properties.

Water, H_2O, has a melting point of 0°C and a normal boiling point of 100°C. The comparable values for H_2S are –85.5°C and –60.7°C, respectively. Thus H_2S is a *gas* at room temperature! Why the difference? We know (Section 2-6) that O–H bonds are more polar than S–H bonds because oxygen is more electronegative than sulfur. The result is that hydrogen bonding is extremely important in liquid and solid water, but essentially absent in H_2S. The presence of this strong attractive interaction between water molecules means that much more thermal energy (i.e., a higher temperature) is required to overcome these interactions and cause a phase change (melting or vaporization).

EXAMPLE 7-21 Explain why the normal boiling point of ethanol is much higher (103°C higher, to be exact) than the normal boiling point of its structural isomer dimethyl ether.

Solution Ethanol (CH_3CH_2OH) has a very polar O–H bond capable of engaging in strong intermolecular hydrogen bonding:

$$CH_3CH_2-\overset{\delta-}{O}\cdots\overset{\delta+}{H}\cdots\overset{\delta-}{O}(H,CH_2CH_3)$$

Dimethyl ether (CH_3OCH_3), on the other hand, has no polar hydrogens, so it cannot engage in hydrogen bonding with itself, though it *could* accept a hydrogen bond from an ethanol molecule:

$$\begin{array}{c}H_3C\\ \\ H_3C\end{array}\overset{\delta-}{O}\cdots\overset{\delta+}{H}-\overset{\delta-}{O}-CH_2CH_3$$

Dimethyl sulfoxide (H_3CSOCH_3, Section 6-5) has a normal boiling point of 189°C, while dimethyl sulfide (H_3CSCH_3) is a gas at room temperature. This is because of the presence of the highly polar S^+-O^- bond in the sulfoxide. Intermolecular attraction between the positive end of one dipole and the negative end of another tends to cause the molecules to associate in the liquid state, making it harder to vaporize them.

$$\begin{array}{c}H_3C\\H_3C\end{array}{}^+S-O^-\cdots {}^+\begin{array}{c}H_3C\\H_3C\end{array}S-O^-$$

B. Steric interactions

Steric effects, which are repulsive in nature, usually *lower* melting points, because the molecules are not as tightly held in the crystal lattice. For example, benzene (Section 4-7) melts at 5.5°C, while *tert*-butylbenzene melts at −57.8°C. While the planar benzene molecules are stacked very efficiently in the crystal, introduction of a bulky *tert*-butyl group interferes with the crystal packing, making the molecules less tightly bound to one another. Interestingly, the boiling point of *tert*-butyl benzene (169°C) is *higher* than that of benzene (80°C). This is because in the liquid phase the attractive forces between the molecules are comparable in both cases, but the mass of benzene is less (MW 78 vs. 134).

The strength of van der Waals attractions between molecules in the liquid state increases as chain length increases but decreases as the molecules become more spherical in shape (thereby presenting less surface area). Thus pentane has a higher normal boiling point (36°C) than its more "spherical" isomer 2,2-dimethylpropane (bp 9.5°C).

$$H_3CCH_2CH_2CH_2CH_3 \qquad \begin{array}{c}H_3C\\H_3C\end{array}\overset{CH_3}{\underset{}{C}}-CH_3$$

pentane

2,2-dimethylbutane

It is possible to make certain generalizations about trends in melting point and boiling point among groups of similar molecules. Of course, as with any generalization there are occasional exceptions. Furthermore, each of these generalizations carries the proviso "all other things being equal":

- The longer the chain length, the higher the boiling point;
- The more polar (or intermolecularly hydrogen bonded) a molecule is, the higher will be its boiling point;
- The more rigid and symmetrical a molecule is, the stronger will be the intermolecular forces in the crystalline state, and the higher will be the melting point.

C. Solubility

In addition to their importance in determining the physical properties of pure compounds, intermolecular forces also play a determining role in the properties of solutions. A **solution** is a homogeneous mixture of two or more substances. Although both gaseous and solid solutions are known, most solutions are liquids composed of one or more **solutes** dissolved in a liquid **solvent** or solvent mixture.

The vast majority of organic reactions take place in liquid solutions and yet, until fairly recently, the effects of solvents on reactions were not well understood or were neglected completely. We'll be discussing such solvent effects in some detail later on, but for now we'll content ourselves with just a little about the dissolving process itself.

One substance (the solute) will dissolve in another (the solvent) when the following condition applies: the energy required to separate solute molecules (i.e., to overcome the intermolecular attractions between solute molecules) plus the energy required to separate solvent molecules must be less than the energy liberated from the intermolecular solute–solvent interactions. In general, polar molecules interact strongly with other polar molecules, nonpolar molecules interact somewhat less strongly with each other, but the interactions of polar molecules with nonpolar molecules are the weakest of all. The upshot of this is that **like dissolves like**. That is, polar solutes tend to dissolve in polar solvents, nonpolar solutes dissolve in nonpolar solvents, but polar solutes do *not* dissolve in nonpolar solvents and vice versa. Common examples include the fact that oil (composed of nonpolar hydrocarbon-like molecules) will not dissolve in water (a polar molecule), while salt (a very polar ionic solid) and table sugar (a molecule with eight hydroxy groups with which to form hydrogen bonds) dissolve readily in water, but not in nonpolar solvents.

Most common organic solvents such as hydrocarbons and ethers are quite nonpolar. Ketones are somewhat more polar, and low-molecular-weight alcohols (e.g., methanol and ethanol) are more polar still. Liquid carboxylic acids are quite polar. There are also a limited number of very special compounds called **dipolar aprotic** solvents, which combine high polarity without the relatively acidic hydrogens (Chapter 9) present in acids and alcohols. Three common examples of such compounds are dimethyl sulfoxide (DMSO, Section 6-5), *N,N*-dimethylformamide (DMF, Example 7-5), and hexamethylphosphoramide (HMPA).

$$(H_3C)_2N-\overset{\overset{\textstyle O}{\|}}{\underset{\underset{\textstyle N(CH_3)_2}{|}}{P}}-N(CH_3)_2$$

HMPA

One of the best measures of solvent polarity is the dielectric constant mentioned in Section 7-1. Tables of dielectric constants can be found in the CRC "Handbook of Chemistry and Physics."

EXAMPLE 7-22   Which do you expect to be more soluble in water: Ethanol or diethyl ether? The dielectric constants of these three compounds are 80, 24, and 4, respectively.

Solution   From the dielectric constant values we confirm our suspicion that water is the most polar and ether is the least. Therefore, ethanol (with its polar hydroxy group) is more soluble in water than is ether in water. In fact, ethanol and water are **miscible,** that is, mutually soluble in all proportions. By contrast, the solubility of ether in water (at 25°C) is limited to a few % by weight. Incidentally, water (as solute) is also only slightly soluble in ether (as solvent).

SUMMARY

1. Each organic compound has a characteristic set of physical properties including melting point, boiling point, index of refraction, density, and dielectric constant. Unknown compounds can often be identified by comparison of their physical properties with those tabulated in the chemical literature.

2. The elemental composition (% by weight) data for a compound can be used to calculate the empirical (simplest) formula of the compound. This empirical formula, along with the compound's molecular weight (from colligative properties or mass spectrometry), can be used to calculate the compound's molecular formula.

3. Spectroscopic techniques are used to deduce structural information about the compound. The four most widely used techniques, along with the type of information available from each, are:
   (a) mass spectrometry: molecular weight, molecular formula, and some information about the sequence of atoms in the molecule
   (b) ultraviolet absorption spectroscopy: information about the presence (or absence) of conjugated multiple bonds in the molecule
   (c) infrared absorption spectroscopy: the types of bonds and functional groups present (or absent) in the molecule
   (d) nuclear magnetic resonance spectroscopy: detailed information about the number, types, and molecular environments of certain nuclei, especially carbon and hydrogen

4. The three types of intramolecular strain are angle strain (departures from ideal bond angles), steric strain (from crowding parts of the molecular structure too close together), and torsional strain (from eclipsing vicinal bonds). Strain *increases* the energy and reactivity of a molecule and *decreases* its stability.

5. Some intramolecular interactions are *attractive* in nature, thereby *stabilizing* the molecule. Examples include certain dipole–dipole interactions such as hydrogen bonding.

6. Conformations of a molecule are different instantaneous arrangements of the atoms, which are rapidly interconverted by rotation around bonds. Staggered conformations are more stable than eclipsed conformations.

7. Newman projections depict the appearance of a molecule when viewed along a particular bond, usually the one undergoing rotation.

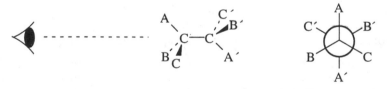

side view   Newman projection (end view)

8. The relative amount (ratio) of two different conformations is a function of the difference in energy between them (ΔE) and the absolute temperature (T) according to the equation

$$\frac{\text{less stable}}{\text{more stable}} = e^{-\Delta E/RT} \qquad \textbf{(7-2)}$$

9. Except for cyclopropane, all cycloalkanes adopt nonplanar conformations in order to relieve eclipsing and angle strain as much as possible.

10. The most stable conformation of cyclohexane is the chair form, which places six hydrogens in axial (a) positions and six in equatorial (e) positions. Substituent groups prefer to occupy equatorial positions to avoid 1,3-diaxial repulsions.

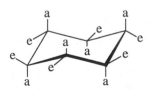

11. Cyclohexane derivatives undergo constant and rapid interconversion of the two possible chair forms by a process called ring flipping, which interchanges all axial positions with equatorial ones.

12. Stereoisomers of disubstituted cyclohexane derivatives have the following arrangement of the substituent groups (e = equatorial, a = axial; most stable conformation is underlined):

substitution pattern	trans	cis
1,4	<u>ee</u>, aa	ea, ae
1,3	ea, ae	<u>ee</u>, aa
1,2	<u>ee</u>, aa	ea,ae

13. *Inter*molecular forces include not only steric repulsion, dipole–dipole interactions, and hydrogen bonding, but also weakly attractive van der Waals (instantaneous dipole) forces. The strength of van der Waals attractions increase with increasing chain length, but decrease as the molecular structure becomes more "spherical."

14. Like dissolves like, that is, polar solutes tend to dissolve in polar solvents, nonpolar solutes dissolve in nonpolar solvents, while nonpolar solutes do not dissolve in polar solvents and vice versa.

RAISE YOUR GRADES

Can you define...?

- ☑ melting point
- ☑ boiling point
- ☑ index of refraction
- ☑ density
- ☑ dielectric constant
- ☑ volatile
- ☑ empirical formula
- ☑ spectrum
- ☑ molecular ion
- ☑ fragment ion
- ☑ exact mass
- ☑ chemical shift
- ☑ inter- and intramolecular
- ☑ angle strain
- ☑ steric strain
- ☑ torsional strain
- ☑ dipole–dipole forces
- ☑ hydrogen bonding
- ☑ conformation
- ☑ eclipsed
- ☑ staggered
- ☑ gauche
- ☑ Newman projection
- ☑ dihedral angle
- ☑ chair form of cyclohexane
- ☑ axial and equatorial positions
- ☑ ring flipping
- ☑ boat form of cyclohexane
- ☑ 1,3-diaxial repulsion
- ☑ van der Waals attraction
- ☑ solution
- ☑ solute
- ☑ solvent
- ☑ dipolar aprotic solvent

Can you explain...?

- ☑ how you would go about identifying an unknown organic compound
- ☑ how to calculate the empirical formula of a compound from its elemental analysis
- ☑ how to calculate the molecular formula of a compound from its empirical formula and its molecular weight
- ☑ what structural information is available from a compound's mass spectrum, UV spectrum, IR spectrum, and NMR spectrum
- ☑ the influence of strain and other intramolecular forces on a molecule's stability and reactivity
- ☑ how the various types of strain present in a cycloalkane can be relieved
- ☑ when hydrogen bonding is likely to be encountered
- ☑ how the energy of a molecule varies with its conformation
- ☑ the difference between stereoisomers and conformations
- ☑ how to calculate the equilibrium ratio of conformations when their relative energies are known
- ☑ the consequences of a cylohexane ring flip
- ☑ why cyclohexane substituents are more stable in an equatorial position than in an axial position
- ☑ how to draw the most stable conformations of cyclic and acyclic molecules
- ☑ how physical properties indicate the strength of intermolecular interactions
- ☑ what factors control the solubility of one compound in another

SOLVED PROBLEMS

PROBLEM 7-1   Deduce the molecular formula of each compound below from the elemental analysis and molecular weight data.

Compound **A**: C, 12.9%; H, 1.1%; Br, 86.0%; MW 185 ± 2.
Compound **B**: C, 90.6%; H, 9.4%; MW 106 ± 1.
Compound **C**: C, 25.7%; H, 6.4%; P, 22.1%; MW 142 ± 2.

Solution   Proceed as in Example 7-2.
Compound **A**: First, we note that all the mass is accounted for (12.9% + 1.1% + 86.0% = 100%). Next, we calculate the empirical formula.

	C	H	Br	
% by weight	12.9	1.1	86.0	empirical formula: $C_1H_1Br_1$ or CHBr
atomic weight	12.0	1.0	79.9	EFW: 12.0 + 1.0 + 79.9 = 92.9
decimal ratio	1.08	1.1	1.08	$n = MW/EFW = 185/92.9 = 2$
integer ratio	1	1	1	molecular formula: $(CHBr)_2 = C_2H_2Br_2$

Compound **B**: Again, we note that all the mass is accounted for (90.6% + 9.4% = 100%). Next, we calculate the empirical formula.

	C	H	
% by weight	90.6	9.4	empirical formula: C_4H_5
atomic weight	12.0	1.0	EFW: 4(12.0) + 5(1.0) = 53
decimal ratio	7.55	9.4	$n = MW/EFW = 106/53 = 2$
integer ratio	1 (or 4)	1.25 (or 5)	molecular formula: $(C_4H_5)_2 = C_8H_{10}$

Compound **C**: In this case there is 100% – (25.7% + 6.4% + 22.1%) = 45.8% of mass missing. We'll tentatively ascribe this to oxygen.

	C	H	P	O	
% by weight	25.7	6.4	22.1	45.8	empirical formula: $C_3H_9PO_4$
atomic weight	12.0	1.0	31.0	16.0	EFW: 3(12.0) + 9(1.0) + 31.0 + 4(16.0) = 140
decimal ratio	2.14	6.4	0.71	2.86	$n = MW/EFW = 142/140 = 1$
integer ratio	3	9	1	4	molecular formula: $(C_3H_9PO_4)_1 = C_3H_9PO_4$

note: The fact that we come out with exactly four atoms of oxygen lends confidence to our assumption that the missing mass is indeed due to oxygen.

PROBLEM 7-2   Using the molecular formulas calculated above, along with the additional spectral information given below, propose a structure for each of the three compounds. Then, for practice, name each compound.

Compound **A**: The hydrogen NMR spectrum exhibits one signal for the two equivalent vinyl hydrogens, and the IR spectrum indicates that these two hydrogens are cis to each other.
Compound **B**: The UV spectrum suggests an aromatic ring, while the IR spectrum indicates a para-disubstituted benzene ring.
Compound **C**: The IR spectrum indicates a phosphoryl group (P=O); the carbon NMR spectrum exhibits one signal for three equivalent sp^3-hybridized carbons, each attached to an oxygen.

Solution
Compound **A**: We begin by calculating the I value [Eq. (6-1)] for the molecular formula $C_2H_2Br_2$:

$$I = \frac{2(2)+2-2-2}{2} = 1$$

Since it is impossible to have a ring with just two carbons and no other polyvalent atoms, the I value must indicate one carbon–carbon double bond. The spectral data indicate a partial structure

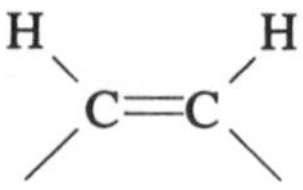

Thus there is only one possibility to complete the structure:

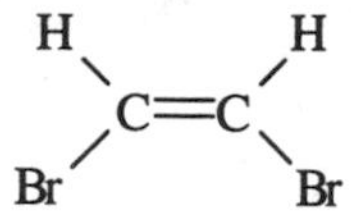

cis-1,2-dibromoethene

Compound B: Here $I = \dfrac{2(8)+2-10}{2} = 4$

The spectral data indicate a benzenoid aromatic ring, which itself requires an I value of four, so there is no additional unsaturation. The spectral data suggest the partial structure below, where the lines indicate as yet unspecified substituents:

(C_6H_4)

There are two carbons and six hydrogen remaining, and these must comprise two methyl groups:

p-xylene (1,4-dimethylbenzene)

Compound C: $I = \dfrac{2(3)+2+1-9}{2} = 0$

Remember (Section 6-6) that a P=O (or P⁺–O⁻) bond is not counted by the I value. The phosphoryl group requires three additional substituents:

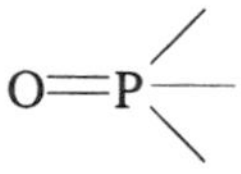

Since the three carbons are equivalent and each is attached to an oxygen, they must occur as three methoxy (CH_3O) groups:

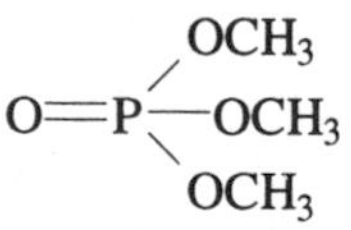

This compound is a phosphate triester, and although you're not expected to be able to name such compounds, you might be able to guess that this compound is trimethyl phosphate.

PROBLEM 7-3 Compound **D** has molecular formula $C_5H_{10}O$, and its IR spectrum indicates a ketone carbonyl group. Draw and name all possible structures that fit this description.

Solution The I value for this molecular formula is

$$I = \frac{2(5)+2-10}{2} = 1$$

so the carbonyl (C=O) is the only unsaturation. There are three possible structural isomers:

$$\underset{\text{2-pentanone}}{CH_3CH_2CH_2\overset{\displaystyle O}{\overset{\displaystyle \|}{C}}CH_3} \qquad \underset{\text{3-pentanone}}{CH_3CH_2\overset{\displaystyle O}{\overset{\displaystyle \|}{C}}CH_2CH_3} \qquad \underset{\text{3-methyl-2-butanone}}{(CH_3)_2CH\overset{\displaystyle O}{\overset{\displaystyle \|}{C}}CH_3}$$

PROBLEM 7-4 The mass spectra of ketones usually show strong peaks for acylium fragment ions:

$$\underset{\text{molecular ion}}{R-\overset{\displaystyle :\overset{+}{O}}{\overset{\displaystyle \|}{C}}-R'} \quad \begin{array}{c} \longrightarrow \quad R-C\equiv O:^+ \;+\; R'\cdot \\[2ex] \longrightarrow \quad R'-C\equiv O:^+ \;+\; R\cdot \end{array}$$

acylium ions

The mass spectrum of compound **D** (Problem 7-3) shows a very intense fragment at m/e 57, but no significant signals at m/e 43 or 71. Propose a structure for the compound.

Solution Both 2-pentanone and 3-methyl-2-butanone are unsymmetrical ketones (i.e., R $\neq$ R'). So each would show two different acylium ion fragments, one at m/e 43 (for $CH_3C\equiv O:^+$) and 71 (for $C_3H_7C\equiv O:^+$). However, 3-butanone is symmetrical (R = R'), so there is only one acylium ion possible, $CH_3CH_2C\equiv O:^+$, with m/e 57. So, compound **D** is 3-butanone.

PROBLEM 7-5 The **heat of combustion** of a substance is the amount of heat energy released when the substance is burned in an excess of oxygen. Each methylene group in a cycloalkane undergoes combustion to yield CO_2 and H_2O according to the equation

$$CH_2 + 3/2\; O_2 \longrightarrow CO_2 + H_2O$$

The heat of combustion of cyclohexane is 3967 kJ/mol, while that of cyclobutane is 2755 kJ/mol. From these data, estimate the total strain energy in cyclobutane. Describe your reasoning.

Solution First of all, we remember that cyclohexane is essentially strain free (Section 7-5). But since there are six CH_2 groups in cyclohexane and only four in cyclobutane, we must first multiply 3967 by 4/6 (or 2/3) to estimate the hypothetical "strain-free" heat of combustion of cyclobutane. The resulting value is 2645 kJ/mol. The actual heat of combustion of cyclobutane is higher than this hypothetical value by 2755 − 2645 = 110 kJ/mol, and this extra energy, released during combustion, can be taken as a measure of the strain present in cyclobutane.

PROBLEM 7-6 How much of the total strain calculated in Problem 7-5 can be ascribed to C–H eclipsing interactions? What is the source of the rest of the strain?

Solution From Example 7-11 we know there are eight C–H eclipsing interactions in cyclobutane, at an energetic cost of 4 kJ/mol each. Thus the total C–H eclipsing strain is 8(4 kJ/mol) = 32 kJ/mol. The remaining strain, 110 − 32 = 78 kJ/mol, results from a combination of angle strain and C–C eclipsing interactions.

PROBLEM 7-7 As we'll see in Chapter 9, a fundamental requirement for a reaction to take place between two molecules (a so-called bimolecular reaction) is that they must be able to collide with each other. When the reaction takes place in solution, as most reactions do, the solvent molecules tend to surround the solute molecules, and the "tightness" of the solute–solvent interaction will depend on the strength of intermolecular forces described in Section 7-7. Do you think that the rate of a bimolecular reaction will be helped or hindered by strong solute–solvent interactions? Why?

Solution If either or both of the reacting molecules are tightly surrounded by a shell of solvent molecules, it may very well prevent them from getting close enough to react. This would be an example of a type of steric effect. On the other hand, if the reaction product is more strongly solvated than the reactants, the reaction will be more favorable.

PROBLEM 7-8   Acetic acid has a molecular weight of 60, yet in the vapor phase or dissolved in nonpolar solvents it often behaves as if it had a molecular weight of 120. Can you suggest a reason for this?

Solution   The fact that the apparent molecular weight of acetic acid in solution (in nonpolar solvents) is exactly double its true molecular weight suggests that it forms some type of **"dimer"** (two molecules somehow stuck together) in solution and in the vapor phase. A little thought should convince you that a carboxylic acid molecule is ideally constituted to form such a dimer through double hydrogen bonding:

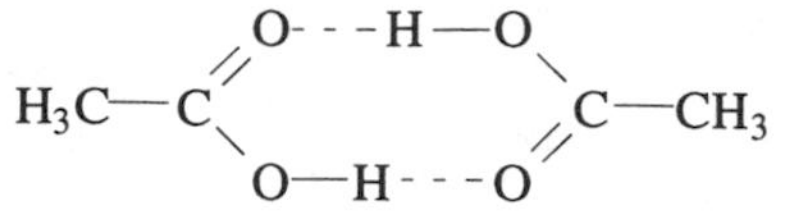

PROBLEM 7-9   The **heat of hydrogenation** of an alkene is the amount of heat energy released when a molecule of hydrogen is added to the bond according to the reaction:

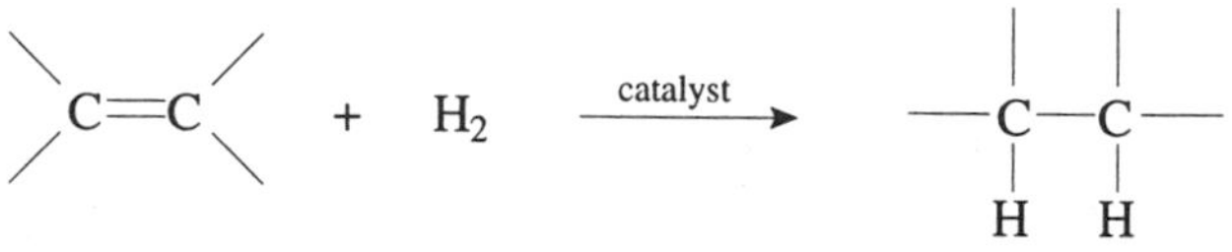

The heat of hydrogenation of *cis*-2-butene is 120 kJ/mol, while that of the trans isomer is 115 kJ/mol. **(a)** What is the hydrogenation product in each case? **(b)** To what do you ascribe the difference in heats of hydrogenation?

Solution
(a) In both cases, the hydrogenation product is butane:

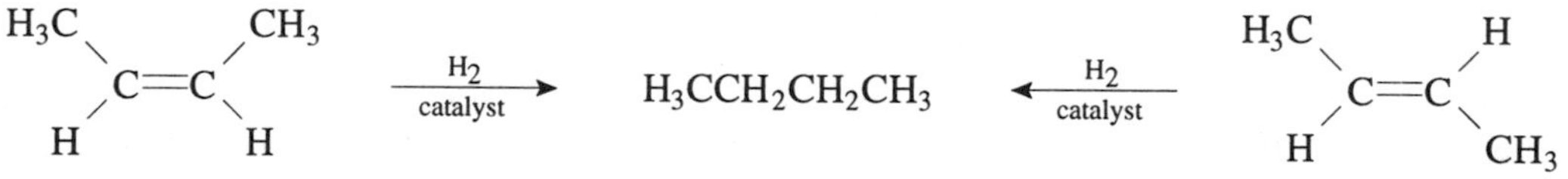

(b) Since both butene isomers give the same final product, the fact that the cis isomer gives off 5 kJ/mol more energy must indicate that it starts out with 5 kJ/mol more internal energy than the trans isomer. That is, the cis isomer is 5 kJ/mol less stable than the trans isomer. A likely cause of this difference is steric repulsion between the two methyl groups in the cis isomer.

PROBLEM 7-10   **(a)** Draw the Newman projection (sighting along the central C–C bond) of butane in the conformation with the methyl groups eclipsed. What is the dihedral angle between the methyl groups in this conformation? **(b)** Draw Newman projections for the other conformations of butane which exhibit the following dihedral angles between the methyl groups: 60°, 120°, 180°, 240°, 300°, and 360°. Label each conformation as either eclipsed or staggered. How would you describe the spatial relationship between the methyl groups in each conformation? **(c)** Calculate the relative energy of each of the above conformations using the following strain energies (all in kJ/mol): H–H eclipsing, 4; H–CH₃ eclipsing, 6; CH₃–CH₃ eclipsing plus steric repulsion, 11; CH₃–CH₃ gauche steric repulsion, 4. **(d)** Plot a graph (similar to Figure 7-3) of relative energy versus dihedral angle for the conformations of butane.

Solution

(a)

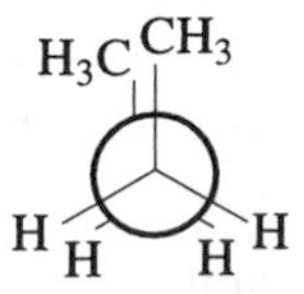

In this conformation the dihedral angle between the methyl groups is 0°. Two such groups are said to be **s-cis** (that is, cis about a sigma bond) or **syn** to each other.

(b and c)

dihedral angle	projection	label	methyls	relative energy (kJ/mol)
60°	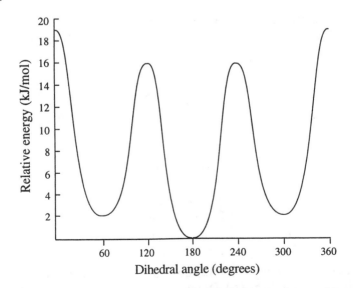	staggered	gauche	4
120°		eclipsed	—	4 + 6 + 6 = 16
180°		staggered	s-trans (anti)	0
240°		eclipsed	—	4 + 6 + 6 = 16
300°		staggered	gauche	4
360°		eclipsed	s-cis	11 + 4 + 4 = 19

(d) See Figure 7-4.

Figure 7-4 Conformational energy of butane as a function of rotation around the central carbon–carbon bond.

PROBLEM 7-11   (a) Draw two chair conformations of phenylcyclohexane, one with the phenyl group equatorial, and the other with the group axial. Which of these is more stable? (b) The energetic preference for a phenyl group to be equatorial rather than axial is 12.6 kJ/mol. At 298 K what is the ratio of chair conformations with an axial phenyl compared to those with an equatorial phenyl? (c) In terms of its equatorial preference, which group is sterically larger, a phenyl or a *tert*-butyl? How do you know?

Solution

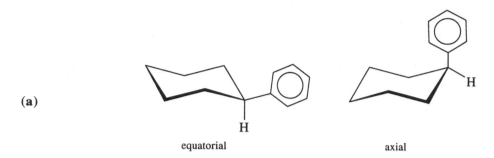

(a)

equatorial   axial

   The equatorial is more stable because it avoids 1,3-diaxial repulsions.
(b) We can use Eq. (7-2):

$$\frac{\text{axial}}{\text{equatorial}} = e^{-\Delta E/RT} = e^{-(12.6 \text{ kJ/mol})/(0.0083 \text{ kJ/mol-K})(298 \text{ K})}$$

$$= 0.0061$$

Thus for every 1000 equatorial phenyls, there will be 6 axial ones.
(c) From Section 7-6, we know that the equatorial preference of a *tert*-butyl group is 23 kJ/mol, nearly twice the value for a phenyl group. So the *tert*-butyl group is sterically larger than a phenyl, at least in the context of 1,3-diaxial repulsions.

PROBLEM 7-12   At room temperature chlorocyclohexane exists as a rapidly interconverting mixture of chair forms with the chlorine either axial or equatorial. However, at very low temperature (approximately –150°C), the interconversion stops, and the more stable equatorial form can actually be separated from the axial form. Under these conditions are the two forms still conformations, or are they now stereoisomers?

Solution   An absolute requirement for two structures to be conformations is that they be *rapidly interconverting*. If this interconversion is suppressed, the two structures are stereoisomers.

PROBLEM 7-13   Complete the Newman projection below of the ea conformation of *cis*-1,2-dichlorocyclohexane (Example 7-19), sighting along the C(Cl)–C(Cl) bond.

Solution

PROBLEM 7-14   (a) Draw the most stable conformation of *trans*-1,4-di-*tert*-butylcyclohexane, and show the result of a ring flip. (b) Draw the most stable conformation of *cis*-1,3-di-*tert*-butylcyclohexane, and show the result of a ring flip. (c) Which is more stable, the cis or trans isomer? Why? **(d)** Is the following structure a conformation of the cis isomer or the trans? Would you expect this conformation to be more or less stable than the one in (b)? Why?

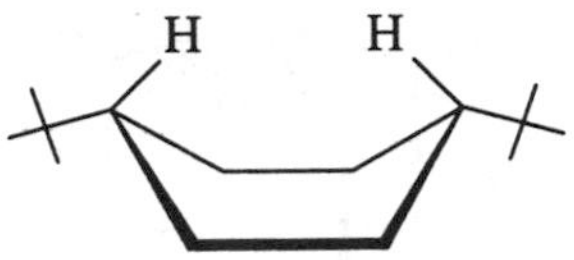

Solution
(a)

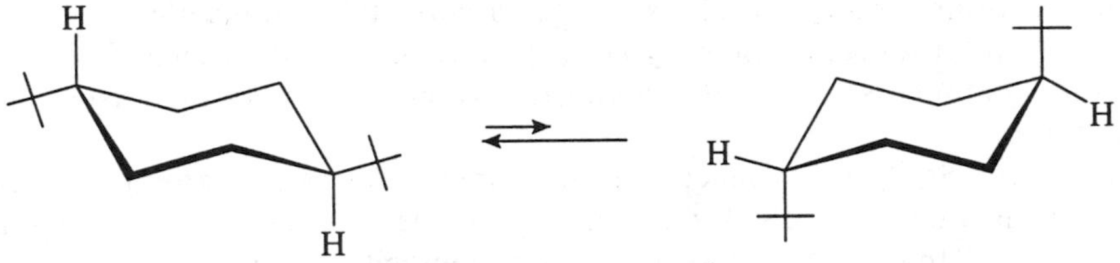

(ee) most stable   (aa) less stable than ea

(b)

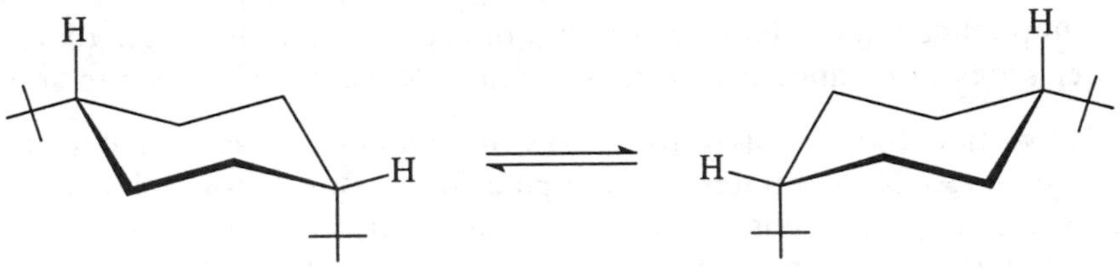

(ea)   equally stable   (ae)

(c) The trans isomer is more stable for it has an accessible conformation that has both *tert*-butyl groups equatorial. The cis isomer (in the chair form) must have one *tert*-butyl group axial, at an energetic cost of 23 kJ/mol.

(d) This is another conformation (boat form) of the cis stereoisomer. The good news is that both *tert*-butyl groups are now able to be free of 1,3-diaxial repulsions by occupying what are called **bowsprit** positions. This saves 23 kJ/mol of energy. The bad news is that this is a boat conformation, which itself costs about 29 kJ/mol. So the cis isomer is stuck between a rock and a hard place, and actually adopts a still different conformation (called a twist boat), which is the best available compromise. It is more stable than the one in (b).

PROBLEM 7-15   The structure shown below is named adamantane. To answer this question, it will help to make a model of the molecule.

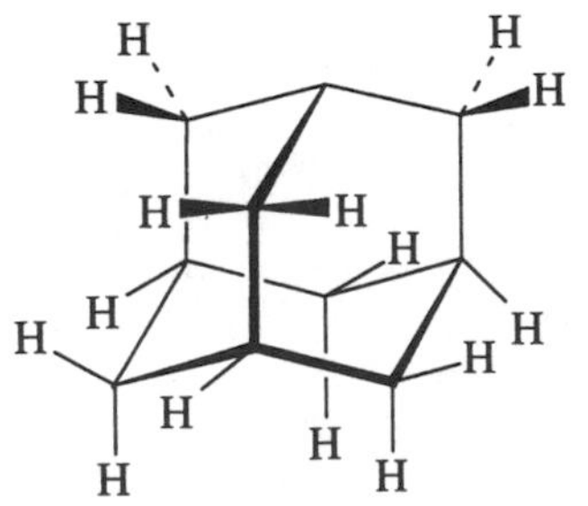

(a) Using the method of Example 5-10, determine the number of rings in adamantane.
(b) The six-membered rings in this structure are in what conformations? Do you expect this to be a stable molecule?
(c) Is this molecule flexible or rigid?
(d) What might you guess about its melting point?

Solution

(a) It is tricyclic, that is, the molecule can be opened up by cleaving a minimum of three bonds:

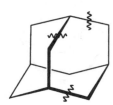

(b) We can regard adamantane as three chair cyclohexanes sharing various sides. There are no eclipsing interactions, or angle strain, so this *is* a very stable molecule.

(c) From either your intuition, or the model you built, you can see that the chair forms are all locked by one another, so there is none of the flexibility present in cyclohexane itself. This is a very rigid molecule.

(d) Because it is both rigid and symmetrical, adamantane forms an extremely tightly ordered crystal, with strong intermolecular forces, making for one of the highest melting points (256°C) of any hydrocarbon. In fact, the carbon atoms in a diamond crystal are arranged in exactly this type of arrangement, extending to "infinity" in all three directions.

PROBLEM 7-16 Among organometallic compounds (Section 6-5), organolithium compounds are much more common than organosodium and organopotassium compounds. This is because the organolithium compounds tend to be more soluble in organic solvents (e.g., alkanes and ethers) than are the other types of compounds. Suggest a reason for this difference in solubility.

Solution We know that "like dissolves like," so that nonpolar organic solvents such as alkanes and ethers will prefer less polar solutes to more polar ones. But we should also remember from Section 2-6 that a carbon–lithium bond, while somewhat polar, will be considerably less polar than a carbon–sodium bond or a carbon–potassium bond. And, though you would have had no way to know this, there is also much evidence that alkyllithium reagents tend to form clusters of molecules (held together by dipole–dipole forces resembling the lithium equivalent of hydrogen bonding), which make them less polar overall and aid in making them soluble in organic solvents.

STEREOCHEMISTRY

THIS CHAPTER IS ABOUT

- ☑ **Isomers Revisited**
- ☑ **Symmetry and Equivalence**
- ☑ **Asymmetry, Dissymmetry, and Enantiomers**
- ☑ **Naming Enantiomers: The R–S System**
- ☑ **Properties of Enantiomers: Polarimetry**
- ☑ **Racemic Mixtures and Optical Purity**
- ☑ **Diastereomers and Multiple Chiral Centers**
- ☑ **Optical Resolutions**
- ☑ **Fischer Projections, Invertomers, and Other Dissymmetric Molecules**
- ☑ **Stereochemistry as a Mechanistic Probe**

8-1. Isomers Revisited

Stereochemistry is the study of the three-dimensional spatial relationships between the atoms in a molecule. As such, it is mainly concerned with the physical and chemical properties of stereo-isomers. Conformational analysis is usually considered to be part of stereochemistry, but we have already discussed that topic in Chapter 7.

It was back in Section 2-4 that we first encountered the concept of *isomers*: Compounds that have the same molecular formula but different structures, and that are (in principle) separable from one another. Structural isomers differ from each other in the *sequence* (connectivity) of atoms and bonds, that is, which atoms are bonded to which other atoms. Stereoisomers (Section 5-5), on the other hand, have the *same* sequence of atoms and bonds, but a different three-dimensional arrangement of the atoms. Furthermore, remember that neither resonance forms (Section 4-3) nor conformations (Section 7-4) constitute isomers. Resonance structures represent various ways of distributing the electrons in a given molecular structure, while conformations are instantaneous arrangements of the atoms in a molecule that are interconverted rapidly by rotations around bonds.

In this chapter we'll explore stereoisomers in somewhat more detail, and we'll find that there are several kinds of stereoisomers. But before we can delve more deeply into stereochemistry, we need to have at least a little familiarity with molecular symmetry.

8-2. Symmetry and Equivalence

note: It is absolutely essential that you construct molecular models of every structure that we discuss in this chapter. Otherwise, it will be almost impossible for you to fully appreciate the spatial relationships being described.

A. Symmetry operations

In Chapter 7 we used two new words without really defining them: *symmetry* and *equivalence*. Most of us would agree, however, that symmetry has something to do with "balance," while equivalent means somehow "indistinguishable." We are now going to formalize our understanding of both terms. Objects (e.g., molecules) can be characterized by their symmetry properties, i.e., the symmetry they possess. To label the symmetry properties of an object, we begin by defining a **symmetry operation** as some actual or imagined manipulation that leaves an object in a new orientation that is indistinguishable from the object in its original orientation.

note: Doing *nothing* to an object certainly leaves it indistinguishable, but we needn't consider this trivial **identity operation** (symbol E) any further.

B. Symmetry elements

Symmetrical objects are said to possess one or more **symmetry elements**, which are axes, planes, or points that help describe symmetry operations. We use these symmetry elements to show how an object's orientation changes when a symmetry operation is performed.

1. Axis of symmetry, C: Imagine, if you will, a perfect glass sphere. Assuming that we leave its center of mass unmoved, how might we manipulate the sphere and still leave it looking exactly as it did before we moved it? For one thing, we could rotate it around any imaginary axis through its center, and the sphere would appear unchanged. Not only that, we could also rotate the sphere any number of degrees around such an axis, and its appearance would still be indistinguishable from its original appearance. Such a **symmetry** (or **rotational**) **axis** is given the symbol C (from *cyclic*). Our sphere has an infinite number of such axes.

2. Plane of symmetry, σ: Here is another manipulation we could perform. Suppose we were to position an imaginary two-sided mirror directly through the center of the sphere, reflecting the sphere onto itself. Because the mirror image of the left half is indistinguishable from the right half, and vice versa, the reflection leaves the sphere indistinguishable from the original. Therefore, a sphere is said to have a **plane of symmetry** (symbol σ) because one half of the object is the mirror image of the other half. In fact, the sphere has an infinite number of such planes.

3. Center of symmetry, i: In addition to axes and planes of symmetry, there is another symmetry element called a **center of symmetry** (symbol i). An object has a center of symmetry (located at its center of mass), when every point (a in the diagram below) on the object has a superimposable counterpart (a') located an equal distance ($d = d'$) from the center along a line from a to a'. Our sphere also has a center of symmetry.

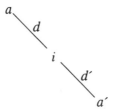

Clearly, the sphere is the most symmetrical of all objects, with its center of symmetry, as well as an infinite number of planes and axes of symmetry. Still, many objects and most small molecules possess at least some symmetry. Let's discuss a few examples.

C. Examples of symmetry in molecules

Consider chloromethane (**1**),

$$\begin{array}{c} Cl \\ | \\ \diagup C \diagdown H \\ H \qquad H \end{array}$$

1

which possesses several types of symmetry (see Figure 8-1). If you rotate the molecule around an imaginary axis passing through the carbon and chlorine atoms (colinear with the C–Cl bond), you will notice that every 120° of rotation brings each hydrogen to a position previously

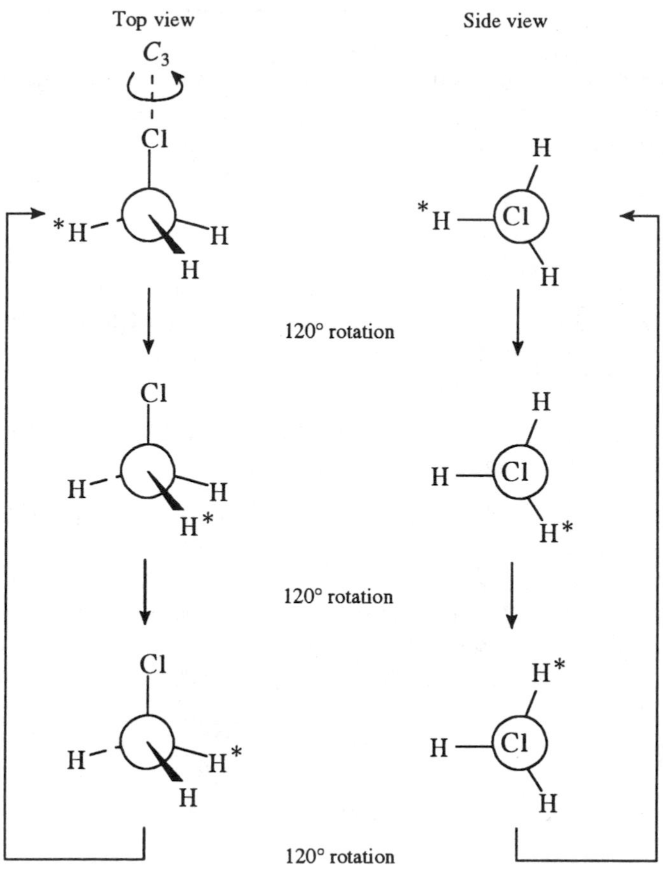

FIGURE 8-1. The C_3 axis of chloromethane. The circle represents the central carbon atom. One hydrogen is labeled (*) to show its position during rotation. (It is actually indistinguishable from any other hydrogen.)

occupied by another hydrogen. Thus each 120° rotation provides an orientation totally indistinguishable from the original. Of course, after the third 120° rotation we would return to where we started. So, we call this a threefold (C_3) axis because the object assumes exactly indistinguishable orientations three different times during one complete (360°) rotation.

- An object with an *n*-fold (C_n) axis becomes indistinguishable from its original orientation *n* times during one 360° rotation, and once every 360/*n* degrees. Furthermore, every object has a C_1 axis. (Why?)

Notice that if we were to rotate the molecule around the C_3 axis by any other angle than an integer multiple of 120°, the resulting orientation *would* be distinguishable from the original. For example, rotation of the upper left structure in Figure 8-1 by 60° or 180° gives structures **1′** and **1″**, respectively. Can you see how these orientations *are* distinguishable from the original?

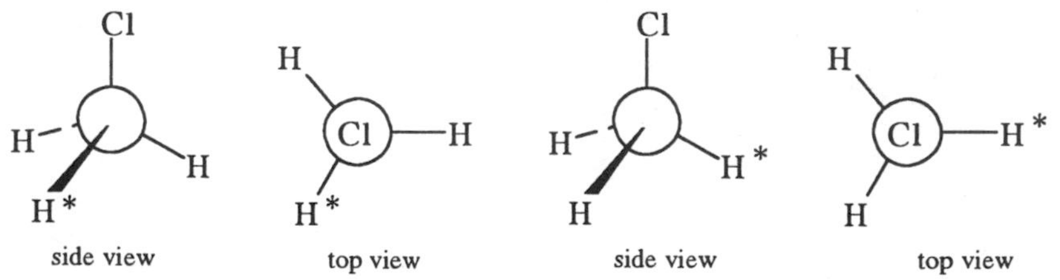

EXAMPLE 8-1

(a) A snowflake has a C_6 axis. How many times does it become indistinguishable from the original orientation during one full rotation around the axis? How many degrees of rotation bring it to an indistinguishable orientation?

(b) What is the highest-order (value of n) axis in a sphere?

Solution

(a) From the definition of a C_n axis, the snowflake presents an indistinguishable orientation six times during one 360° rotation around the axis, once each 360°/6 = 60°.

(b) Since the sphere assumes indistinguishable orientations an infinite number of times during one rotation, the highest-order axis is a C_∞.

- Atoms in a molecule that are related by virtue of a symmetry operation are said to be (symmetry-) **equivalent,** and are indistinguishable.

Thus the three hydrogens in chloromethane are equivalent by virtue of the C_3 axis and as such are chemically indistinguishable.

Reflection is the symmetry operation described by a plane of symmetry. Can you find the three planes of symmetry in chloromethane? There is one containing each C–H bond together with the chlorine atom, as shown in Figure 8-2. Reflection in any one of these planes relates the two hydrogens *not* in the plane, so the three hydrogens are also equivalent by virtue of these planes of symmetry.

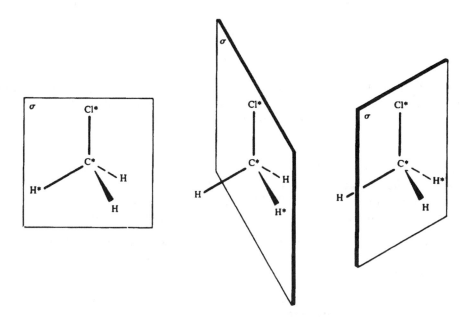

FIGURE 8-2. The mirror planes of chloromethane. The asterisk indicates which atoms lie *in* the plane; the remaining hydrogens are related *by* the plane.

Can you now find a center of symmetry in chloromethane? I hope not, because it doesn't have one. The easiest way to see this is that there is no second chlorine to balance the first one.

Next, let's consider the three isomers of dichloroethylene, *cis*-1,2-(**2**), *trans*-1,2-(**3**), and 1,1-(**4**).

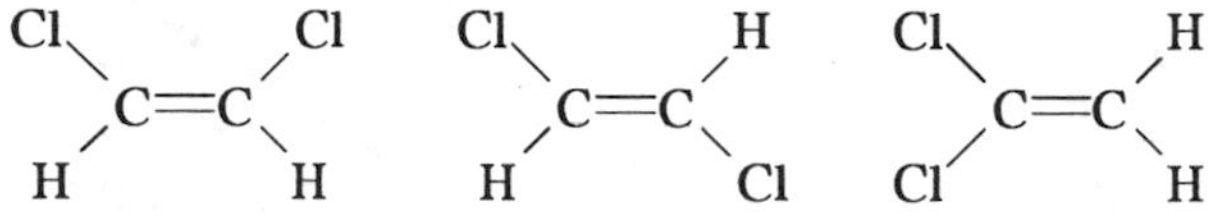

| **2** | **3** | **4** |

Each of these structures is planar, with all six atoms in each structure occupying the same plane (the molecular plane). What symmetry relationships can you find in each of these structures? Figure 8-3 shows that in all three cases the pairs of like atoms (H, C, and Cl) are equivalent by virtue of C_2 axes and/or σ planes. Note also that for each of these structures the molecular plane itself constitutes a plane of symmetry. But this symmetry element doesn't relate any of the atoms together, because all of the atoms are *in* the plane. Of the three, only the trans isomer (**3**) has a center of symmetry.

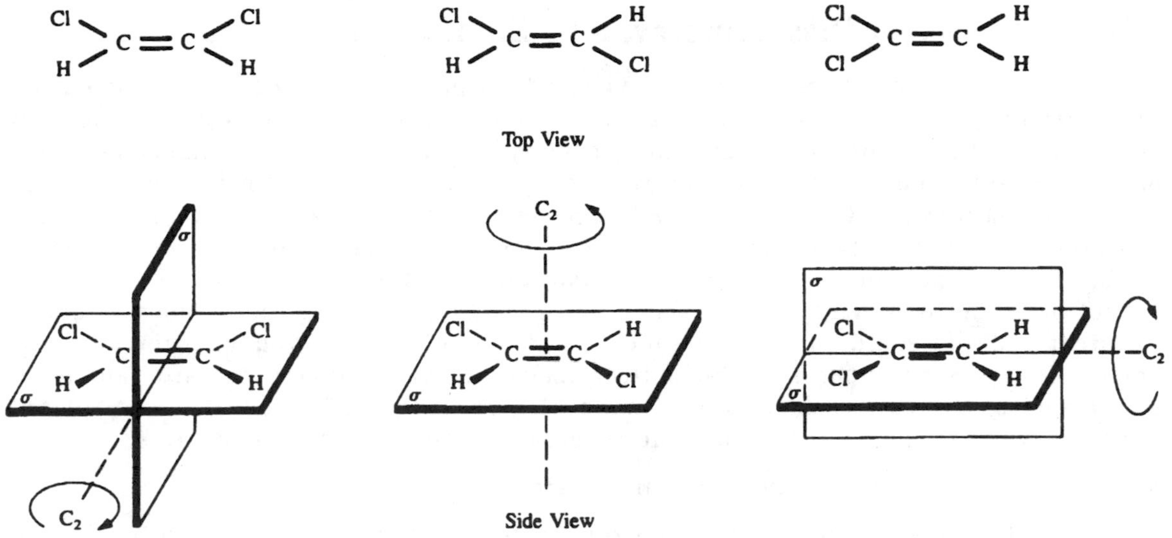

FIGURE 8-3. Symmetry properties of the dichloroethylenes.

EXAMPLE 8-2 What symmetry elements can you discover for benzene (structure **5**)?

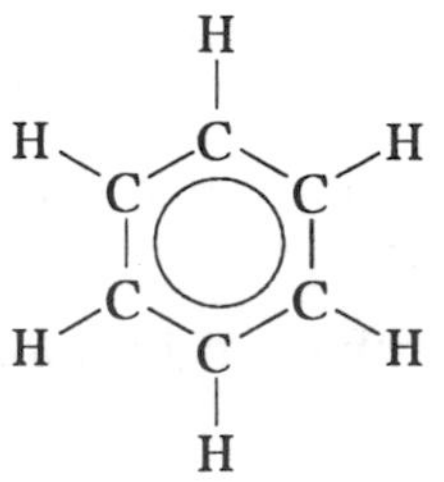

5

Solution Benzene has a C_6 axis perpendicular to the ring and passing through its center. In addition, there are six C_2 axes *in* the plane of the ring. There are seven planes of symmetry, the molecular plane and six others perpendicular to the ring (one containing each C_2 axis). Finally, there is a center of symmetry. Some of these are shown in Figure 8-4. Can you locate the rest of them?

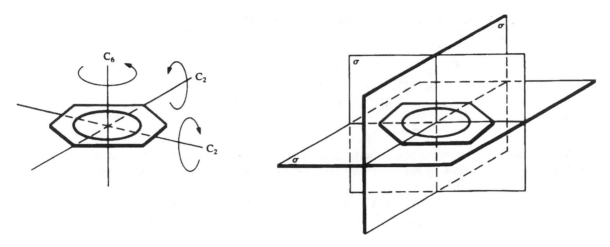

FIGURE 8-4. Some symmetry elements of benzene.

8-3. Asymmetry, Dissymmetry, and Enantiomers

Take a moment to look at your shoes. Can you find any planes, axes, or centers of symmetry for either of them? Unless your shoes are out of the ordinary, there will be no symmetry elements for them. Objects that have no symmetry (other than trivial elements such as E or C_1) are said to be **asymmetric**. But even though your shoes are both asymmetric, they *do* bear a relationship: They are **mirror images** of each other. That is, if you look at your right shoe in a mirror, it looks like your left one, and vice versa (neglecting scuffs, broken laces, etc.). Furthermore, your shoes are **nonsuperimposable**; they are not congruent and cannot be made to coincide exactly. (If they *were* superimposable, both shoes would be identical and you could wear either shoe on either foot!) Chemists have a term for two objects that are nonsuperimposable mirror images of each other; they are said to be **enantiomers**. Notice that enantiomeric molecules *do* fit the definition of stereoisomers: same sequence of atoms and bonds, but different three-dimensional structures.

- Like shoes, enantiomers *always* occur in pairs.

Now, of course, *all* objects have mirror images. In many cases the mirror image *is* superimposable with the original object. For example, all the molecules mentioned so far in this chapter have mirror images that *are* superimposable on the original molecule. In Figure 8-5 we show each of these structures, and in the parallelogram are shown the mirror image structures. Remember that when you're testing to see if the mirror image is superimposable on the original, you can reorient it in any way you want, so long as you don't break any bonds. And once again, it's best to make models if you have any doubts. Don't continue on until you are convinced that in each example in Figure 8-5, the mirror image *is* superimposable on the original structure. Notice also that the structures in Figure 8-5 have something in common: They all have at least one plane of symmetry.

- Objects with a plane or center of symmetry (which obviously includes planar molecules) are *not* asymmetric, and do *not* have enantiomers.

But now, carefully consider the two structures below. Make a model of each, then describe the relationship between them.

$$
\underset{\textbf{6}}{\overset{\displaystyle \text{Cl} \diagdown \diagup \text{H}}{\underset{\displaystyle \text{I} \diagup \diagdown \text{Br}}{\text{C}}}}
\qquad\qquad
\underset{\textbf{7}}{\overset{\displaystyle \text{H} \diagdown \diagup \text{Cl}}{\underset{\displaystyle \text{Br} \diagup \diagdown \text{I}}{\text{C}}}}
$$

If you said they're identical, look again! They *are* mirror images of each other, but they are *not* superimposable, so they are *enantiomers* of each other. The fact that the two are nonidentical is

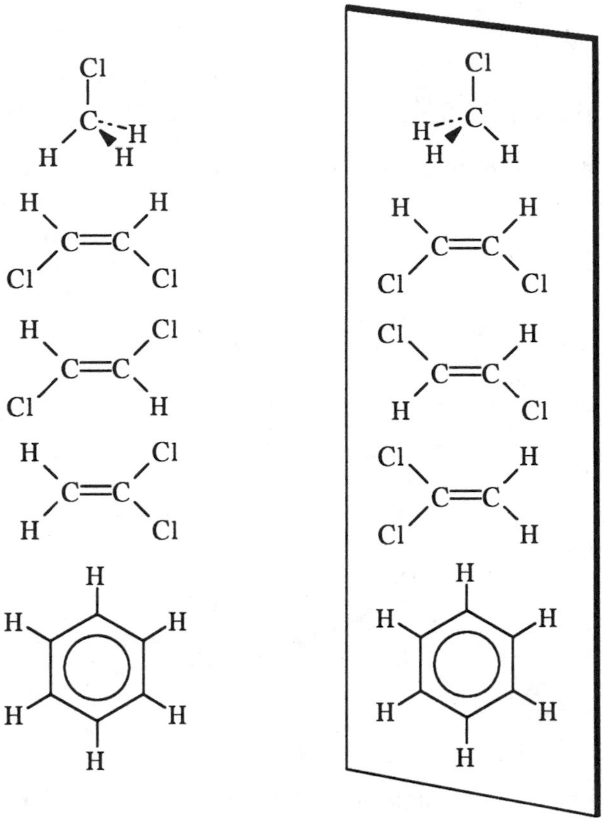

FIGURE 8-5. Five structures and their *superimposable* mirror images.

perhaps more easily appreciated if you rotate **7** through an angle of 180° to get **7'**, and compare the positions of the iodine and chlorine atoms in the two structures. They are in opposite positions!

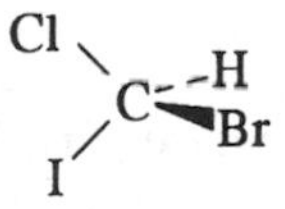

      6 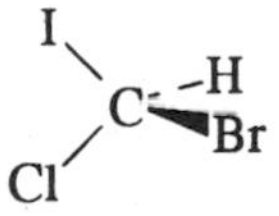   **7'**

- One enantiomer can be converted to its mirror image by interchanging the positions of any two of the groups. (Verify this with your models.)

Can you find any symmetry in either structure? There is none: No planes, no axes, no centers. These two structures are both asymmetric.

- A tetrahedrally substituted atom with four different groups attached (such as the carbons in **6** and **7**) is called an **asymmetric center** and is sometimes denoted by an asterisk (∗).

Any object that has a nonsuperimposable mirror image is said to be **chiral**, which means "to have handedness," i.e., left or right. Just as you have left and right feet, hands, and shoes, chiral molecules exist as left- and right-handed enantiomers. An asymmetric center is also called a **chiral center.**

- All asymmetric objects are chiral.

EXAMPLE 8-3 Which of the structures below is chiral? For each chiral structure, draw the other enantiomer.

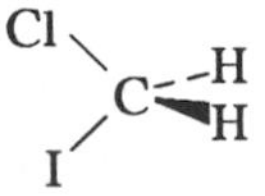

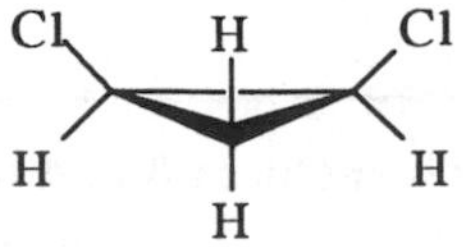

(c) 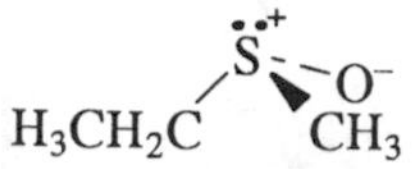   (d)

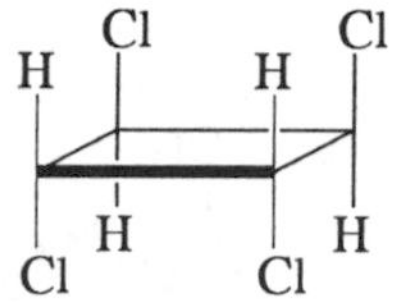

Solution   **(a and b)** Each of these structures has a plane of symmetry, so each is **achiral** (not chiral). **(c)** The sulfoxide structure has a tetrahedral *sulfur* (Problem 6-7) with four different groups attached (the nonbonding pair counts as one group; Section 3-2). So this structure *is* chiral, but see Section 8-9. The other enantiomer is

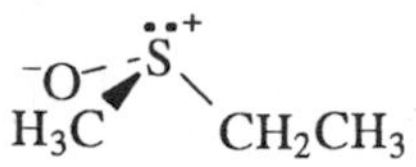

(d) This molecule has both a plane of symmetry and a center of symmetry (as well as a C_2 axis), so it is achiral.

Now here's a tough one; be careful! What symmetry elements, if any, can you find in *trans*-1,2-dichlorocyclopropane (**8**)?

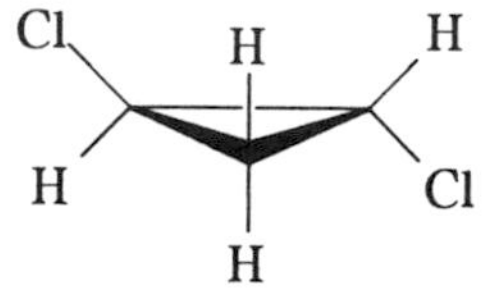

8

The answer is, a C_2 axis passing through the CH_2 carbon and a point midway between the other two carbons:

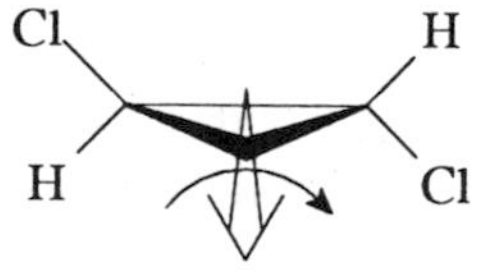

C_2

So, this is *not* an asymmetric molecule. Now comes the hard part. What is the relationship between **8** and its mirror image (**9**)?

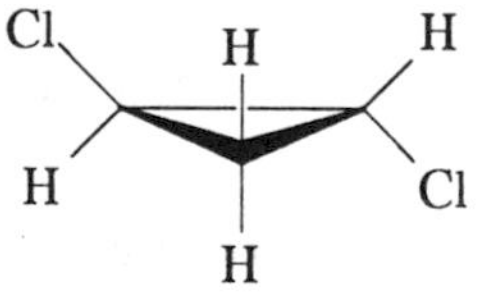

8 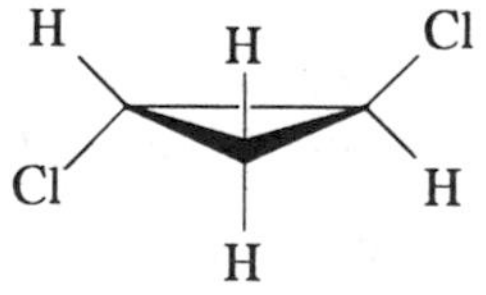   **9**

Surprise! They are *not* superimposable; they are enantiomers of each other. If you have trouble seeing this, make models of both structures.

The fact that *trans*-1,2-dichlorocyclopropane is chiral demonstrates that, although all asymmetric structures are chiral, not all chiral structures are necessarily asymmetric. It turns out that, while planes and centers of symmetry *do* preclude chirality, structures that have *only* axes of symmetry *can* be chiral. Another term you will often see in this context is **dissymmetric**, which is synonymous with chiral, i.e., having a nonsuperimposable mirror image. Notice that *dissymmetric* structures can have axial symmetry, while asymmetric structures cannot have any nontrivial symmetry.

● All asymmetric structures are dissymmetric, but not all dissymmetric structures are asymmetric.

EXAMPLE 8-4   In structures **8** and **9** which atoms, if any, are chiral centers?

Solution   We're looking for tetrahedrally substituted atoms with four different groups attached.

Therefore, both chlorine-containing carbons are chiral (asymmetric) centers. The four different groups attached to each one are H, Cl, CH_2, and CHCl.

8-4. Naming Enantiomers: The R–S System

Although all of us know the difference between our right and left hands, let's play a little mind game. Imagine trying to explain the difference between your left hand and your right hand to a person who cannot see you, and whom you cannot see. At first, you might say that your right hand is the one you write with. But suppose your friend is left-handed. Or perhaps you'd say that the left hand is the one with the wedding ring on it. But maybe your friend comes from a culture where wedding rings are worn on the right hand. Finally, you hit on an idea. You say your left hand is on the same side of your body as your heart. Okay, but is there any way to differentiate right from left *without* making reference to some other dissymmetric object such as the position of your heart in your body? Go ahead. See if you can find a way.

In fact, it is impossible to differentiate right from left without making reference to some universally accepted dissymmetric standard. The only reason you know right from left is that somebody told you which was which, and someone told them. Left and right are terms that have meaning only when we all agree what dissymmetric object we're comparing them with (e.g., the human body). A physicist might say "the physical laws of the universe are invariant on reflection." But how does this impact on our study of stereochemistry?

Suppose you were asked to name structures **6** and **7**. Clearly, both are bromochloroiodomethane, and yet we've already agreed that the two structures are enantiomers, and therefore different from each other. We might be tempted to call one "right" and the other "left," but which is which? Correct! We need some universally agreed-on dissymmetric standard. Chemists have agreed to use the concept of "clockwise and counterclockwise." (Pity a generation raised solely on digital clocks. How are we going to define clockwise and counterclockwise for them? Oh well, that's another problem.) To use this system you may want to review the CIP priority system we described in Section 5-5 for labeling E–Z isomers.

Here is our three-step system for labeling enantiomers:

1. Identify the chiral center, and the relative priority of the four different groups attached to it (1 = highest, 4 = lowest). For example,

10

2. Draw the molecule as it would appear if viewed along the bond *from* the chiral center *toward* the group of *lowest* priority. Thus, group 4 will be directly behind the chiral center, and therefore hidden from view. Be very careful making this drawing; do not transpose any groups! Making a molecular model helps. (To view the model properly, take hold of group 4 and its bond to the chiral center in one hand and hold the model so that the other three groups are pointed toward you.) The above example would now look something like this:

10

3. Draw a curved arrow from group 1 through group 2 to group 3. If the path of the arrow is clockwise, the **absolute configuration** of the asymmetric center is labeled **R** (from the Latin *rectus*, right). If the curved arrow defines a counterclockwise path, the asymmetric center is said to have the **S** absolute configuration (from the Latin *sinister*, left). In our example, the curved arrow is counterclockwise, so we have the S configuration.

Just for practice, let's examine the mirror image of **10**.

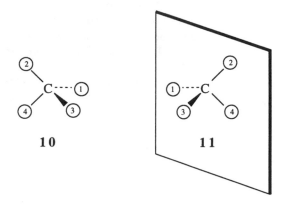

10   **11**

The mirror image, **11**, has the same relative priority of its four groups, and would look like this when viewed with group 4 behind the carbon:

11

Here the curved arrow describes a clockwise path, so **11** has the R configuration.

- The enantiomer of a molecule with the S configuration always has the R configuration, and vice versa.

EXAMPLE 8-5   Specify the absolute configurations of enantiomers **6** and **7**.

Solution   Recalling that a group's relative priority is based on the atomic number of its first atom, the four different groups in **6** have the relative priorities I > Br > Cl > H.

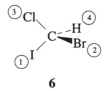

6

Viewing along the C–H bond, the structure would appear like this:

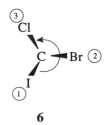

6

The counterclockwise arrow indicates an S configuration, so the complete name for this compound is (*S*)-bromochloroiodomethane. Structure **7**, the enantiomer of **6**, must therefore have the R configuration. This can be confirmed by looking at the curved arrow in either **7** or **7′**. (Think of **7′** as derived from **6** by interchanging Cl and I, thus converting **6** to its enantiomer.)

8-5. Properties of Enantiomers: Polarimetry

A. Physical and chemical properties

In Chapter 7 we described how physical and spectroscopic properties can be used to identify the structure of a compound. We know, for example, that two compounds that are structural isomers of each other will have different physical and spectroscopic properties. In general, the more similar the structures, the more similar are the properties. Cis–trans stereoisomers also have different properties, but in many cases these differences are not very dramatic. However, because all physical laws are invariant on reflection,

- Enantiomers have *identical* physical, chemical, and spectroscopic properties under normal conditions. (The proviso "under normal conditions" means in the absence of an asymmetric perturbation in the environment; more about this in Section 8-8.)

Thus, the melting point, boiling point, IR spectrum, NMR spectrum, etc., of the R enantiomer of a chiral compound will be identical with those of the S enantiomer of the compound. But if enantiomers have identical physical, chemical, and spectroscopic properties, how can they be experimentally differentiated?

B. Differentiation by X-ray crystallography

If the compound is crystalline and meets certain other criteria, it is possible (in principle) to use X-ray crystallography (Sections 3-1 and 7-2) to determine the exact positions of each atom in the molecule, and hence the absolute configuration at each asymmetric center. Although this is the most unambiguous way to establish absolute configurations, it can be expensive, time consuming, and not universally applicable. Fortunately, there *is* another property that can be used to differentiate enantiomers. But to describe this technique, we need to talk a little about plane-polarized light.

C. Differentiation by polarized light

White light is composed of electromagnetic waves of all different wavelengths (colors) in the visible spectrum. These waves are oscillating in random directions perpendicular to the beam's direction of travel. When white light is passed through a **monochromator**, one specific wavelength is selected, but the oscillations are still in random directions. If the monochromatic beam is then passed through a **polarizing filter**, only the waves oscillating parallel to the filter's axis emerge, as shown in Figure 8-6. Such monochromatic light is said to be **plane polarized.**

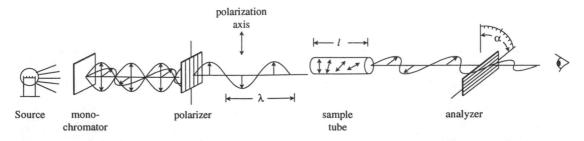

FIGURE 8-6. Essential features of a polarimeter, showing positive (clockwise) optical rotation.

Now, if this beam of plane-polarized monochromatic light is allowed to pass through a solution of a pure enantiomer, a strange and wonderful thing occurs. As it passes through the solution, the orientation of the polarization plane begins to twist either clockwise or counterclockwise. The amount of twist, measured in degrees using a second polarizing filter called the **analyzer,** is called the **optical** (or **observed**) **rotation** (symbol α) of the solution. The *sign* of the optical rotation is considered to be positive (+) if the beam was twisted clockwise (viewed as shown in Figure 8-6), and negative (−) if it was twisted counterclockwise. Any sample that has a nonzero optical rotation is said to be **optically active.** This is why, in the older literature, enantiomers were often called **optical isomers.** The instrument used to measure optical rotations is called a **polarimeter** and has all the essential features shown in Figure 8-6.

D. The specific rotation

The *magnitude* of the optical rotation is dependent on the concentration of the solution (expressed in grams of solute per milliliter of solution), the length of the sample tube (expressed in decimeters), the wavelength of the monochromatic light (the so-called D line emission of sodium vapor is often used), and the temperature at which the measurement was conducted. Normally, the dependence on concentration (c) and path length (l) can be removed by calculating a **specific rotation** ($[\alpha]$) by the relationship

$$[\alpha]_\lambda^T = \frac{\alpha}{lc} \tag{8-1}$$

The wavelength (λ) and temperature (T) of the measurement are recorded as sub- and superscripts on the $[\alpha]$ value, so that comparisons can be made between measurements under the same set of conditions. The *specific* rotation is a characteristic physical property of a pure, optically active compound. But here's the most important thing to remember:

- The specific rotation of the pure R enantiomer of a given chiral compound will be *equal in magnitude but opposite in sign* to that of the pure S enantiomer.

EXAMPLE 8-6   A certain pure chiral compound with the R configuration is being prepared for determination of its optical rotation. The temperature of the measurement is 25°C; sodium D line light (sodium vapor lamp) will be used. Using a solution with a concentration of 0.120 g/mL and a sample tube length of 1.00 dm, the polarization plane twists 2.70° counterclockwise.
(a) What is the *specific* rotation of this R enantiomer?
(b) What would be the *specific* rotation of the S enantiomer under the same conditions?
(c) What would be the *optical* rotation of the R enantiomer (at the same wavelength and temperature) if the concentration and path length were 0.041 g/mL and 2.00 dm, respectively?

Solution
(a) Use Eq. (8-1); the value of α is negative since it was counterclockwise.

$$[\alpha]_D^{25} = \frac{\alpha}{lc} = \frac{-2.70°}{(1.00)(0.120)} = -22.5°$$

(b) The specific rotation of the S enantiomer will be equal and opposite to that of the R enantiomer. Thus the value would be +22.5°.
(c) Solve Eq. (8-1) for α:

$$\alpha = [\alpha]_D^{25} lc = (-22.5°)(2.00)(0.041) = -1.8°$$

E. Rotation vs. configuration

At this point we must emphasize something that confuses many students. Although the sign of the optical rotation is determined by whether the *polarization plane* twists clockwise (+) or counterclockwise (−), and the absolute configuration is labeled by the clockwise (R) or counterclockwise (S) *pattern of group priorities,*

- Without additional information there is no universal way to predict for a given chiral compound whether the R enantiomer will have a (+) or (−) specific rotation.

All we can be sure of is that the R and S enantiomers will have equal but opposite specific rotations (under the same conditions). For this reason, it has become a common practice to label to a compound with a (+) rotation; it is called a **dextrorotatory** (or ***d***) compound. A compound with a (−) rotation is called **levorotatory** (or ***l***). But remember that these two new terms are simply words meaning (+) and (−) optical rotations.

EXAMPLE 8-7 Suppose you are the first person to isolate and purify a new, optically active compound. You determine its specific rotation to be +101°. **(a)** Is this new compound the *d* or *l* isomer? Why? **(b)** Is this new compound the R or S isomer? Why?

Solution **(a)** Because the compound has a *positive* specific rotation, it is by definition the *d* enantiomer. **(b)** Without some additional information, we have no way to determine whether this isomer has the R or S configuration. (But see Problem 8-8.)

8-6. Racemic Mixtures and Optical Purity

Now let's consider what happens when we have a 50/50 mixture of the *d* and *l* isomers of a given chiral compound. What would be the optical rotation of such a mixture? The answer is *zero*, because the positive rotation of the *d* isomer will be exactly balanced by the equally negative rotation of the *l* isomer. Such a mixture is said to be **optically inactive** and is called a **racemic modification.** The term **racemic mixture** has a slightly different meaning: a mixture of equal amounts of *crystals* of the two enantiomers.

A little thought should convince you that the net specific rotation of any mixture of the *d* and *l* isomers of a given chiral compound is equal to the weighted average of the rotations due to both isomers. Expressed mathematically,

$$[\alpha]_{net} = f_d[\alpha_d] + f_l[\alpha_l] \tag{8-2}$$

where f_d and f_l are the fractions of *d* and *l* isomer, respectively, while $[\alpha_d]$ and $[\alpha_l]$ are the specific rotations of the pure *d* and *l* isomers, respectively. Since $f_l = 1 - f_d$, and $[\alpha_l] = -[\alpha_d]$, we can revise Eq. (8-2):

$$\begin{aligned}[\alpha]_{net} &= f_d[\alpha_d] + (1 - f_d)(-[\alpha_d]) \\ &= f_d[\alpha_d] - [\alpha_d] + f_d[\alpha_d] \\ &= [\alpha_d](2f_d - 1) = [\alpha_l](2f_l - 1)\end{aligned} \tag{8-2'}$$

Notice how, for a racemic mixture, $f_d = f_l = 0.5$, so the term $(2f - 1)$ equals zero, and hence so does the net rotation. Any other *dl* mixture will have a specific rotation whose *sign* is the same as that of the predominant isomer. The term $2f_d - 1$ (or $2f_l - 1$ if the *l* isomer predominates) is called the **enantiomeric excess** (*ee*), or **optical purity,** of the mixture.

EXAMPLE 8-8 The pure *d* isomer of a certain chiral compound has $[\alpha]_D^{25} = +55°$. A nonracemic *dl* mixture of this compound has a net $[\alpha]_D^{25}$ of −11°. **(a)** Which enantiomer predominates in this mixture? **(b)** What is the enantiomeric excess (optical purity) of this isomer? **(c)** What is the fraction of both isomers in the mixture?

Solution
(a) Since the net rotation of the mixture is negative, the *l* isomer is the predominant one.
(b) We know that $ee = 2f_l - 1$, so we can solve Eq. (8-2′) for this term:

$$ee = (2f_l - 1) = \frac{[\alpha]_{net}}{[\alpha_l]} = \frac{-11°}{-55°} = 0.20 \quad \text{(or 20\%)}$$

(c) Now we can solve for f_l:

$$(2f_l - 1) = 0.20$$

$$f_l = \frac{(0.20 + 1)}{2} = 0.60 \quad \text{(or 60\%)}$$

and

$$f_d = 1 - f_l = 1 - 0.60 = 0.40 \quad \text{(or 40\%)}$$

8-7. Diastereomers and Multiple Chiral Centers

A. How many isomers?

So far we have encountered two kinds of stereoisomers: Cis–trans (and E–Z) isomers and enantiomers. It turns out that cis–trans (and E–Z) isomers are but one example of a broad class of stereoisomers called **diastereomers** (or **diastereoisomers**). The definition of diastereomers is simply "stereoisomers that are *not* enantiomers," as shown with the "isomer tree" below:

```
                         isomers
                        /       \
          structural isomers    stereoisomers
                                      |
                                     / \
                        enantiomers    diastereomers
```

We know what stereoisomers are (same sequence of atoms, different three-dimensional structure) and what enantiomers are (nonsuperimposable mirror images). So, whenever we encounter two *stereoisomers* that are *not* enantiomerically related, they must be diastereomers of each other. And remember:

- Diastereomers (unlike enantiomers) have *different* physical, chemical, and spectroscopic properties.

Although cis–trans (and E–Z) isomers represent one kind of diastereomer, diastereomers are also encountered with structures that have multiple (two or more) chiral centers. Consider, for example, 2-bromo-3-chlorobutane (**12**). How many different stereoisomers can you draw for this compound, and what are the relationships between the structures?

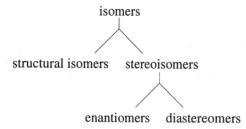

12

A good way to approach this question is to recognize that the two middle carbons (C-2 and C-3) in **12** are chiral centers because each has four different groups attached (halogen, haloethyl, methyl, and hydrogen, in that order of priority). Since each chiral center can have either the R or S configuration, there are four possible structures: (2R,3R), (2R,3S), (2S,3R), and (2S,3S). Here is what each of them looks like:

(2R,3R)-**12**   (2R,3S)-**12**

Br⟍   H
H◣C—C◢CH₃
H₃C   Cl

(2S,3R)-12

Br⟍   CH₃
H◣C—C◢H
H₃C   Cl

(2S,3S)-12

Thus the complete name of the first structure is (2R,3R)-2-bromo-3-chlorobutane, and the complete names of the other three include the appropriate stereochemical designators. Furthermore, notice how the configuration at any chiral center can be **inverted** (changed from R to S, or S to R) by simply interchanging any *two* of the four groups around that center. Just for practice, you might want to confirm the R or S configuration at each carbon in these structures. [Actually, you would have to do this for only one of the four structures, because once you are sure about one of them, you can generate the other three by exchanging pairs of groups. For example, if you have definitely established the first one as (2R,3R), you would know that the (2R,3S) would be like the first structure but with the configuration of C-3 inverted by exchanging the CH₃ and H, etc.]

Now, carefully compare the (2R,3R) and (2S,3S) structures. What is the relationship between them? The answer is that they are enantiomers of each other. Similarly, the (2R,3S) and (2S,3R) structures are enantiomers of each other. To generalize, if two stereoisomers have the opposite configuration at *each and every* chiral center, they are *enantiomers* of each other.

But what is the relationship between the (2R,3R) structure and the (2R,3S) structure? They have the *same* configuration at C-2, but opposite configuration at C-3. So, they cannot be enantiomers. Yet, they *are* stereoisomers (same sequence of atoms, different three-dimensional arrangement). So, they are...diastereomers! The various relationships between all four structures can be shown pictorially.

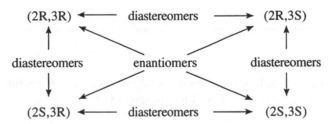

As the example above shows, the presence of two different chiral centers in a molecule gives rise to four possible stereoisomers, two enantiomeric pairs. With three different chiral centers, there can be eight stereoisomers (four enantiomeric pairs), and with *n* different chiral centers there can be a maximum of 2^n stereoisomers (2^{n-1} enantiomeric pairs).

EXAMPLE 8-9 A certain compound has four different chiral centers. How many stereoisomers are possible? How many enantiomeric pairs? What are the stereochemical designators for these isomers?

Solution There are $2^4 = 16$ stereoisomers, 8 enantiomeric pairs. Below, the enantiomers are paired horizontally; all other relationships are diastereomeric.

configuration at chiral center:

1	2	3	4		1	2	3	4
R	R	R	R		S	S	S	S
R	R	R	S		S	S	S	R
R	R	S	R		S	S	R	S
R	S	R	R		S	R	S	S
S	R	R	R		R	S	S	S
R	R	S	S		S	S	R	R
R	S	R	S		S	R	S	R
S	R	R	S		R	S	S	R

EXAMPLE 8-10 The female sex hormone estradiol, the structure of which is shown below, has several chiral centers. Identify each one with an asterisk. How many stereoisomers of estradiol are possible?

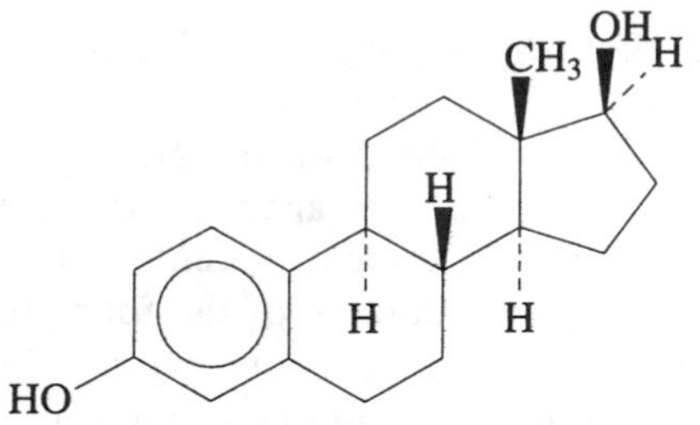

Solution Remember, we're looking for tetrahedral atoms with four different groups attached. Further, remember that different groups can have the same first atom, with differences appearing in the second or even third atoms out (Section 5-5). There are five chiral centers in the structure:

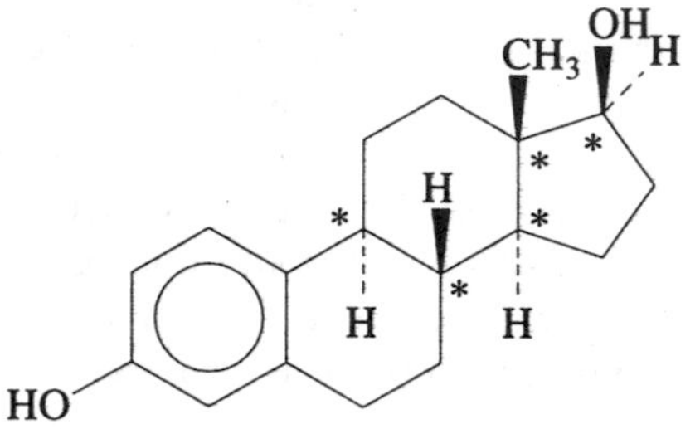

The five chiral centers give rise to $2^5 = 32$ stereoisomers, 16 *dl* pairs.

B. The meso isomer

We encounter a slight problem when two chiral centers have the same *set* of four different groups. Consider, for example, 2,3-dichlorobutane (**13**).

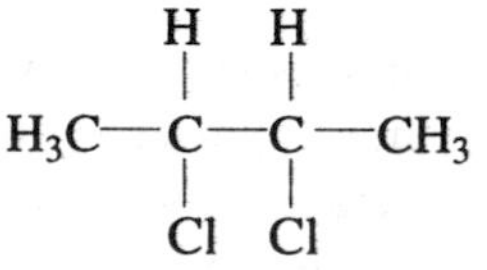

13

As with 2-bromo-3-chlorobutane (**12**), we still have two chiral centers. So, let's redraw the four structures of **12**, replacing the Br atoms with Cl atoms.

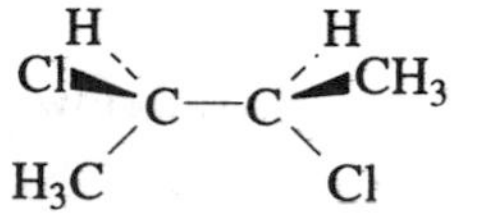

(2R,3R)-**13**

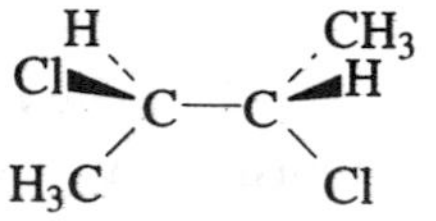

(2R,3S)-**13**

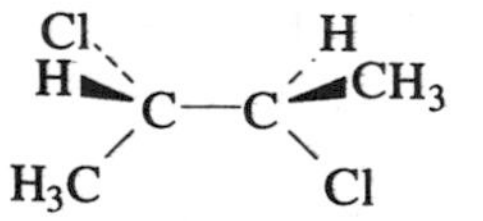

(2S,3R)-**13**

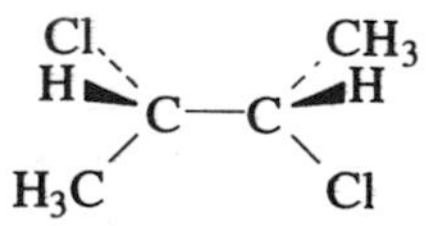

(2S,3S)-**13**

Now look very carefully at the (2R,3S) and (2S,3R) structures. Are they diastereomers? The answer is NO! They represent identical configurations because both chiral centers have the same set of four different groups (chloro, chloroethyl, methyl, and hydrogen). They are, however, in different *conformations*, as is best seen by rotating the central bond until the chlorines are s-cis:

$$(2R,3S)\text{-}\mathbf{13} \longrightarrow \qquad\qquad \equiv \qquad\qquad \longleftarrow (2S,3R)\text{-}\mathbf{13}$$

Furthermore, because the configuration at one chiral center is the exact mirror image of the other center, the molecule has a plane of symmetry and is therefore achiral and *optically inactive* (having a specific rotation of zero). Such a stereoisomer with multiple asymmetric centers but a plane of symmetry (in at least one conformation) is called a **meso** structure. It would be a good idea for you to make models of the ones we've labeled as (2R,3S)-**13** and (2S,3R)-**13** and see for yourself that they are indeed identical and that each can be rotated into a conformation with a plane of symmetry. [Compare this with models of (2R,3S)-**12** and (2S,3R)-**12**.] Thus there are only *three* stereoisomers of 2,3-dichlorobutane, two of which are optically active (enantiomers), and one of which is optically inactive (meso). Therefore,

- A molecule with *n* chiral centers can have a *maximum* of 2^n stereoisomers, but fewer if pairs of chiral centers in the structure have the same set of four different groups and a possible plane of symmetry.

EXAMPLE 8-11 We have already established that 1,2-dichlorocyclopropane has two chiral centers. Draw and name all possible stereoisomers. Is any of these a meso form?

Solution There are just *three* stereoisomers, trans-(1R,2R), trans-(1S,2S), and cis. The cis isomer could be called either (1R,2S) or (1S,2R), but in either case it is a meso form and has the requisite plane of symmetry.

trans-(1R,2R)   trans-(1S,2S)   cis

8-8. Optical Resolutions

The process of separating enantiomers is called **optical resolution**, or just **resolution** for short. However, if enantiomers have identical physical and chemical properties under normal conditions, how can they be separated? The answer is that we provide an asymmetric environment that allows us to differentiate them.

Suppose we had a box containing ten right and ten left feet made of plastic. We are not allowed to see or touch them, yet we wish to separate out the right feet. Our lab assistant volunteers his help, but unfortunately he has never learned right from left. But there *is* a way. We give him ten right shoes of the appropriate size and ask him to bring us all the feet that fit the shoes. This he is able to do, because even though he doesn't know right from left, he does recognize a good fit from a bad one.

The resolution of enantiomers follows a similar strategy. To a racemic mixture of two enantiomers we add a **resolving reagent** with the following characteristics: (1) It must itself be a pure enantiomer, and (2) it must react with one or both of the enantiomers to form the diastereomers of a new compound or complex from which the original enantiomer(s) can be easily regenerated. If the resolving reagent reacts with just one of the enantiomers, it is usually a simple matter to separate the unreacted enantiomers from the complex, since they are expected to have grossly different physical and chemical properties. But even if both enantiomers react with the resolving agent, the separation can be accomplished, because the resulting stereoisomers of the complex are *diastereomerically* related. For example, if the resolving agent has the R configuration, the two stereoisomers of the complex will be **RR** and **SR** (where the boldface designator describes the configuration of the enantiomers being separated). These two are *diastereomers*, with different properties that permit their separation. Once separated, the two diastereomers of the complex are decomposed back into the now separate enantiomers, along with the reusable resolving agent. The techniques for optical resolution have become quite sophisticated. It is now possible, for example, to pass the racemic mixture through a chromatography column (Section 7-1) charged with an asymmetric stationary phase. In principle, one enantiomer will pass through the column more rapidly than the other, obviating the need to make, separate, and decompose the diastereomers of the complex. But the overall principle is the same: The enantiomers (with identical physical and chemical properties) are temporarily converted by use of a resolving agent into diastereomeric complexes with different properties. In any event, the resolution–purification–regeneration steps are continued until the increasingly pure enantiomer reaches a maximum specific rotation that does not increase on further purification.

EXAMPLE 8-12   Most NMR spectra (Section 7-2) are determined with the sample dissolved in some solvent. Do you think it would be possible to differentiate two enantiomers by NMR, using a solvent that itself was a pure enantiomer? Why?

Solution   Yes, it would be possible. In this case, the chiral solvent serves as the resolving reagent. The solvation complexes it forms with the two enantiomers will be diastereomerically related, and hence (in principle) will have different spectroscopic properties. This technique can be used to confirm the enantiomeric purity of a compound directly, as well as the enantiomeric excess in any mixture.

8-9. Fischer Projections, Invertomers, and Other Dissymmetric Molecules

A. Fischer projections

Ever since Section 3-1, we have been using perspective drawings such as **6** (below) to indicate the relative positions of groups around a tetrahedral atom. In such a structure, the normal lines indicate bonds lying *in* the page, the wedge indicates a bond protruding toward the reader, while the dashed line shows a bond behind the page.

6

However, such drawings may become unwieldy in molecules with multiple asymmetric centers. To portray stereochemical relationships in such molecules more easily, a Nobel Prize-winning chemist named Emil Fischer developed a way to depict chiral structures. To be able to interpret and use such **Fischer projections**, you must first understand how such a drawing is constructed.

Taking structure **6** as an example, we begin by re-orienting the molecule as it would appear if viewed along the line bisecting any pair of bonds, say the C–Cl and C–Br bonds. Once again, it is perhaps easiest to see this with a model. At any rate, the structure now look like this:

The Fischer projection of this structure would look like this:

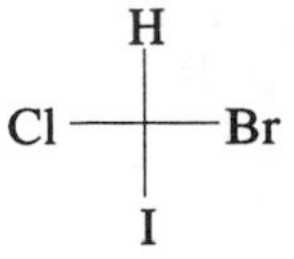

$$
\begin{array}{c}
H \\
| \\
Cl\!-\!\!\!-\!\!\!-\!Br \\
| \\
I
\end{array}
$$

Notice that all four bonds are now shown as normal lines, and the carbon is understood to be at their intersection. But, most importantly, *the horizontal bonds are understood to be coming out of the page toward the reader, while the vertical bonds are behind the page.* One way to remember this is to view the horizontal bonds as resembling a bow-tie (worn by someone else, not by you!).

It is very important to realize that rotating a given Fischer projection 90° around the viewing axis results in *not* the same structure, but rather its enantiomer. Thus if we redraw the above Fischer projection of **6** rotated 90°, the new projection is really **7**, the enantiomer of **6**.

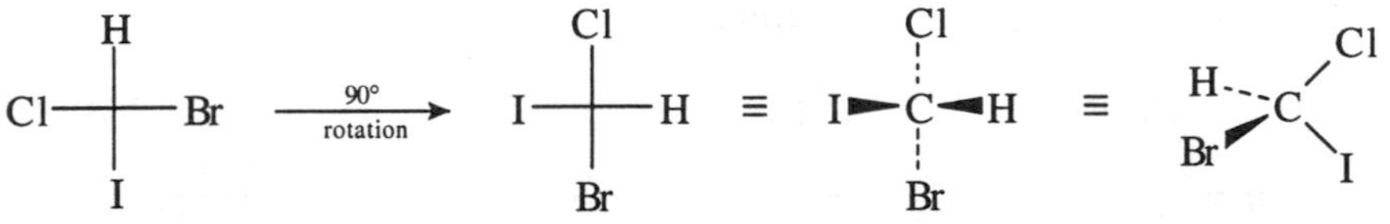

This is because a 90° rotation interchanges front groups with back ones, and thereby inverts the configuration. Also, as we've seen before, interchanging any two groups around the chiral center causes an inversion of configuration. Each of the structures below results from an interchange of a pair of groups in the projection of **6**, and each is therefore a Fischer projection of **7**:

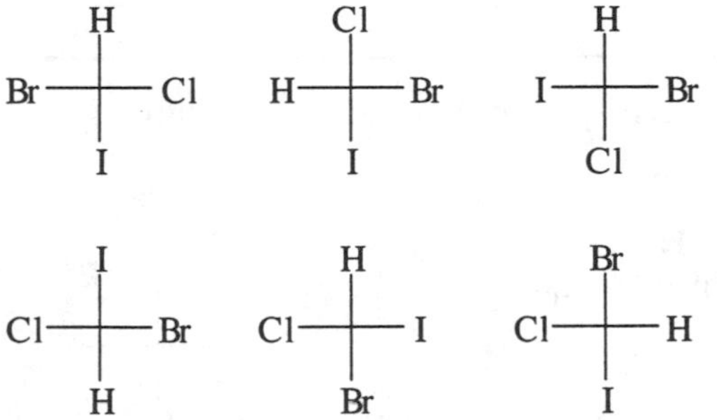

Whenever you're comparing two Fischer projections, trying to decide if they are identical or enantiomers, count how many interchanges of two groups are necessary to convert one structure into the other. An odd number of interchanges indicates enantiomers, while an even number indicates identical structures.

EXAMPLE 8-13 By counting the number of interchanges necessary to convert the Fischer projection of **6** to the structure below, decide if the two structures are the same or enantiomers.

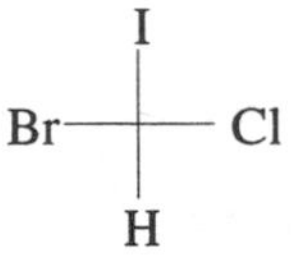

$$
\begin{array}{c}
I \\
| \\
Br\!-\!\!\!-\!\!\!-\!Cl \\
| \\
H
\end{array}
$$

Solution Two (an even number) interchanges are required, first interchanging the vertical groups, then interchanging the horizontal groups. So, the two structures are identical. You may have noticed also that the two structures are related by a 180° rotation around the viewing axis. While a 90° rotation of a Fischer projection gives a projection of the other enantiomer, a 180° rotation ($2 \times 90°$) gives another projection of the same molecule.

To draw the Fischer projection of a molecule with multiple asymmetric centers, it is first necessary to visualize the structure in a conformation that has the carbon chain vertical and the horizontal groups coming out toward the viewer. Thus one Fischer projection of (2R,3R)-**12** can be constructed as shown below:

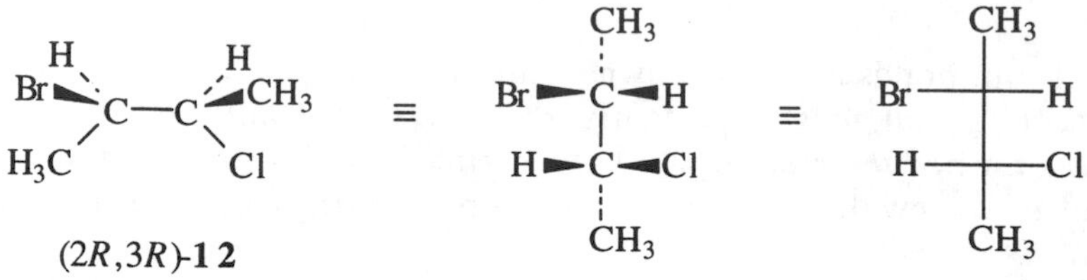

(2R,3R)-**12**

The conventional way to draw Fischer projection formulas of chiral molecules in which there is an extended carbon chain is to show this chain as the vertical line, with C-1 at the top. Substituents are then oriented right or left; hydrogens are often not printed. Thus the four stereoisomers corresponding to **12** become:

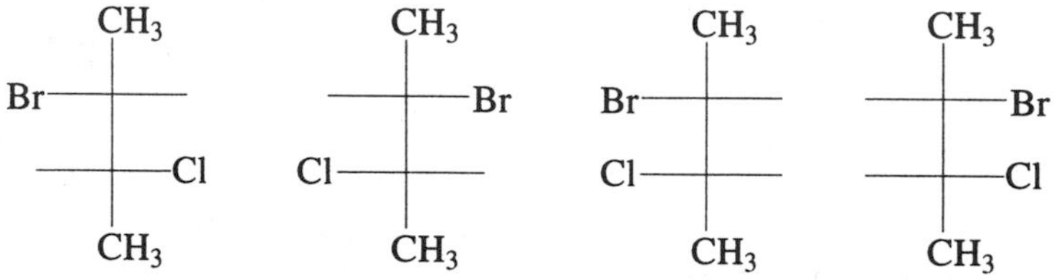

(2R,3R)-**12**   (2S,3S)-**12**   (2R,3S)-**12**   (2S,3R)-**12**

Notice it is easy to see that the first two constitute an enantiomeric pair, as do the second two.

EXAMPLE 8-14   Complete the Fischer projection of structure **14**.

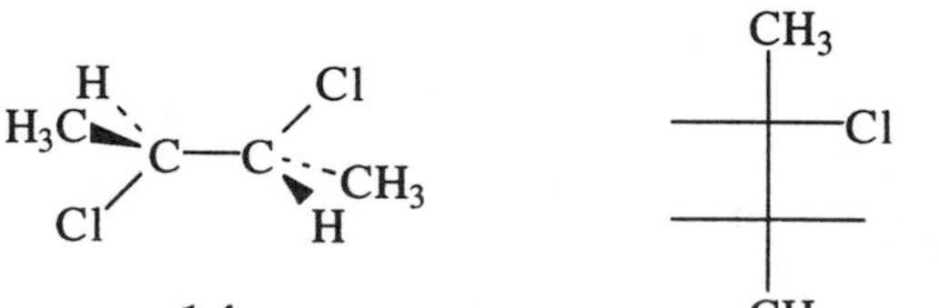

14

Solution   First, we must rotate the central C–C bond to generate a conformation where the two end carbons are s-cis. When we redraw the structure vertically, as viewed with the end carbons behind the page, we see that the two chlorines are also s-cis.

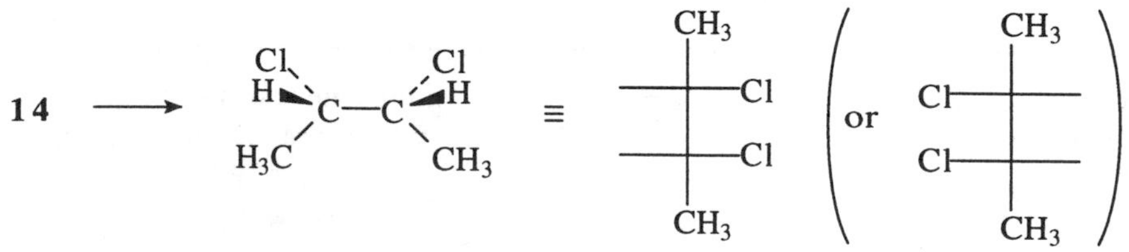

It makes no difference in this case whether we orient both chlorines right or left. This is because **14** is a meso form, with a plane of symmetry. If you rotate one of the projection formulas of **14** through an angle of 180° in the plane of the paper, you'll see that the two formulas are identical.

B. Invertomers

In Example 8-3c we decided that sulfoxides with two different R groups are chiral, because the nonbonding pair on the tetrahedral sulfur constituted one of the four necessary different groups. Based on this precedent, we might expect that simple amines (Section 6-2) with three different groups attached to the nitrogen would also be chiral. Thus ethylmethylamine could exist in two enantiomeric forms:

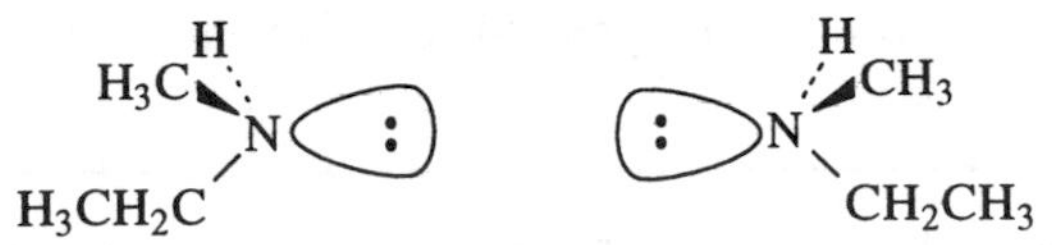

15   **16**

However, no one has ever succeeded in separating the enantiomers of simple amines. That is, although structures **15** and **16** are indeed nonsuperimposable mirror images of each other, they undergo interconversion extremely rapidly at room temperature. Because of this rapid interconversion, the two structures are *not* isomers. Instead, they are called **invertomers** (or inversion isomers). The process by which they are interconverted involves temporary rehybridization at nitrogen ($sp^3 \rightarrow sp^2 \rightarrow sp^3$), "pushing" the nonbonding pair through to the opposite side, as the three other groups reposition themselves.

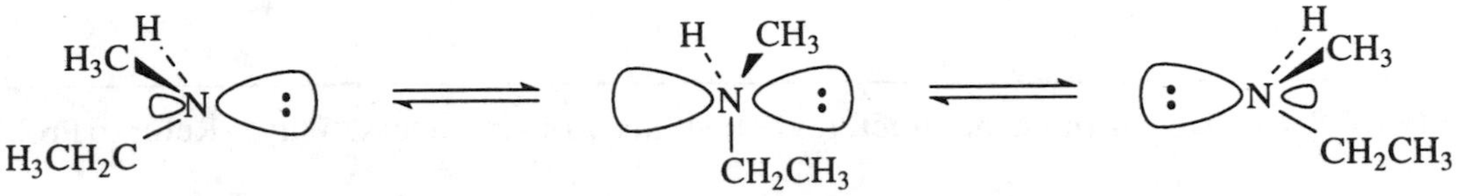

Notice that the middle structure has a plane of symmetry—defined by the nitrogen with its three groups—and is therefore achiral. The occupied *p* orbital is perpendicular to this plane. The probabilities that this structure will proceed to the structure at the right or left are exactly equal. This process has been likened to what happens to an umbrella in a strong wind, and is another example of inversion of configuration (Section 8-7).

It turns out that this "umbrella inversion" of nonbonding pairs is very facile in the case of second period elements such as nitrogen but very much slower for third-period elements such as sulfur and phosphorus. Thus while simple amines cannot be separated into their enantiomers, sulfoxides and phosphines (PRR′R″) *can* be. The latter compounds are therefore described as **optically stable,** because they retain their chirality and do not undergo spontaneous inversion at room temperature.

C. **Other dissymmetric molecules**

It may surprise you to learn that there are certain chiral molecules that possess no chiral atoms! For example, below are two structures of 2,3-pentadiene; what is the relationship between them? (Remember from Chapter 3 and Problem 3-4 that the sets of groups at the ends of a C=C=C linkage must occupy mutually perpendicular planes.)

17   **18**

If you think they are identical, look again at the Newman projections of the two structures, sighting down the C=C=C linkage.

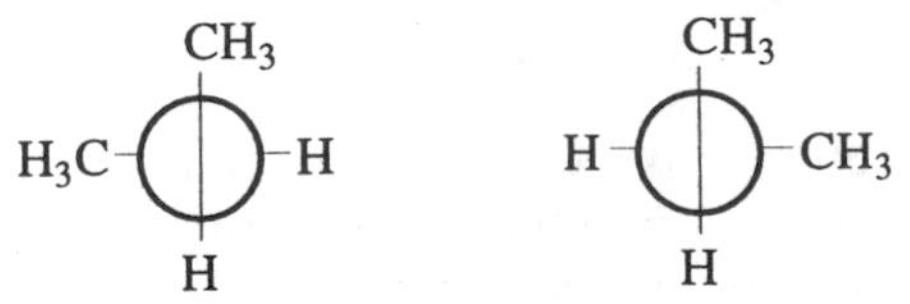

17   **18**

I hope you can see that these two are indeed enantiomers. But there are *no* tetrahedral atoms with four different groups attached. You may have noticed the similarity of these Newman projections to the Fischer projections encountered earlier in this section. Here is the key to this kind of dissymmetry: The atom to which the horizontal groups are attached is different from the atom to which the vertical groups are attached. Whenever you encounter this type of molecular architecture, be aware of the possibility of enantiomers.

EXAMPLE 8-15 Can you find any symmetry in structures **17** and **18**?

Solution Yes! There is a C_2 axis, but it's hard to find. It goes through the central carbon and bisects the CH_3–CH_3 dihedral angle. Can you find the C_2 axis in **18**?

17

EXAMPLE 8-16 Do you think the structures below exist as enantiomers? Why? (Refer to Problem 4-13.)

(a) $H_3CCH=C=C=CHCH_3$ (b)

Solution

(a) No! For this "cumulatriene," the terminal CH_3 and H groups all lie in the same plane. Therefore, this *planar* structure can exist in two diastereomeric forms, cis and trans, but no enantiomeric forms are possible:

cis   trans

(b) It depends! We know from Problem 4-13 that such biphenyl derivatives prefer to adopt a nonplanar conformation (phenyl rings perpendicular) to avoid steric congestion of the R groups.

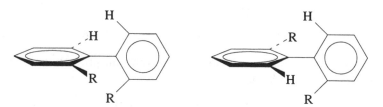

But if R ≠ H these two structures, like **17** and **18**, are enantiomeric. If the R groups are large enough, the rotation of the bond connecting the two rings is restricted, and so the molecules cannot reach the planar conformation necessary to interconvert them. In this case the molecules would be separable, optically stable enantiomers. If, on the other hand, the two groups were small enough, rotation around the central C–C bond will not be completely suppressed, interconversion *will* take place, and the two enantiomers will not be optically stable.

8-10. Stereochemistry as a Mechanistic Probe

One of the most important applications of our new stereochemical knowledge will be to trace the three-dimensional consequences of chemical reactions. This type of information can often be crucial when we attempt to unravel the mechanisms of organic reactions in the following chapters. For example, consider the reaction below:

$$H_3CH_2C \diagdown C - Br + N_3^- \longrightarrow N_3 - C \diagup CH_2CH_3 + Br^-$$

(S)-19   (R)-20

Note that (S)-**19** gives *only* (R)-**20** (the product of stereochemical **inversion** of configuration) and none of its enantiomer (S)-**20** (the product of **retention** of configuration). Similarly, we find that (R)-**19** gives only (S)-**20**:

(R)-**19**   (S)-**20**

Such a reaction is said to involve 100% **stereospecific** inversion. For this to be true, the incoming azide ion (N_3^-) must become attached to the *opposite* side of the carbon from which the bromide ion departs, while the three other groups undergo umbrella inversion.

$$N_3^- \longrightarrow C - Br$$

But this essential detail becomes apparent only by studying the stereochemical outcome of the reaction.

SUMMARY

1. Stereochemistry is the description of the three-dimensional structure of molecules, as well as the physical, chemical, and spectroscopic properties of stereoisomers, conformations, and invertomers.
2. A symmetry operation is the manipulation of an object in such a way that its new orientation is indistinguishable from the original orientation. There are three main types of symmetry elements that help describe such symmetry operations: Axes (C_n), planes (σ), and centers (i).
3. An *n*-fold axis of symmetry (C_n) is an imaginary axis passing through the center of an object. Rotation of the object around this axis by $360°/n$ leaves the orientation of the object indistinguishable from its original orientation.
4. A plane of symmetry is an imaginary (mirror) plane through the center of an object that cuts it in half. One half of the object is the mirror image of the other half.
5. An object has a center of symmetry when every point in its structure has a superimposable counterpart located an equal distance from the center along an extension of the line from the point to the center.
6. Atoms that are related by one or more symmetry elements are said to be (symmetry-) equivalent, and are therefore indistinguishable.
7. A molecule that has a nonsuperimposable mirror image (enantiomer) is said to be chiral (or dissymmetric). An asymmetric object has no (nontrivial) elements of symmetry, and *is* chiral. An object that has a plane and/or a center of symmetry is *achiral*.

8. A chiral (asymmetric) center is a tetrahedral atom with four different groups attached.

9. Enantiomers have *identical* physical, chemical, and spectroscopic properties under symmetrical conditions.

10. The absolute configuration label (R or S) is assigned to a chiral center based on the relative positions and priorities of the four groups attached. If, when the structure is viewed along the bond *from* the chiral center *toward* the lowest priority group, the pattern from highest priority group to second to third is *clockwise*, the correct label is R. If the pattern is counterclockwise, the correct label is S.

11. The separate, pure enantiomers of a chiral compound are capable of twisting the polarization plane (axis) of plane-polarized light by equal amounts but in opposite directions. The observed value of this twist (α) is called the optical rotation of the sample, and the value is usually converted to a specific rotation ($[\alpha]$) by Eq. (8-1):

$$[\alpha] = \frac{\alpha}{lc} \quad \textbf{(8-1)}$$

This specific rotation is a characteristic physical property of an enantiomer.

12. A mixture of the *d* and *l* (+ and −) enantiomers of a chiral compound has a net specific rotation given by Eq. (8-2′):

$$[\alpha]_{net} = [\alpha_d](2f_d - 1) = [\alpha_l](2f_l - 1) \quad \textbf{(8-2′)}$$

where the term ($2f - 1$) is called the enantiomeric excess (or optical purity) of the mixture. A racemic modification is a 50/50 *dl* mixture, which is optically inactive (zero optical rotation).

13. Diastereomers are stereoisomers that are *not* enantiomers.

14. The maximum number of stereoisomers of a compound with *n* chiral centers is 2^n (2^{n-1} *dl* pairs).

15. A meso structure is a molecule with multiple chiral centers and at least one conformation with a plane of symmetry.

16. A Fischer projection is a way to depict a chiral molecule. The structure is drawn as shown below; the horizontal groups (1 and 3) are understood to be in front of the page, while the vertical groups (2 and 4) are understood to be behind the page.

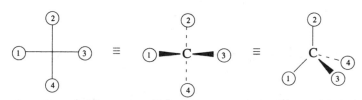

17. Invertomers are two enantiomerically related structures, each with a nonbonding pair as one of the groups around the chiral center, which are rapidly interconverted by "umbrella" inversion.

18. Certain structures such as allenes ($>$C$=$C$=$C$<$) and biphenyls can exist as enantiomers even though they have no chirally substituted atoms. In such molecules the four groups are divided into two pairs attached to different atoms and occupying mutually perpendicular planes. The only requirement is that the two groups in any pair must be different.

RAISE YOUR GRADES

Can you define...?

- ☑ stereochemistry
- ☑ symmetry operation
- ☑ symmetry element (C$_n$, σ, i)
- ☑ (symmetry-) equivalent
- ☑ asymmetric
- ☑ nonsuperimposable mirror image
- ☑ enantiomer
- ☑ chiral (dissymmetric)
- ☑ achiral
- ☑ chiral (asymmetric) center

- ☑ polarimeter
- ☑ optical isomer
- ☑ *d* (dextrorotatory) and *l* (levorotatory)
- ☑ optically inactive
- ☑ racemic modification
- ☑ enantiomeric excess (optical purity)
- ☑ diastereomer
- ☑ meso structure
- ☑ (optical) resolution
- ☑ Fischer projection

☑ optical activity
☑ optical rotation
☑ specific rotation
☑ R–S labels for absolute configuration
☑ stereospecific

☑ inversion (of configuration)
☑ invertomer
☑ "umbrella" inversion
☑ retention (of configuration)

Can you explain...?

☑ how to determine if an object has a C_n, σ, or i element
☑ how to determine when two or more atoms in a molecule are equivalent
☑ when to expect an object to have a nonsuperimposable mirror image, and when not
☑ how to label the absolute configuration of a chiral center either R or S
☑ the relationship between observed optical rotation and specific rotation
☑ how a polarimeter works
☑ how the specific rotation of any *dl* mixture varies with the enantiomeric excess of that mixture
☑ how to determine when two structures are diastereomerically related
☑ how to generate stereochemical labels for structures with multiple chiral centers
☑ how to identify each chiral center in a complex molecule
☑ how to identify a meso structure
☑ how enantiomers can be resolved (separated)
☑ how to recognize and construct Fischer projections of a chiral molecule
☑ how invertomers are interconverted
☑ how some molecules with no chiral atoms can still exist as enantiomers

SOLVED PROBLEMS

PROBLEM 8-1   List all axes, planes, and centers of symmetry present in the two structures below. Which of the hydrogens in each structure are equivalent?

(a)

(b)

Solution

(a) The boat conformation of cyclohexane has a C_2 axis perpendicular to the center of the ring. There are two mirror planes that intersect at the C_2 axis, one parallel to the page ($\sigma_{\parallel}$), the other perpendicular to the page ($\sigma_{\perp}$).

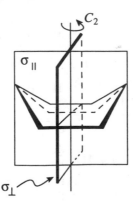

There are four sets of equivalent hydrogens in the structure:

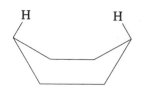

equivalent by virtue of the C_2 or $\sigma_\perp$

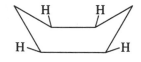

equivalent by virtue of the C_2 or $\sigma_\perp$

equivalent by virtue of the C_2 and $\sigma_\perp$ or $\sigma_\parallel$

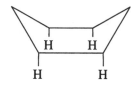

equivalent by virtue of the C_2 and $\sigma_\perp$ or $\sigma_\parallel$

(b) The staggered conformation of ethane has a C_3 axis along the C–C bond and three C_2's perpendicular to it (120° from each other). There are three mirror planes intersecting at the C_3 axis, one containing each C_2 axis. There is also a center of symmetry.

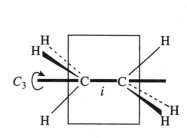

All six hydrogens are equivalent by virtue of the C_3 axis and the center of symmetry.

PROBLEM 8-2   Which of the isomeric alcohols below are chiral, and which are achiral? Indicate all chiral centers with an asterisk (∗).

(a)
$$\begin{array}{ccccc} \text{H} & \text{H} & \text{OH} & \text{H} & \text{H} \\ | & | & | & | & | \\ \text{H—C—} & \text{C—} & \text{C—} & \text{C—} & \text{C—H} \\ | & | & | & | & | \\ \text{H} & \text{H} & \text{H} & \text{H} & \text{H} \end{array}$$

(b)
$$\begin{array}{ccccc} \text{H} & \text{OH} & \text{H} & \text{H} & \text{H} \\ | & | & | & | & | \\ \text{H—C—} & \text{C—} & \text{C—} & \text{C—} & \text{C—H} \\ | & | & | & | & | \\ \text{H} & \text{H} & \text{H} & \text{H} & \text{H} \end{array}$$

(c)
$$\begin{array}{ccccc} \text{OH} & \text{H} & \text{H} & \text{H} & \text{H} \\ | & | & | & | & | \\ \text{H—C—} & \text{C—} & \text{C—} & \text{C—} & \text{C—H} \\ | & | & | & | & | \\ \text{H} & \text{H} & \text{H} & \text{H} & \text{H} \end{array}$$

(d)
$$\begin{array}{ccc} \text{H} & \text{H} & \text{OH} \\ | & | & | \\ \text{H—C—} & \text{C—} & \text{C—CH}_3 \\ | & | & | \\ \text{H} & \text{H} & \text{CH}_3 \end{array}$$

(e)
$$\begin{array}{ccc} \text{H} & \text{OH} & \text{H} \\ | & | & | \\ \text{H—C—} & \text{C—} & \text{C—CH}_3 \\ | & | & | \\ \text{H} & \text{H} & \text{CH}_3 \end{array}$$

(f)
$$\begin{array}{ccc} \text{H} & \text{H} & \text{H} \\ | & | & | \\ \text{H—C—} & \text{C—} & \text{C—CH}_2\text{OH} \\ | & | & | \\ \text{H} & \text{H} & \text{CH}_3 \end{array}$$

Solution Structures (a), (c), and (d) are achiral by virtue of a plane of symmetry—none has a chiral center. Structures (b), (e), and (f) each have one chiral center.

$$
\textbf{(b)} \quad H-\underset{\underset{H}{|}}{\overset{\overset{H}{|}}{C}}-\underset{\underset{H}{|}}{\overset{\overset{OH}{|}}{\overset{*}{C}}}-\underset{\underset{H}{|}}{\overset{\overset{H}{|}}{C}}-\underset{\underset{H}{|}}{\overset{\overset{H}{|}}{C}}-\underset{\underset{H}{|}}{\overset{\overset{H}{|}}{C}}-H
\qquad
\textbf{(e)} \quad H-\underset{\underset{H}{|}}{\overset{\overset{H}{|}}{C}}-\underset{\underset{H}{|}}{\overset{\overset{OH}{|}}{\overset{*}{C}}}-\underset{\underset{CH_3}{|}}{\overset{\overset{H}{|}}{C}}-CH_3
$$

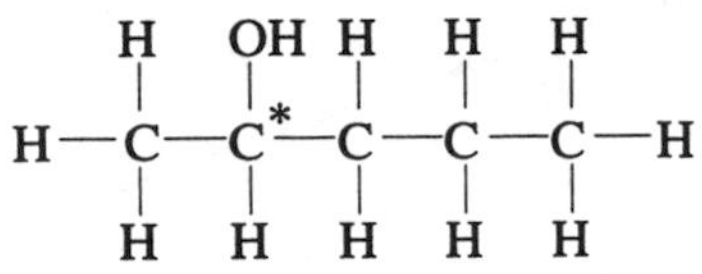

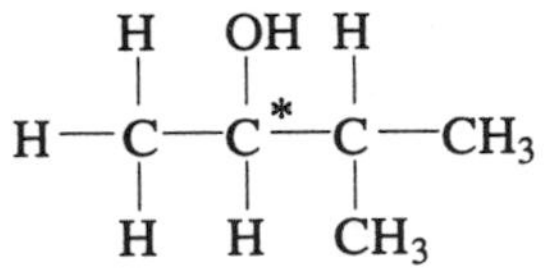

$$
\textbf{(f)} \quad H-\underset{\underset{H}{|}}{\overset{\overset{H}{|}}{C}}-\underset{\underset{H}{|}}{\overset{\overset{H}{|}}{C}}-\underset{\underset{CH_3}{|}}{\overset{\overset{H}{|}}{\overset{*}{C}}}-CH_2OH
$$

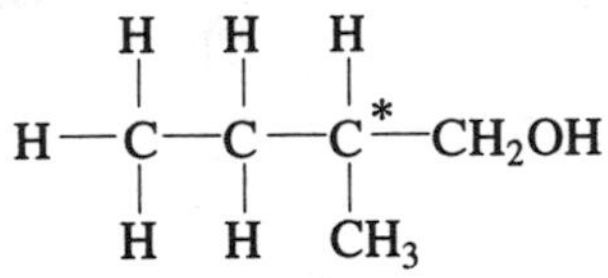

PROBLEM 8-3 For each pair of structures below, indicate whether they are enantiomers or identical.

(a)

$$
\underset{H_3CH_2C}{\overset{Br}{\diagdown}}\overset{|}{\underset{H}{C}}\text{---}CH_3
\qquad
\underset{H}{\overset{Br}{\diagdown}}\overset{|}{\underset{CH_3}{C}}\text{---}CH_2CH_3
$$

(b)

$$
\underset{H_3C}{\overset{HO}{\diagdown}}\overset{}{\underset{H}{C}}\text{---}CH_2CH_3
\qquad
\underset{H_3C}{\overset{H}{\diagdown}}\overset{}{\underset{OH}{C}}\text{---}CH_2CH_3
$$

(c)

$$
H-\overset{\overset{Cl}{|}}{\underset{\underset{CH_2CH_3}{|}}{C}}-CH_3
\qquad
H_3C-\overset{\overset{Cl}{|}}{\underset{\underset{H}{|}}{C}}-CH_2CH_3
$$

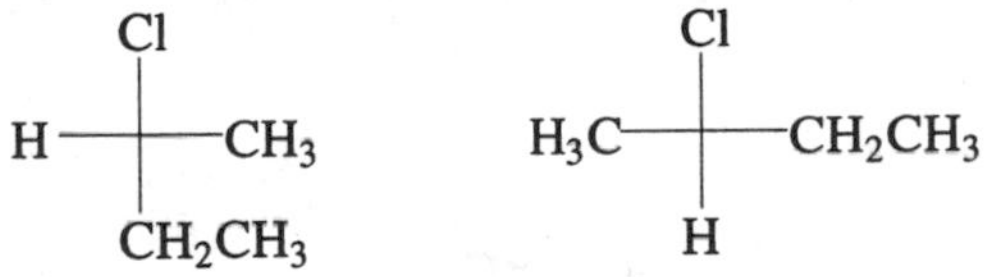

(d)

$$
H-\overset{\overset{Cl}{|}}{\underset{\underset{CH_2CH_3}{|}}{C}}-CH_3
\qquad
H_3CH_2C-\overset{\overset{H}{|}}{\underset{\underset{CH_3}{|}}{C}}-Cl
$$

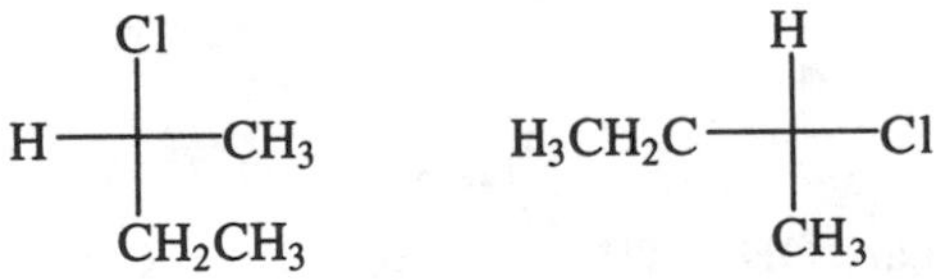

Solution **(a)** identical, resulting from a 120° rotation around the C–Br bond; **(b)** enantiomers, interconverted by one group interchange (HO and H); **(c)** identical, resulting from two group interchanges; **(d)** enantiomers, resulting from either a 90° rotation or three group interchanges.

PROBLEM 8-4 Explain why *trans*-1,3-dichlorocyclobutane is achiral, while *trans*-1,3-dichlorocyclopentane is chiral.

Solution *trans*-1,3-Dichlorocyclobutane has a plane of symmetry (containing the two chlorines and the carbons to which they're bonded) and therefore cannot be chiral.

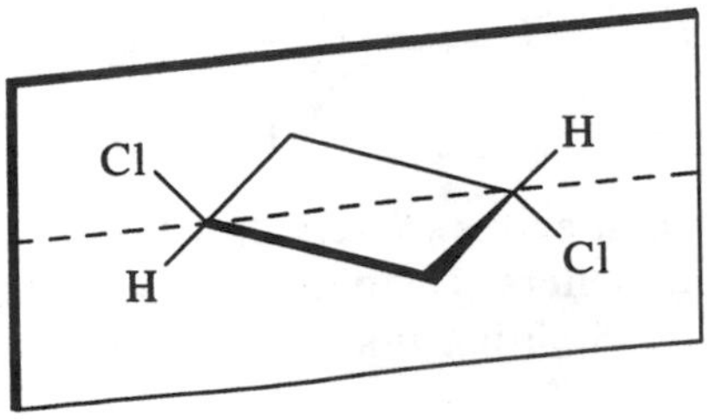

trans-1,3-Dichlorocyclopentane, on the other hand, lacks the mirror plane (though it does have a C_2 axis), and has a nonsuperimposable mirror image.

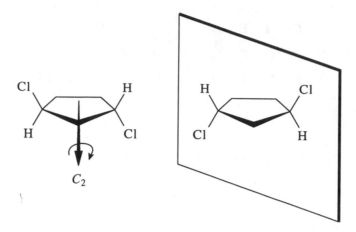

PROBLEM 8-5   Specify the absolute configuration (R or S) at each chiral center in the structures below, then provide a complete name for each structure.

(a)
H
    \ * —OH
     C
    / ⁀CH₂CH₃
H₃C

(b)
CHO
|*
H———OH
|
CH₂OH

(c)
Cl ╲ ╱ H
  *   *
H ╱ ╲ Cl

Solution

(a) The group priorities are OH > CH_2CH_3 > CH_3 > H. Redrawing the structure in the appropriate perspective (as viewed down the C–H bond), it is seen to have the S configuration. Name: (*S*)-2-butanol.

CH₂CH₃ ②
③ H₃C ➤ C
           ↘
           OH ①

(b) The group priorities here are OH > CHO $\left(= \begin{matrix} O \\ \diagdown \\ -C-H \\ \diagup \\ O \end{matrix} \right)$ > CH_2OH > H. The Fischer projection depicts the structure below:

CHO
|
H ➤ C ◀ OH
|
CH₂OH

When viewed down the C–H bond, the structure would look like this:

CHO ②
① HO ➤ C
          ↘
          CH₂OH ③

Thus it has the R configuration, and is (*R*)-2,3-dihydroxypropanal.

(c) This structure has two chiral centers. The one on the left has the following group priorities: Cl > CH_2CHCl > CH_2CH_2 > H. Sighting along the C–H bond gives this view, which indicates the S configuration:

Similarly, the chiral center on the right would look like this:

This one also has the S configuration. So, the complete name is (1S,3S)-1,3-dichlorocyclopentane.

PROBLEM 8-6   Explain why two enantiomers of a given chiral compound might have different fragrances.

Solution   Although enantiomers have identical physical, chemical, and spectroscopic properties under normal (i.e., nondissymmetric) conditions, a compound's fragrance is based on the complicated way it interacts with the cells in the nasal membrane. These cells are made up of many complex biomolecules, most of which are themselves asymmetric. Therefore, these biomolecules can serve as resolving agents that discriminate between enantiomers, giving rise to a different perceived fragrance for each one.

PROBLEM 8-7   The optical rotation of a solution of pure natural camphor is found to be +5.76° under the following conditions: $c = 0.130$ g/mL, $l = 1.00$ dm, wavelength = sodium D line, $T = 25°C$.
(a) Calculate the specific rotation of camphor under these conditions.
(b) Is natural camphor the d or l isomer?
(c) A sample of synthetic camphor contains both the d and l isomers in unequal amounts. The *specific* rotation of the sample is found to be −13.9°. What is the enantiomeric excess of this sample?
(d) What is the percentage of each enantiomer in the sample?

Solution
(a) Use Eq. (8-1):

$$[\alpha]_D^{25} = \frac{\alpha}{lc} = \frac{+5.76°}{(1.00)(0.130)} = +44.3°$$

(b) The positive rotation indicates the d isomer.
(c) The negative rotation of this sample indicates that the l isomer predominates. Solving Eq. (8-2′) for the enantiomeric excess $(2f_l - 1)$,

$$ee = (2f_l - 1) = \frac{[\alpha]_{net}}{[\alpha_l]} = \frac{-13.9°}{-44.3°} = 0.314$$

(d) Isolate f_l and f_d:

$$2f_l - 1 = 0.314$$
$$f_l = \frac{1.314}{2}$$
$$= 0.657 \quad (66\%)$$
$$f_d = 1 - f_l$$
$$= 1 - 0.657 = 0.34 \quad (34\%)$$

PROBLEM 8-8   Refer back to Example 8-7, and assume the structure has a single chiral center. Furthermore, suppose that it is later found that the *l* isomer of this compound has the R configuration. Is this additional information sufficient to determine the configuration of your *d* isomer?

Solution   Certainly! If the *l* (−) isomer of the chiral compound is found to have the R configuration, its enantiomer, the *d* (+) isomer, must have the S configuration.

PROBLEM 8-9   Review the essential features of a polarimeter as shown in Figure 8-6. A little thought should convince you that an optical rotation of, say, +40° would appear the same to a polarimeter as a rotation of −320°.

How could you differentiate between these two optical rotations?

Solution   You would have to redetermine the rotation of the sample using either a different concentration or a different sample path length. For example, by using a sample tube one-fourth as long, or a concentration one-fourth as great, the optical rotation would decrease to one-fourth its original value. Thus, a +40° value would become +10°, while a −320° rotation would become −80°, and these two outcomes could be easily differentiated.

PROBLEM 8-10   Using terms such as identical, enantiomers, different conformation, diastereomers, or structural isomers, describe the relationship between structure **21** and each structure in (a)–(f) below.

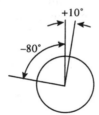

21

Solution (a) different conformation of the same molecule; (b) identical (different views of the same structure); (c) identical; (d) diastereomers (same configuration at one chiral center, opposite configuration at the other); (e) structural isomers; (f) enantiomers. If you had difficulty with any of these comparisons, be sure to make the appropriate molecular models.

PROBLEM 8-11 Indicate each chiral center in prostaglandin E$_1$ (below) with an asterisk (*). Neglecting the configuration of the C=C double bond, how many stereoisomers of prostaglandin E$_1$ are possible?

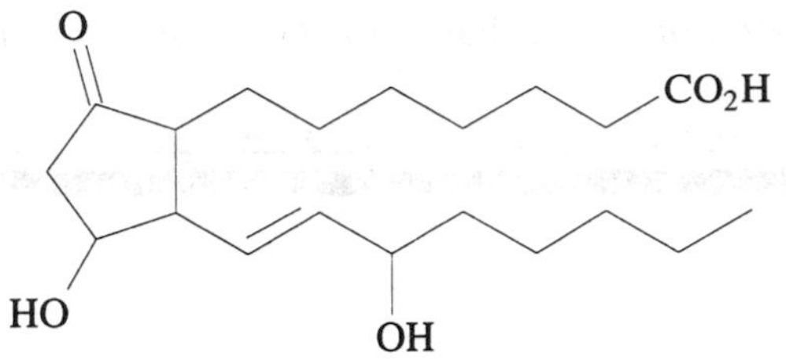

Solution There are four chiral centers, giving $2^4 = 16$ stereoisomers (eight *dl* pairs).

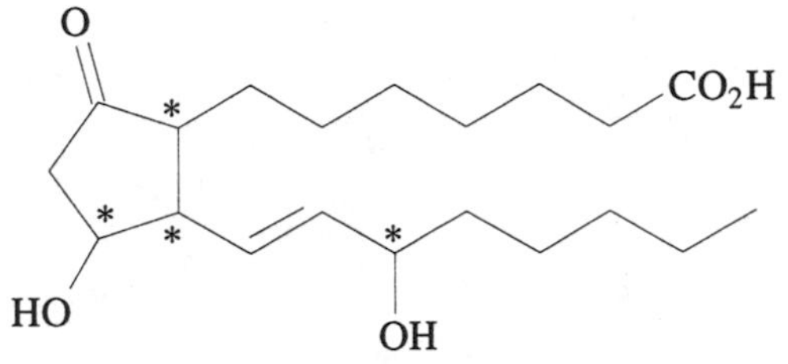

It's interesting that, as is the case with virtually all chiral compounds that have physiological activity, each stereoisomer has different physiological properties. This is another manifestation of the dissymmetric nature of our biological systems (see Problem 8-6).

PROBLEM 8-12 Complete the Fischer projection of structure **22**.

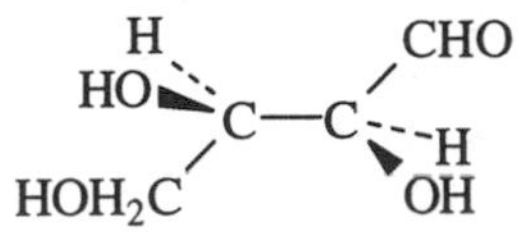

22

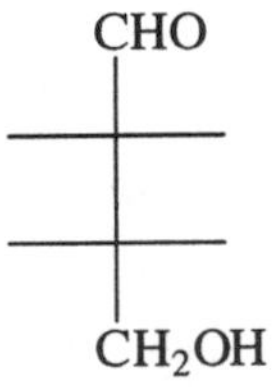

Solution Start by redrawing the molecule in the appropriate conformation, then locate the various groups in the correct relative positions.

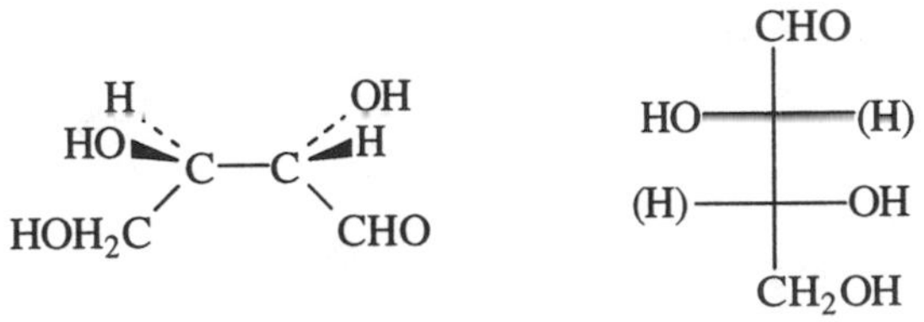

PROBLEM 8-13 Draw the structure that would result from "umbrella" inversion at the nitrogen atom in structure **23**. What is the relationship between these two "invertomers"? Are they identical, enantiomers, or diastereomers?

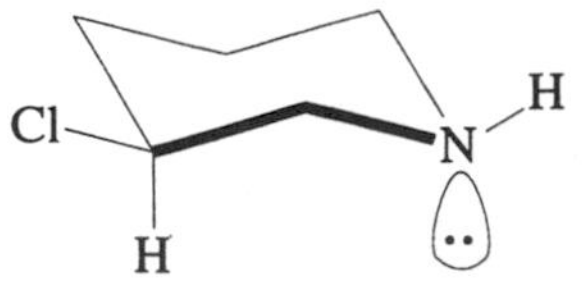

23

Solution

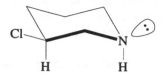

Although the configuration around the nitrogen has been inverted, the other chiral center remains unchanged. So, these two structures are diastereomers—but remember, they're interconverting rapidly.

EXAM 2
(Chapters 5–8)

1. **(a)** Provide a complete systematic name for each hydrocarbon below:

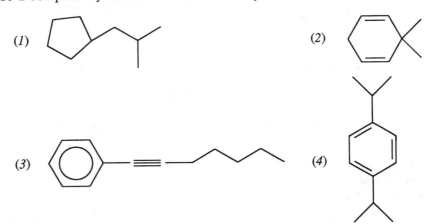

(1)   (2)   (3)   (4)

(b) For structure (1) above, how many of each of the following are there in the molecule: methyl groups, methylene groups, methine groups, quaternary carbons, 1° hydrogens, 2° hydrogens, and 3° hydrogens?

2. For each structure below indicate whether the stereochemical label cis, trans, E, Z, or none of these, is appropriate.

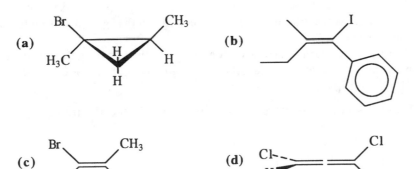

(a)   **(b)**

(c)   **(d)**

3. Provide a systematic name for each functionalized compound below:

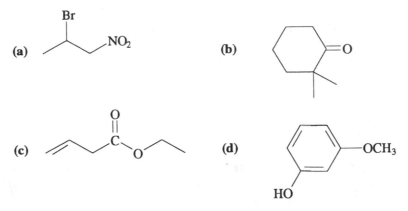

(a)   **(b)**

(c)   **(d)**

4. (a) Indicate with an asterisk each chiral center in structures **I** and **II**.

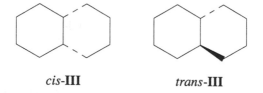

I (estrone)   II (a nucleotide)

(b) Fill in the table below with the appropriate data for structures **I** and **II**.

	I	II
Molecular formula		
I value		
Number of rings		
Number of π bonds		
Number of nonbonding pairs		
Functional group class(es) represented (indicate 1°, 2°, or 3°, as appropriate)		

5. Write the structural formula that corresponds to each common name below:

   (a) acetone   **(b)** acetonitrile   **(c)** formaldehyde   **(d)** acetic acid

6. Describe how the solubility of linear primary alcohols [$H(CH_2)_nOH$] in water varies with the chain length (*n*).

7. (a) Draw the eclipsed and staggered conformations of propane, sighting along one of the C–C bonds. What is the dihedral angle between the methyl group and each vicinal hydrogen in each conformation?

   (b) Using the torsional strain data below, construct a graph showing the conformational energy of propane as a function of the dihedral angle from 0° to 360°. H–H eclipsing: 4 kJ mol⁻¹; H–CH₃ eclipsing: 6 kJ mol⁻¹.

8. Decalin (**III**) exists in two diastereomeric forms, the cis and the trans.

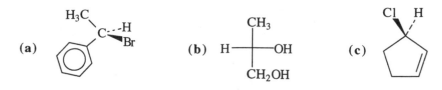

cis-**III**   *trans*-**III**

   (a) Redraw the two structures showing how each six-membered ring can adopt a chair conformation.
   (b) Which diastereomer of decalin is more stable, and why?
   (c) Which of the two diastereomers is more conformationally flexible, and why?
   (d) Which of the two diastereomers is chiral?

9. List all centers, planes, and axes of symmetry present in staggered ethane.

10. Specify the absolute configuration (R or S) at each chiral center in the structures below:

11. One pure enantiomer of 2-methyl-1-butanol has a specific rotation of +5.90° (25°, D line). An unequal mixture of the *d* and *l* enantiomers of this compound has a net specific rotation of +1.80° (25°, D line).

(a) Which enantiomer predominates in the mixture, *d* or *l*?

(b) What is the enantiomeric excess in the above mixture? What is the fraction of each enantiomer in the mixture?

(c) Does *d*-2-methyl-1-butanol have the R or S configuration? Explain.

12. Suggest a unique structure consistent with each set of data below. Describe your reasoning.

Compound **IV**: Mass spectrum indicates a molecular formula of $C_3H_6O_2$. Its IR spectrum suggests the presence of an aldehyde carbonyl and a hydroxy group. Polarimetric examination indicates that the molecule is optically active.

Compound **V**: Mass spectrum indicates a molecular formula of C_6H_8. The UV spectrum suggests two conjugated double bonds. The carbon NMR spectrum indicates that the molecule is symmetrical; there are just three signals representing two carbons each. One signal can be assigned to two equivalent methylene groups. The other two signals represent vinyl carbons (two of each), and each vinyl carbon has one hydrogen attached to it.

Answers to Exam 2

1. **(a)** (*1*) isobutylcyclopentane [(2-methylpropyl)cyclopentane]
   (*2*) 3,3-dimethyl-1,4-cyclohexadiene

   (*3*) 1-phenyl-1-heptyne
   (*4*) 1,4-diisopropylbenzene (*p*-diisopropylbenzene)
   (b) methyl groups, 2; methylene groups, 5; methine groups, 2; quaternary carbons, 0; 1° H, 6; 2° H, 10; 3° H, 2

2. Where appropriate, matching groups are circled:

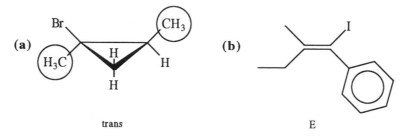

   (a) trans   (b) E

   Note: The structure in part (a) could also be labeled Z (Br vs. CH₃).

   (c) Both groups attached to the vinyl carbon on the right are the same (CH₃), so there are no stereoisomers of this structure.
   (d) Because the two groups at each end of the allene system lie in mutually perpendicular planes, this molecule is chiral, and diastereomers do not exist.

3. **(a)** 2-bromo-1-nitropropane
   (b) 2,2-dimethylcyclohexanone
   (c) ethyl 3-butenoate
   (d) 3-methoxyphenol (*m*-methoxyphenol)

4. **(a)** Chiral centers are shown by an asterisk (∗); nonbonding pairs are also shown.

(b)

	I	II
Molecular formula	$C_{18}H_{22}O_2$	$C_9H_{14}N_3O_7P$
I value (see Note 1)	8	5
Number of rings	4	2
Number of π bonds	4 (1 C=O, 3 C=C)	3 (1 C=C, 1 C=O, 1 C=N)
Number of nonbonding pairs	4	17

Functional group class(es)	phenol	phosphate monoester
	ketone	ether
		2° alcohol
		3° amide (see Note 2)
		disubstituted C=C
		imine (see Note 2)

note 1: Notice that the sum of rings + π bonds equals the *I* value and that P=O bonds are not counted as π bonds.

note 2: In actuality, an $\underset{\underset{O}{\|}}{NCN}$ functional group belongs to the urea class.

5. (a) $\underset{\underset{O}{\|}}{H_3CCCH_3}$　　(b) $CH_3C{\equiv}N$　　(c) $H_2C{=}O$　　(d) $\underset{\underset{O}{\|}}{CH_3COH}$

6. The hydroxyl group, which can hydrogen-bond to water molecules, is **hydrophilic** (water-loving) and therefore contributes to water solubility. The alkyl group residue is nonpolar and **hydrophobic** (water-hating) and lowers water solubility. Thus water solubility should decrease as *n* increases. As expected, the first three alcohols in the series, methanol (*n* =1), ethanol (*n* = 2), and 1-propanol (*n* = 3), are all miscible (soluble in all proportions) with water. The water solubility of the next five alcohols decreases from 7.9 g/100 g water for 1-butanol (*n* = 4) to 0.05 g/100 g water for 1-octanol (*n* = 8).

7. (a)

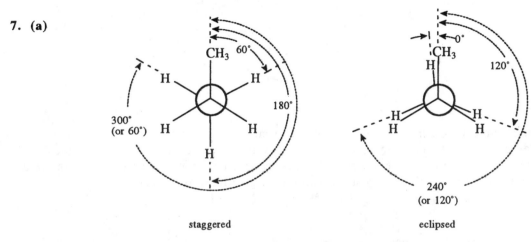

staggered　　　　　　　　　　　　eclipsed

(b) The conformational energy of the staggered form is zero relative to that of the eclipsed form, which is 2(4 kJ/mol) + 6 kJ/mol = 14 kJ/mol.

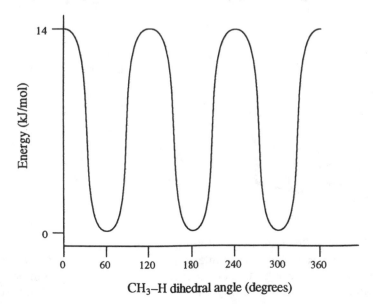

8. (a)

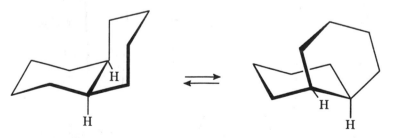

trans   cis

(b) The trans diastereomer has no axial carbon branches, so it avoids all 1,3-diaxial steric repulsions. The cis isomer has one axial carbon branch off each ring, causing unfavorable 1,3-diaxial interactions:

Thus, the cis isomer is less stable (by ~8 kJ/mol) than the trans.

(c) Neglecting highly unstable boat conformations, the trans isomer is "locked" in the double chair conformation shown in part (a). This is because a chair–chair flip of either ring would require the second ring to span the 1,2-diaxial positions, an impossibility.

The cis isomer, on the other hand, *can* undergo a double ring flip to give an energetically equivalent cis conformation:

Thus the cis isomer *is* much more flexible than the trans.

(d) The trans isomer has a plane of symmetry [most easily seen in the Newman projection in part (a)] and is therefore achiral. The cis isomer lacks a plane of symmetry (having instead a C_2 axis), and *is* chiral. Interestingly, the chair–chair flip described in part (c) interconverts the two enantiomers of *cis*-decalin. Because this interconversion is extremely rapid, the two structures are best considered to be enantiomeric *conformations*, rather than enantiomers (enantiomeric *isomers*).

9.

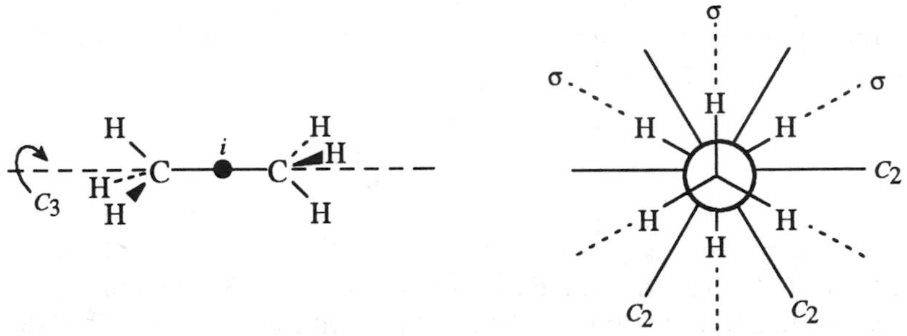

The highest-order axis is a C_3 along the C–C bond. There are three σ planes, each defined by the C–C bond and a pair of s-trans vicinal hydrogens. There are three C_2 axes, each one perpendicular to a σ plane and through the center of the C–C bond. Finally, there is a center of symmetry.

10. (a)

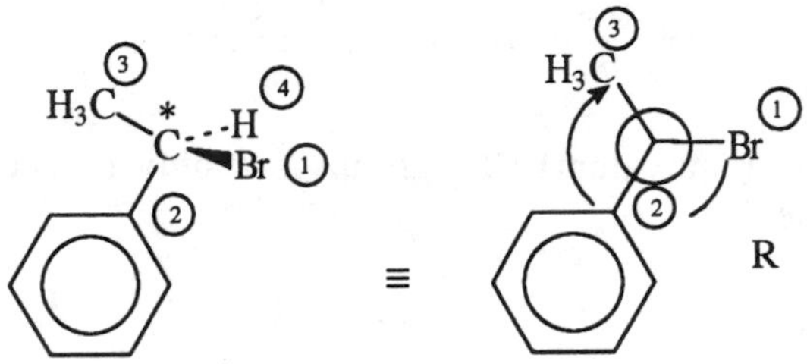

(b)

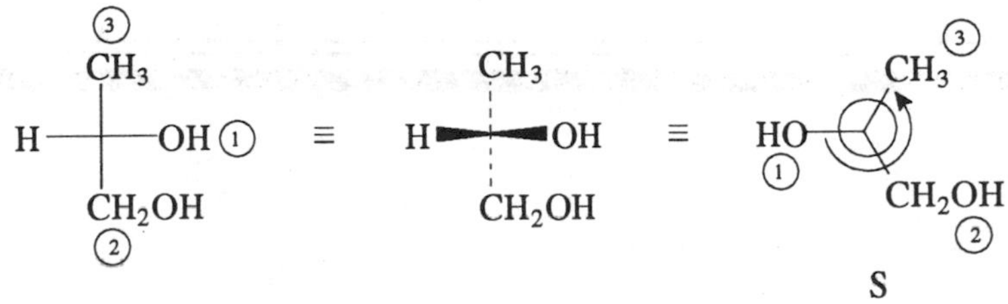

(c)

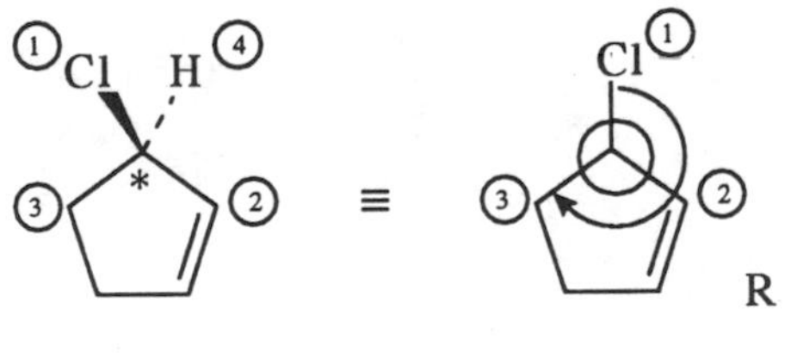

11. (a) The net rotation is positive, so the *d* isomer predominates.

  (b) Use Eq. (8-2′):

$$[\alpha]_{net} = [\alpha](2f_d - 1)$$

$$ee = 2f_d - 1 = \frac{[\alpha]_{net}}{[\alpha]} = \frac{+1.80}{+5.90} = 0.305$$

$$f_d = \frac{0.305 + 1}{2} = 0.653$$

$$f_l = 1 - f_d = 1 - 0.653 = 0.347$$

  (c) Without additional information there is no way to know whether the *d* isomer has the R or S configuration.

12. Compound **IV**: $C_3H_6O_2$ has an *I* value of $[2(3) + 2 - 6]/2 = 1$. This must be the π bond of the aldehyde's carbonyl (C=O). The indicated fragments are

$$-C \overset{O}{\underset{H}{\diagdown}} , \quad -OH, \quad \overset{*}{\diagup}C\overset{\diagup}{\diagdown}$$

plus a carbon and four hydrogens. The only structure that can accommodate all these fragments is either enantiomer of 2-hydroxypropanal:

$$\underset{H_3C}{\overset{H}{\diagdown}}C\overset{-CHO}{\underset{OH}{\diagdown}}$$

Compound **V**: C_6H_8 has an *I* value of $[2(6) + 2 - 8]/2 = 3$. The conjugated diene indicated by the UV spectrum accounts for two of these; the third may be a ring. The NMR spectrum suggests this partial structure, where the dotted line indicates some sort of symmetry relationship (axis or plane):

$$\underset{H-C}{\overset{H}{\diagdown}}\overset{}{\underset{\diagup}{C}}\overset{}{\overset{\diagdown}{}}\overset{H}{\underset{\diagup}{C}}\overset{\diagdown}{\underset{C-H}{}}$$

The remaining two (equivalent) CH_2 groups can only be accommodated if the structure is 1,3-cyclohexadiene:

9 AN INTRODUCTION TO PHYSICAL ORGANIC CHEMISTRY

9-1. Why Do Chemical Reactions Occur? The Bond-Strength Criterion

So far in this Outline we've had the soup and salad course. Our focus has been on the *structure* of organic molecules: The nature of atoms and chemical bonds, molecular shape and stereochemistry, nomenclature, and physical properties. Now we begin the main course—organic reactions, the conversion of one organic molecule into another. Certainly, it is the ability to synthesize new compounds with wonderfully varied properties that has elevated organic chemistry to its central position in today's highly industrialized world.

The detailed chemical reactivity of each functional group and the integration of this information into strategies for organic synthesis are the topics that fill most of Volume II of this Outline. In this chapter, however, we'll describe certain characteristics of *all* chemical reactions, and learn how and why some reactions occur readily, while others do not occur at all. This subdiscipline of organic chemistry is called **physical organic chemistry** because it represents application of the principles of physics (mainly thermodynamics) and physical chemistry to organic reactions.

Consider the conversion of 1-propen-2-ol (**1**, an example of an **enol**) to acetone (**2**):

1	**2**

Under most conditions this reaction takes place very rapidly, and it goes essentially "to completion," that is, virtually all of the reactant (**1**) is converted to product (**2**). Incidentally, this reaction is an example of a **structural isomerization** because **2** is a structural isomer of **1**.

A. Bond dissociation energy

The *first* important criterion for whether or not a chemical reaction is favorable (i.e., whether it will occur spontaneously) is this:

- The bonds that are made must be stronger than the bonds that are broken.

The strength of a chemical bond is measured by the **bond dissociation energy**, the enthalpy change (heat energy) required in the gas phase to cleave the bond *homolytically* (Section 4-2). The symbol for bond dissociation energy is ΔH_d, where ΔH implies enthalpy change (heat absorbed at constant pressure), and the subscript d is for dissociation. The units of ΔH_d are typically kJ mol^{-1}. All bond dissociations are **endothermic** ("uphill," requiring the absorption of heat energy), and by convention this is reflected by the fact that all ΔH_d values are *positive*. Table 9-1 lists ΔH_d values of the bonds most commonly encountered in organic molecules. Because many of these values represent averages over several compounds, the exact dissociation energy of a specific bond in a specific molecule may be somewhat different than the value

TABLE 9-1.   Representative Bond Dissociation Energies[a,b]

Bond	ΔH_d	Bond	ΔH_d	Bond	ΔH_d
C–H		**C–O**		**N–O**	
C(sp^3)–H	410	C(sp^3)–OH	383	N–O	180
C(sp^2)–H	431	C(sp^2)–OH	431	N=O	481
C(sp)–H	523	O=C–OH	456	N–Cl	201
C–C		C(sp^3)–OC(sp^3)	335	**H–O**	
		C(sp^2)–OC(sp^3)	366		
C(sp^3)–C(sp^3)	368	C(sp^3)–OC=O	406	H–OC(sp^3)	427
C(sp^3)–C(sp^2)	372	C=O	732	H–OC(sp^2)	356
C(sp^3)–C(sp)	490			H–OC=O	469
C(sp^2)–C(sp^2)	418	**C-halogen**		**O–O**	
C(sp)–C(sp)	628	C(sp^3)–OH	383	O–O	138
C=C	619	C(sp^2)–OH	431	O=O	402
C≡C	812	O=C–OH	456	H–H	436
C–C π	≈ 251				
C–N		**N–H**		**homonuclear diatomics**	
		N–H	389		
C–N	289			H–H	436
C=N	619	**N–N**		F–F	159
C≡N	891			Cl–Cl	243
		N–N	159	Br–Br	192
		N=N	418	I–I	151
		N≡N	946		

[a] Values are expressed in kJ mol^{-1}.
[b] Data taken from T. H. Lowry and K. S. Richardson, *Mechanism and Theory in Organic Chemistry* (3rd Ed.), Harper and Row, New York, 1987.

quoted in Table 9-1. Nonetheless, these values are very useful for predicting whether or not a reaction meets the bond-strength criterion. To make this determination, we compare the total bond strength of the product(s) with that of the reactant(s), using Eq. (9-1):

ENTHALPY CHANGE FROM BOND STRENGTHS
$$\Delta H = \sum \Delta H_{d\,(\text{reactants})} - \sum \Delta H_{d\,(\text{products})} \tag{9-1}$$

In this equation ΔH is the enthalpy change for the overall reaction, and Σ means "sum of." If ΔH is *negative* the *products* have stronger bonds than the reactants, and the reaction is **exothermic** (liberating heat), favorable from the standpoint of bond strengths. If, on the other hand, ΔH is *positive* (an **endothermic** reaction), heat energy is *absorbed* and the reaction is unfavorable from the standpoint of bond strength. Remember that Eq. (9-1) applies only when all reactants and products are in the gas phase. This interconversion of heat energy and chemical potential energy is the subject of the **First Law of Thermodynamics**, which says that energy can neither be created nor destroyed, only changed in form.

EXAMPLE 9-1 Using the data in Table 9-1, estimate ΔH for the conversion of enol (**1**) to acetone (**2**), and determine whether the reaction is favorable from the standpoint of bond strengths.

Solution Reactant **1** and product **2** have the following bond energies (in kJ mol^{-1}):

1		2	
H–OC(sp^2)	356	C=O	732
C(sp^2)–OH	431	2 C(sp^3)–C(sp^2)	2(372)
C=C	619	6 C(sp^3)–H	6(410)
C(sp^2)–C(sp^3)	372	Total	3936
3 C(sp^3)–H	3(410)		
2 C(sp^2)–H	2(431)		
Total	3870		

Thus the bonds in the product are stronger than the bonds in the reactant and, from Eq. (9-1), we can calculate the value of ΔH:

$$\Delta H = \sum H_{d(\text{reactants})} - \sum H_{d(\text{products})}$$
$$= 3870 - 3936$$
$$= -66 \text{ kJ mol}^{-1}$$

The *negative* sign of ΔH indicates that this process *releases* 66 kJ mol^{-1} of energy. Thus the reaction *is* favorable from the standpoint of enthalpy (bond strengths).

B. Enthalpy diagrams

One way to visualize the energetic relationships in this reaction is to draw an **enthalpy diagram** (Figure 9-1), which shows the relative enthalpies of reactant **1**, product **2**, and separated atoms. Remember that all species are assumed to be in the gas phase.

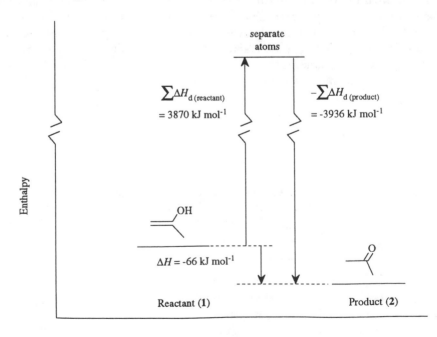

FIGURE 9-1. Enthalpy diagram for the reaction **1** → **2**.

EXAMPLE 9-2 (a) Is the reaction **2** → **1** (the reverse of **1** → **2**) favorable or unfavorable from the standpoint of bond strengths? (b) Draw the enthalpy diagram for this reaction.

Solution (a) In comparing the reaction **2** → **1** with the reaction **1** → **2**, all we have done is to interchange the reactant and product. Therefore, it is now the *reactant* (**2** in this case) that is more stable, and $\Delta H = +66$ kJ mol^{-1}. This reaction is *endothermic* and unfavorable. Of course, unfavorable reactions can be made to occur by the investment of an appropriate amount of energy. (b) See Figure 9-2. See also the discussion of microscopic reversibility in Section 9-5.

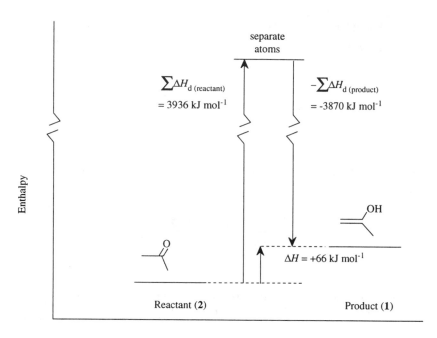

FIGURE 9-2. Enthalpy diagram for the reaction **2** → **1**.

There is an even more accurate way to predict the value of ΔH for a given reaction if you have a table of **standard heats of formation** (ΔH_f°) that includes your reactants and products. Such tables can be found, for example, in the *CRC Handbook of Chemistry and Physics*. The standard heat of formation of a compound is the value of ΔH when that compound is prepared from its constituent elements (not atoms, but *elements*), with all species in their **standard state** (most stable state at 25°C and 1 atm pressure). This definition thereby defines the heat of formation of any element in its standard state to be zero, because such a number would measure the ΔH for making that element from its constituent elements, a nonreaction!

At any rate, ΔH° is calculated from heats of formation, using Eq. (9-2):

ENTHALPY CHANGE FROM HEATS OF FORMATION

$$\Delta H^\circ = \sum \Delta H_{f(products)}^\circ - \sum \Delta H_{f(reactants)}^\circ \qquad (9\text{-}2)$$

Notice that the product and reactant terms have been reversed, compared to Eq. (9-1). This "products minus reactants" order will appear in all thermodynamic calculations except those involving bond strengths. Furthermore, the ΔH° value calculated from heats of formation applies to a more realistic set of conditions (25°C, 1 atm) than those used in bond-strength calculations (the gas phase). The superscript "°" reminds us of these standard conditions.

EXAMPLE 9-3 (a) Calculate ΔH for the hydrogenation of ethylene ($H_2C=CH_2 + H_2 \rightarrow H_3C-CH_3$) in the gas phase, using bond–strength data. (b) The standard heats of formation of ethylene and ethane are +52.4 and −84.6 kJ mol^{-1}, respectively. Use these data to calculate ΔH° for the hydrogenation of ethylene. (c) Account for any differences between the values you calculate in parts (a) and (b).

Solution

(a) Proceed as in Example 9-1 to find the total bond energies of reactants and products:

4 C(sp^2)–H	4(431)	6 C(sp^3)–H	6(410)
1 C=C	619	1 C(sp^3)–C(sp^3)	368
1 H–H	436	Total	2828
Total	2779		

Using Eq. (9-1), we calculate ΔH:

$$\Delta H = \sum \Delta H_{d(\text{reactants})} - \sum \Delta H_{d(\text{products})} = 2779 - 2828 = -49 \text{ kJ mol}^{-1}$$

Alternatively, you can approach this calculation more quickly (and approximately) by looking at the minimum number of bonds broken and formed. In this reaction we're breaking a C–C π bond (251 kJ mol^{-1}, Table 9-1) and an H–H bond (436 kJ mol^{-1}), and making two C(sp^3)–H bonds (2 × 410 kJ). By this method,

$$\Delta H = (251 + 436) - 2(410) \approx -133 \text{ kJ mol}^{-1}$$

(b) Next, we'll use heats of formation to calculate $\Delta H°$. Remember that H_2, being an element in its most stable form, has a heat of formation of zero.

$$\Delta H° = \sum \Delta H°_{f(\text{products})} - \sum \Delta H°_{f(\text{reactants})}$$

$$= \Delta H°_{f(\text{ethane})} - \left[\Delta H°_{f(\text{ethylene})} + \Delta H°_{f(\text{hydrogen})} \right]$$

$$= -84.6 - [52.4 + 0]$$

$$= -137 \text{ kJ mol}^{-1}$$

(c) Although both calculations predict that the hydrogenation of ethylene is exothermic (favorable from enthalpy considerations), the value of $\Delta H°$ calculated from heats of formation is more than twice the value calculated from bond strengths. (It is fortuitous that the more approximate method agrees better.) Remember that the bond-strength calculation involves *average* bond strengths and therefore gives a result that is only approximate and applies only when all species are in the gas phase (not a limitation in this case because the reactants and products *are* gases). On the other hand, the $\Delta H°$ value calculated from heats of formation is very accurate because it uses $\Delta H°_f$ values for the specific compounds of interest, under a specific set of conditions (25°C, 1 atm). Of course, if our table of heats of formation did not include one or more of our reactants or products, we would have to resort to a bond-strength-based calculation.

9-2. The Entropy Criterion

There are some chemical reactions that occur spontaneously even though they are **thermoneutral** ($\Delta H° = 0$) or even endothermic. One example is the racemization (Section 8-6) of certain optically active compounds. For example, when treated with base or acid, optically active (+)-2-methylcyclohexanone [(+)-**3**] undergoes racemization to produce racemic (±)-2-methylcyclohexanone:

This reaction, which is a type of **stereo-isomerization**, must have a $\Delta H°$ of zero because the reactant and product have the same sequence of atoms and bonds, and therefore the same bond strength. And yet, this is a spontaneous process. Why?

There is a second criterion for whether or not a reaction will occur spontaneously, and it relates to changes in the **entropy** (disorder or randomness) of the molecules during the reaction. All other things being equal, a process that occurs with an *increase* in disorder will be favorable, while one that occurs with a *decrease* in disorder (or an increase in order) will be unfavorable. Expressed symbolically, a favorable process is one where $\Delta S° > 0$, with S being the symbol for entropy. The "> 0" is a mathematical way of saying "positive."

The concept of entropy and the natural tendency for things to become disordered (more random) is the subject of the **Second Law of Thermodynamics**. And we can cite numerous examples from our everyday experience to support it. Consider what happens to the apparatus in Figure 9-3, where one of the bulbs initially contains a gas under pressure, while the other bulb is evacuated.

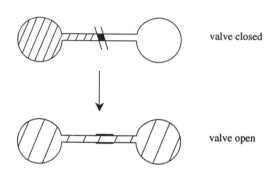

valve closed

valve open

FIGURE 9-3. Spontaneous expansion of a gas into an evacuated bulb ($\Delta S > 0$).

We all know that when the valve is opened, gas molecules will migrate (rapidly!) from the filled bulb to the empty one until the pressure in both bulbs is equal. Yet, no heat energy is absorbed or evolved during this process, so ΔH is zero. The reason this process occurs spontaneously is that having the gas molecules spread out over two bulbs (rather than just one) is a more disordered, or more random, arrangement, so $\Delta S > 0$.

The value of $\Delta S°$ of a chemical reaction is most easily calculated from tabulated **standard entropy values** ($S°$) for each reactant and product, according to Eq. (9-3). Such data can be found in the CRC handbook along with the heat of formation data previously described.

ENTROPY CHANGE
$$\Delta S° = \sum S°_{products} - \sum S°_{reactants} \qquad (9\text{-}3)$$

Note that $S°$ and $\Delta S°$ values are usually expressed in J K^{-1} mol^{-1}, where K refers to the absolute temperature scale in degrees Kelvin. The significance of this will be discussed more fully in Section 9-3. Also, notice that the standard entropy of *any* substance is *positive* above absolute zero; in fact, this is one way to state the **Third Law of Thermodynamics**. As a result, even elements have a positive entropy though their heats of formation were arbitrarily defined as zero.

EXAMPLE 9-4 Calculate $\Delta S°$ for the hydrogenation of ethylene (Example 9-3), given the standard entropy values below (J K^{-1} mol^{-1} at 25°C):

ethylene, 220; hydrogen, 131; ethane, 229

Solution Use the above data in Eq. (9-3):

$$\Delta S° = \sum S°_{products} - \sum S°_{reactants}$$
$$= S°_{ethane} - \left[S°_{ethylene} + S°_{hydrogen} \right]$$
$$= 229 - (220 + 131) = -122 \text{ J K}^{-1} \text{ mol}^{-1} \text{ at } 25°C$$

Notice that $\Delta S°$ in the above example is *negative*, indicating an entropically *un*favorable process where the product (ethane) is *less* disordered than the reactants (ethylene plus hydrogen). A little thought might have enabled you to predict the sign of $\Delta S°$ in this case, because *two* moles of gaseous reactants have been assembled to form *one* mole of gaseous product, a less random (more ordered) arrangement. It is a useful generalization (though not without exception) that reactions that produce fewer moles of product than the number of moles of reactant will often have negative $\Delta S°$ values and be unfavorable from the standpoint of entropy. Conversely, reactions that produce more moles of product than there were moles of reactant will generally have positive $\Delta S°$ values and be favorable from the standpoint of entropy.

EXAMPLE 9-5   For each combination of $\Delta H°$ and $\Delta S°$ below, specify whether the process will be favorable or unfavorable, or explain why you can't tell:

(a) $\Delta H° < 0; \Delta S° > 0$
(b) $\Delta H° > 0; \Delta S° < 0$
(c) $\Delta H° < 0; \Delta S° < 0$
(d) $\Delta H° > 0; \Delta S° > 0$

Solution   Recall that a *negative* $\Delta H°$ corresponds to a favorable process, as does a *positive* $\Delta S°$ Therefore, (a) represents a process that is favorable by both criteria. Case (b), on the other hand, is *un*favorable according to both criteria. In case (c) we have a process that is favorable from the standpoint of enthalpy but unfavorable entropically, while case (d) is just the converse. In these last two cases we cannot be sure whether the overall process is favorable until we know something more about the relative magnitudes of $\Delta H°$ versus $\Delta S°$. This is the topic of the next section.

9-3. $\Delta G°$: The Ultimate Criterion of Favorability

So far, we've discussed two criteria that bear on the favorability of a chemical reaction: Enthalpy change (related to bond strengths) and entropy changes (related to disorder). The ultimate criterion for the favorability (spontaneity) of a reaction is a combination of both enthalpy and entropy changes. It is called the **Gibbs free energy** change, with symbol $\Delta G°$ ($\Delta F°$ in older books). The relationship between $\Delta G°$, $\Delta H°$, and $\Delta S°$ is given in Eq. (9-4):

GIBBS FREE ENERGY CHANGE $\Delta G° = \Delta H° - T\Delta S°$ (9-4)

You will notice that $\Delta S°$ is multiplied by the *absolute* temperature, that is, the *effect* of the entropy change is controlled by the temperature at which the reaction is conducted. From Eq. (9-4) we can see that a favorable (spontaneous) process is one where $\Delta G°$ is negative ("< 0"), while a positive $\Delta G°$ corresponds to an unfavorable process.

EXAMPLE 9-6   Calculate the value of $\Delta G°$ for the hydrogenation of ethylene at 25°C (298 K). See Examples 9-3 and 9-4 for relevant data. Will this reaction proceed spontaneously?

Solution   Use Eq. (9-4) with $\Delta H° = -137$ kJ mol^{-1} and $\Delta S° = -0.122$ kJ K^{-1} mol^{-1} (-122 J K^{-1} mol^{-1}):

$$\Delta G° = \Delta H° - T\Delta S°$$
$$= -137 - (298 \text{ K})(-0.122)$$
$$= -100 \text{ kJ mol}^{-1}$$

Because $\Delta G°$ is *negative*, this reaction *is* favorable and will proceed spontaneously.

The hydrogenation of ethylene, discussed in Examples 9-3, 9-4, and 9-6, is an example of case (c) in Example 9-5: Favorable with regard to enthalpy, but unfavorable with regard to entropy. Yet, overall the reaction *is* favorable because the $\Delta H°$ term outweighs the $-T\Delta S°$ term. This is often, though certainly not always, true: The enthalpy term is usually dominant, with the entropy term being relatively less important. Nonetheless, don't forget that the ultimate test for the favorability of a reaction is $\Delta G°$, which combines *both* enthalpy and entropy terms.

EXAMPLE 9-7   Assuming for the moment that the values of $\Delta H°$ and $\Delta S°$ are independent of temperature (i.e., they don't change even if the temperature does), above what temperature would the hydrogenation of ethylene become *unfavorable*?

Solution   Although $\Delta S°$ is assumed to be independent of temperature, $T\Delta S°$ certainly *is* dependent on temperature. Because $\Delta S°$ is *negative* (unfavorable with regard to entropy), the $-T\Delta S°$ term will make $G°$ increasingly *positive* as the temperature increases. The temperature at which $T\Delta S°$ equals $\Delta H°$ can be found by setting $\Delta G° = 0$ in Eq. (9-4), and solving for T:

$$0 = \Delta G°$$
$$= \Delta H° - T\Delta S°$$
$$T = \frac{\Delta H°}{\Delta S°}$$
$$= \frac{-137 \text{ kJ mol}^{-1}}{-0.122 \text{ kJ K}^{-1} \text{ mol}^{-1}}$$
$$= 1120 \text{ K}$$

Thus, at temperatures above 1120 K (847°C) the entropy term will outweigh the enthalpy term, making $\Delta G°$ *positive*, and the reaction will become unfavorable.

The result described in Example 9-7 can be generalized as follows:

- The effect of the entropy term, favorable or unfavorable, increases with temperature, and

- When the number of molecules of product(s) is *less* than the number of molecules of reactant(s), $\Delta S°$ is usually *negative* and the reaction becomes increasingly *unfavorable* as the temperature increases. Conversely, when there are more molecules of product than reactant, $\Delta S°$ is usually positive and the reaction becomes *more* favorable as the temperature increases.

You can use an alternative way to calculate $\Delta G°$ for a reaction if you can find tabulated values for the **free energies of formation** ($\Delta G_f°$) of each reactant and product. As was true for standard heats of formation and standard entropies, such data are available in the CRC handbook. The value of $\Delta G°$ is found using Eq. (9-5), which is identical to Eq. (9-2) except that G's are substituted for H's.

$\Delta G°$ FROM FREE ENERGIES OF FORMATION
$$\Delta G° = \sum \Delta G_{f(\text{products})}° - \sum \Delta G_{f(\text{reactants})}° \tag{9-5}$$

EXAMPLE 9-8   (a) Given the data below, calculate the values for $\Delta H°$, $\Delta G°$, and $\Delta S°$ for the isomerization of methylcyclopentane to cyclohexane.

$\Delta H_f°$	-107	-123
$\Delta G_f°$	35.8	31.8

(These values are expressed in units of kJ mol^{-1}, at 25°C.)

(b) Is this a favorable reaction from the standpoint of enthalpy? entropy? free energy? Will this reaction proceed spontaneously?

Solution

(a) Use Eqs. (9-2) and (9-5) to calculate ΔH° and ΔG°, respectively.

$$\Delta H^\circ = \sum \Delta H^\circ_{f(products)} - \sum \Delta H^\circ_{f(reactants)}$$
$$= -123 - (-107)$$
$$= -16 \text{ kJ mol}^{-1}$$

$$\Delta G^\circ = \sum \Delta G^\circ_{f(products)} - \sum \Delta G^\circ_{f(reactants)}$$
$$= 31.8 - 35.8$$
$$= -4.0 \text{ kJ mol}^{-1}$$

We calculate ΔS° by solving Eq. (9-4):

$$\Delta G^\circ = \Delta H^\circ - T\Delta S^\circ$$
$$\Delta S^\circ = \frac{\Delta H^\circ - \Delta G^\circ}{298 \text{ K}}$$
$$= \frac{-16 - (-4.0)}{298 \text{ K}}$$
$$= -0.040 \text{ kJ K}^{-1} \text{ mol}^{-1}$$
$$= -40 \text{ J K}^{-1} \text{ mol}^{-1}$$

(b) This reaction is *favorable* from the standpoints of ΔH° and ΔG° (both negative), but *unfavorable* from the standpoint of ΔS° (negative, more ordered). But remember that ΔG° is the ultimate measure of favorability, so the reaction will be spontaneous. Notice also that, once again, the favorable enthalpy term was dominant over an unfavorable entropy term. As to why ΔS° is negative for this reaction, we can only infer that cyclohexane (the product) is somewhat more ordered than methylcyclopentane (the reactant). The detailed reasons for this are beyond the scope of this book.

Before leaving this section we must emphasize once again that the values of ΔH°, ΔS°, and ΔG° (the so-called thermodynamic parameters of a reaction) give us information only about whether or not a reaction is favorable and capable of proceeding spontaneously. Thermodynamic considerations tell us nothing about how *fast* the reaction will proceed, just *if* it will proceed spontaneously.

9-4. The Relationship between ΔG° and the Equilibrium Constant

In the laboratory, we usually describe as useful a reaction in which virtually all of the reactant(s) are spontaneously converted to product(s). However, even for a very favorable reaction we still find a small (sometimes *exceedingly* small) amount of reactant(s) left over when the reaction has gone "to completion." We say that once the amounts of reactant(s) and product(s), as well as other observable variables such as temperature and pressure, are no longer changing with time, the system has come to **equilibrium**. Even though you might think that, at equilibrium, everything has stopped, that is definitely not the case.

Equilibrium is dynamic; *both* the forward *and* reverse reactions are still occurring, but they balance each other in such a way that the amount of product(s) and reactant(s) remains constant with time.

It might come as a surprise to you that even an *unfavorable* reaction (one with a ΔG° value that is positive) will occur spontaneously to some small extent in order to establish equilibrium. The important question is: Just how far does a reaction, favorable or unfavorable, have to go to reach equilibrium? We answer this question with a number called the **equilibrium constant**, *K* (not to be confused with the unit of absolute temperature), which is a ratio of product amounts to reactant amounts. Specifically, for the generalized reaction

$$aA + bB + \cdots \rightarrow xX + yY + \cdots$$

the equilibrium constant is given by

EQUILIBRIUM CONSTANT $K = \dfrac{[X]^x [Y]^y \cdots}{[A]^a [B]^b \cdots}$ (9-6)

where [Y] indicates the equilibrium concentration (or, more accurately, activity) of Y, and the lower-case letters are the **stoichiometric coefficients** in the original balanced equation for the reaction. The "concentration" (activity) of a compound is a dimensionless ratio of its concentration (if it is a dissolved material) to 1.00 M (moles per liter), or the ratio of its pressure (if it is a gas) to 1.00 atm. Thus we normally use the molar concentration of Y *without its units* for [Y]. The expression on the right side of Eq. (9-6) is sometimes called the **mass action expression** for the reaction.

Clearly, a favorable reaction is one in which, at equilibrium, there's a large excess of product(s) with very little of the reactant(s) remaining. In such a case [X] and [Y] in Eq. (9-6) will be much larger than [A] and [B], so K will be a large number ($\gg 1$, which reads "much greater than one"). Conversely, for an unfavorable reaction there will be more reactant(s) than product(s) at equilibrium, so K will be very small ($\ll 1$, "much less than one"). Of course, K can never be a negative number, since concentrations can never be negative.

EXAMPLE 9-9   (a) Write the mass action expression for the hydrogenation of ethylene (Example 9-3). (b) At equilibrium the mass action expression will be equal to what number?

Solution

(a) Referring to the balanced equation in Example 9-3, we note that all stoichiometric coefficients are one. Therefore, the mass action expression is

$$\frac{[H_3CCH_3]}{[H_2CCH_2][H_2]}$$

(b) At equilibrium the mass action expression will equal the equilibrium constant K.

Now, if a favorable reaction is one with a large value of K, and also one with a negative value of $\Delta G°$, is there a relationship between the two quantities? You bet there is!

$$K = e^{-\Delta G°/RT}$$ (9-7)

K/ΔG° RELATIONSHIP   or

$$\Delta G° = -RT \ln K = -2.3 RT \log K$$ (9-7′)

where $R = 8.31 \times 10^{-3}$ kJ K^{-1} mol^{-1}, T is in degrees Kelvin, ln represents the natural (base e) logarithm, and log is the common (base 10) logarithm. Careful inspection of the dimensions of R, T, and $\Delta G°$ should convince you that K must be a dimensionless quantity in order for $\Delta G°$ to have the correct dimensions. Further, note how a *negative* $\Delta G°$ correlates with a K value greater than one, while a *positive* $\Delta G°$ yields a K value less than one. (Remember that log 1 = ln 1 = 0.) By the way, if this equation looks familiar, it's because you've already seen it (in disguise) back in Eq. (7-2). Only now, we know that the "ratio" is called the equilibrium constant, and it is the *free* energy that we must use in the calculation.

EXAMPLE 9-10   (a) Calculate the value of K (at 298 K) for the isomerization of methylcyclopentane to cyclohexane (Example 9-8). (b) How much methylcyclopentane would remain at equilibrium if its initial concentration were 1.0 M?

Solution

(a) Use Eq. (9-7) and the $\Delta G°$ value calculated in Example 9-8 (−4.0 kJ mol^{-1}):

$$K = e^{-\Delta G°/RT}$$

$$= e^{-(-4.0)/[(8.31\times10^{-3})(298)]}$$

$$= e^{1.62}$$

$$= 5.0$$

or, using Eq. (9-7′),

$$\Delta G° = -2.3RT \log K$$

$$\log K = -\frac{\Delta G°}{2.3RT}$$

$$= -\frac{-4.0 \text{ kJ mol}^{-1}}{(2.3)(8.31\times10^{-3} \text{ kJ K}^{-1} \text{ mol}^{-1})(298 \text{ K})}$$

$$= 0.702$$

$$K = 5.0$$

(b) The mass action expression for this reaction looks like this:

$$\frac{[\text{cyclohexane}]}{[\text{methylcyclopentane}]} = 5.0 \quad \text{(at equilibrium)}$$

Suppose that x moles per liter of methylcyclopentane had been converted to cyclohexane, leaving $(1.0 - x)$ moles per liter unreacted. We can substitute these literal values in the above equation and solve for x.

$$\frac{x}{(1.0 - x)} = 5.0$$

$$x = 5.0(1.0 - x) = 5.0 - 5.0x$$

$$6.0x = 5.0$$

$$x = 0.83 \text{ M}$$

$$1.0 - x = 0.17 \text{ M}$$

Thus the K value of 5.0 tells us that only 17% of the methylcyclopropane remains at equilibrium.

9-5. How Reactions Occur: The Mechanism

A. The reaction rate

By now you might be thinking that all we need to know about a reaction is the value (sign and magnitude) of $\Delta G°$. With that alone we can predict whether or not the reaction will proceed spontaneously, as well as which side of the equation (reactants or products) will be favored at equilibrium. Unfortunately, just because a reaction is thermodynamically favorable doesn't necessarily mean that it will proceed at a useful rate. For example, although both reactions are favorable, the isomerization of enol (**1**) to acetone (**2**) (Section 9-1) is fast under most conditions, while the hydrogenation of ethylene requires a catalyst to proceed at even a moderate rate. And the isomerization of methylcyclopentane to cyclohexane, though favorable thermodynamically, is virtually impossible to carry out in the laboratory. These facts are true because the rate at which a reaction approaches completion (equilibrium) is *not* determined by the thermodynamic factors discussed so far. To begin to understand why this is so, we must first learn something about the **activation barrier** of a reaction.

B. Free energy of activation

For even a favorable reaction there is an energetic barrier separating reactants from products, as shown in Figure 9-4.

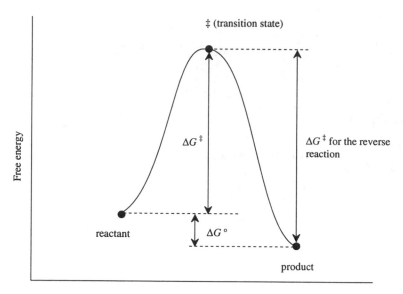

FIGURE 9-4. A typical reaction profile diagram.

The height of this barrier is a measure of the **free energy of activation** ($\Delta G^{\ddagger}$), if the ordinate scale is free energy. In some contexts this barrier is called the **(Arrhenius) activation energy** (E_a). Although E_a and $\Delta G^{\ddagger}$ are not exactly the same, they are basically similar and the differences between them can be neglected for the purposes of this discussion. Regardless of which way you measure the activation barrier,

● The higher the activation barrier, the slower the reaction.

C. Path of a reaction

The curved line in Figure 9-4 connecting reactants with products depicts the energetic path that a reactant molecule must follow to be converted to a product molecule. The highest energy point along this path (which defines the height of the activation barrier) is called the **transition state** (or **activated complex**), denoted by the symbol ‡. Notice that $\Delta G^{\ddagger}$ for the forward reaction differs from $\Delta G^{\ddagger}$ of the reverse reaction by exactly $\Delta G°$.

The abscissa (horizontal axis) in Figure 9-4 is labeled **reaction coordinate**, which represents some (usually unspecified) measure of the progress of one reactant molecule's conversion to product. We might choose the reaction coordinate to be a critical bond length, a bond angle, the hybridization at a certain atom, or any other molecular parameter that changes uniformly as a reactant molecule is converted to a product molecule.

The graph in Figure 9-4 is known as a **reaction** (or **energy**) **profile diagram**. It is a representation of the **mechanism** of a reaction, a complete description of the exact molecular path by which the reactant is converted to product. For most reactions there are two or more reaction coordinates, so the "path" leading from reactant to product is actually a curved surface in two or more dimensions (i.e., saddle-shaped). We'll assume that our reaction profile diagrams, with just a one-dimensional reaction coordinate, represent the lowest energy path from reactant to product, that is, the lowest path over the saddle from one stirrup to the other. At any rate, our goal when studying the mechanism of a given reaction is to know as much as possible about its energy profile diagram. Notice that this reaction profile not only represents the lowest energy path for the *forward* reaction (reactant to product), but also the lowest energy path for the *reverse* reaction (product to reactant). Thus if you know the mechanism of the forward reaction, you also know the mechanism of the reverse reaction. This is known as the **principle of microscopic reversibility**.

Here's something that may be bothering you: If all the reactant molecules start out at a certain free energy, [i.e., $\Delta G°_{f(reactants)}$], how do any of them get over the activation barrier? Ah, good question! It turns out that whenever we have a collection of molecules at a fixed temperature, there is a **distribution** (range) of energies among the molecules, even though the *average* energy is ΔG_f°. This distribution is shown in graphical form in Figure 9-5 and can be thought of as resembling a distribution of exam scores vs. number of students.

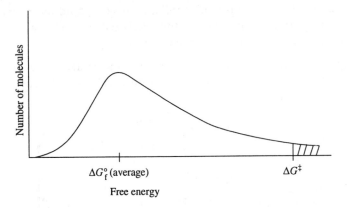

FIGURE 9-5. The distribution of free energy among a
collection of molecules at a given temperature.

Again, although the average energy is ΔG_f°, most molecules have either more or less free energy. And some small fraction (designated by the cross-hatched area) has enough energy ($\Delta G^\ddagger$) to surmount the activation barrier. The reason that the rate of a chemical reaction increases with temperature can also be understood in this context. As the temperature goes up, the distribution broadens to give not only a higher average energy but also a greater fraction with the required free energy of activation (see Figure 9-6).

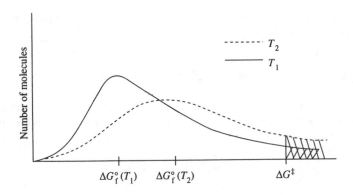

FIGURE 9-6. The effect of increasing temperature
on the free-energy distribution ($T_2 > T_1$).

Furthermore, in an exothermic reaction, the heat evolved can be used to raise the temperature, helping the remaining reactant molecules over the activation barrier.

D. The catalyst

Before we look in more detail at factors that control reaction rates, there is one more thing we might point out. Suppose we have a reaction such as the hydrogenation of ethylene, which is infinitesimally slow under most conditions, even at elevated temperature. Is there any way to accelerate its rate to take advantage of its thermodynamic favorability? Often, the answer is *yes*, by changing the mechanism (and hence, energy profile) to one with a lower activation barrier. This is most often done by finding a **catalyst**, a chemical substance capable of providing a different mechanistic path for the reaction, one with a lower barrier, without itself being permanently changed by the reaction or becoming part of the product. The catalyst is the nearest we come in chemistry to getting something (a faster rate) for nothing (the catalyst is not consumed, at least in principle).

EXAMPLE 9-11 (a) Assume that $\Delta G^{\ddagger}$ for the uncatalyzed hydrogenation of ethylene is 400 kJ mol^{-1}. Draw an accurately scaled free-energy profile diagram for the reaction. (b) When a platinum catalyst is added to the above reaction mixture, a relatively rapid hydrogenation takes place, with a $\Delta G^{\ddagger}$ of 100 kJ mol^{-1}. Draw the profile for the catalyzed reaction on the same graph. Be sure to label all axes and other relevant points on each graph.

Solution (a) and (b) As can be seen from Figure 9-7, the catalyzed reaction (dashed line) follows a different mechanism, involving a lower activation barrier—and hence a faster reaction.

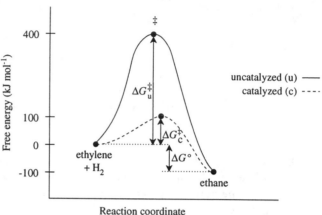

FIGURE 9-7. Reaction profile diagrams for an uncatalyzed reaction and the corresponding catalyzed reaction.

Notice from Example 9-11 that

- A catalyst does *not* affect the magnitude of ΔG° (or the value of K); it affects only the rate of approach to equilibrium.

9-6. Multistep Reactions and Intermediates

The reaction profile diagrams we discussed in the previous section all had one thing in common: In each there was only one transition state. Suppose you encountered a reaction profile like the one in Figure 9-8.

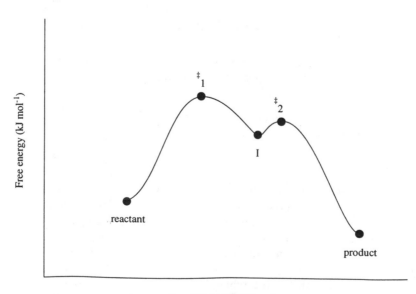

FIGURE 9-8. Reaction profile diagram for a two-step reaction.

In this case the conversion of reactants to products proceeds over *two* activation barriers, through *two* transition states (‡). We call this a two-step reaction (two-step mechanism).

● The number of steps in a reaction mechanism equals the number of transition states in the reaction profile.

The "dip" (minimum), labeled I, in the reaction profile represents the free energy of an **intermediate** in the reaction, a substance that is formed in one step, then consumed in a subsequent step. Some intermediates are compounds that are relatively stable, having relatively low ΔG_f° values and surrounded by high activation barriers. Such intermediates can usually be isolated and characterized if the reaction is interrupted before completion. Other intermediates have high free energies, low activation barriers, and are very unstable; examples include the carbocations, carbanions, and radicals (see Section 4-3). Unstable intermediates can rarely be isolated because their lifetimes are too short. Nonetheless, their occurrence on the reaction profile and their composition and structure can often be confirmed spectroscopically (Section 7-2), or by inference from other reaction characteristics (discussed later). Just remember this important distinction:

● An intermediate has a finite lifetime, can be isolated (in principle), and is always represented by a local *minimum* (dip) in the reaction profile, while a transition state has no finite lifetime, cannot ever be isolated, and is represented by a local **maximum** (peak) in the reaction profile.

EXAMPLE 9-12 **(a)** For the reaction profile diagram in Figure 9-9, label the axes, reactant(s), product(s), transition states, and intermediates. **(b)** How many steps are there in this mechanism? **(c)** Which intermediate(s) are likely to be unstable, and which are relatively stable?

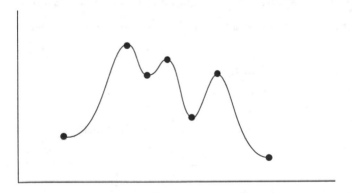

FIGURE 9-9. Incomplete reaction profile diagram for Example 9-12.

Solution
(a) See Figure 9-10.

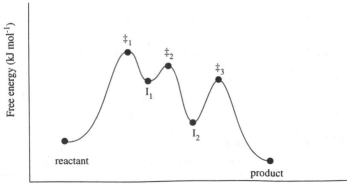

FIGURE 9-10. Completed reaction profile diagram for Example 9-12.

(b) Three, the same as the number of transition states.
(c) The first intermediate (I_1) has a relatively high free energy and is surrounded by low activation barriers; it is therefore unstable. I_2, by contrast, has a lower free energy and higher barriers, so it is relatively stable.

As a preview of Section 9-7, where we discuss the factors that control reaction rates, here is another thing to remember about multistep mechanisms:

- The step with the highest transition state free energy will be the *slowest* step. Because no reaction can go faster than its slowest step, this step is often called the **rate-limiting step** of the reaction.

EXAMPLE 9-13 What is the rate-limiting step of the mechanism pictured in Figure 9-10?

Solution The *first* step is rate-limiting since it has the transition state (labeled $\ddagger_1$) with the highest free energy.

The chemical reaction profiled in Figure 9-10 can be written in several ways, depending on which features we wish to emphasize. The equation

$$\text{Reactant(s)} \longrightarrow \text{product(s)}$$

when balanced, is called the **net reaction**. It tells us what reactant(s) is (are) converted to what product(s), as well as the stoichiometry (how many moles of each is consumed or formed). But,

- The net reaction tells us nothing about the mechanism of the reaction.

Alternatively, we might write the reaction *steps* in order to provide mechanistic information, like this:

$$\text{Reactant(s)} \xrightarrow{\text{slow}} [I_1]$$
$$[I_1] \longrightarrow I_2$$
$$I_2 \longrightarrow \text{product(s)}$$

or

$$\text{Reactant(s)} \xrightarrow{\text{slow}} [I_1] \longrightarrow I_2 \longrightarrow \text{product(s)}$$

Such an equation not only gives the information contained in the net reaction equation, but it also gives us qualitative information about the number of steps, which one is rate-limiting (the one labeled "slow"), and which intermediates are unstable (those shown in brackets). Of course, a complete reaction profile diagram gives us all of this information plus quantitative data that allow us to calculate equilibrium constants and, as we shall see in the next section, even the rates of reactions.

9-7. Chemical Kinetics: Parameters that Describe Reaction Rates

The branch of chemistry that deals with reaction rates is called **chemical kinetics**. The rate of a chemical reaction is the speed at which the products are formed or, alternatively, the speed at which the reactants are consumed. The dimensions (units) of rate are typically $M\ s^{-1}$ (moles per liter per second), provided that all species are in the same phase (e.g., in solution). When describing the *formation* of a product, the rate is taken to be positive, while for the *disappearance* of a reactant the rate is negative.

The rate of a reaction depends on the mechanism of the reaction, the temperature at which the reaction is conducted, and the concentrations of reactants. Although a detailed foray into chemical kinetics is beyond the scope of this book, a few important facts are summarized below.

A. The rate law

The concentration dependence of a reaction rate is usually expressed as a **rate law** (or **rate equation**) of the form

RATE LAW $$Rate = k[A]^n[B]^m \cdots$$ (9-8)

where the quantities in brackets are the various reactant concentrations (in moles per liter), exponents n and m are the **order** of the reaction in reactant A and B, respectively, and k is the **rate constant** (or **specific rate**) of the reaction at a given temperature. The **overall order** of the reaction is the sum of the exponents ($n + m$). Note that, since rate has units of M s^{-1}, the units of k depend on the overall order of the reaction. A first-order reaction will have a k value with units of s^{-1}, a second-order one will have units M^{-1}s^{-1}, etc.

Rate laws range in complexity from simple to complex, *depending on the mechanism* of the reaction. Thus some processes that have simple net reactions turn out to have complicated-looking rate laws, while some very complex mechanisms have relatively simple rate laws.

- The form of the rate law depends on the *mechanism* of the reaction, not the net reaction.

B. Relation between rate law and mechanism

In the case of reactions with relatively simple mechanisms, only those reactants required to reach the rate-limiting transition state appear in the rate law. Reactants involved *after* the rate-limiting step (provided they are not limiting reagents) do not appear in the rate law for the simple reason that the steps in which they're involved are so fast that they do not affect the overall rate. The order of each reactant tells us something about how many molecules of that reactant are involved in reaching the rate-limiting transition state. For example, suppose that the reaction A + B → C goes by a simple one-step mechanism, where one molecule of A collides with one molecule of B to produce one molecule of C. This step, involving the collision of two molecules, is said to be **bimolecular**. In this case the rate law would take the form

$$Rate = k[A][B]$$ (9-9)

since one molecule of A and one molecule of B are involved in the rate-limiting transition state (the *only* transition state in this case!). Notice that the order (exponent) for both A and B is 1. The overall order is 1 + 1 = 2.

Suppose, instead, that the mechanism of the reaction was

$$A \xrightarrow{\text{slow}} [I]$$

$$[I] + B \longrightarrow C$$

which can also be written

$$A \xrightarrow{\text{slow}} [I] \xrightarrow{B} C$$

In this case, A is converted to unstable intermediate I in a *unimolecular* slow step, while B becomes involved only *after* the rate-limiting step. Here, the rate law would be

$$Rate = k[A]$$ (9-10)

which is first order in A, zero order in B, and first order overall.

Finally, consider the mechanism

$$2A \xrightarrow{\text{slow}} [I]$$

$$[I] + B \xrightarrow{\text{fast}} C + A$$

The *net* reaction is still A + B → C, but now the rate law would be

$$Rate = k[A]^2$$ (9-11)

which is second order in A and second order overall, because two molecules of A are involved in a bimolecular process in the first step.

The point of these examples is to show that the rate law depends on the mechanism and the rate law must, therefore, be *experimentally determined*. Its form cannot be predicted from the net reaction alone. By the same token, once the rate law is known, it gives valuable information about the mechanism, though it does not *prove* the mechanism. Of this you can be sure: Whenever the rate law does *not* include one or more of the reactants, the reaction must involve a multistep mechanism, with those reactants involved *after* the rate-limiting step. Moreover, just because the rate law *does* happen to match the stoichiometry of the net reaction, it doesn't prove the reaction involves a one-step mechanism.

EXAMPLE 9-14   Consider the net reaction

$$RCl + OH^- \longrightarrow ROH + Cl^-$$

where R represents an alkyl group. We'll assume for the purposes of this discussion that $\Delta G°$ for this process is sufficiently negative that the reaction is quite favorable. When R is CH_3 the rate law for the reaction is

$$Rate = k[CH_3Cl][OH^-]$$

indicating that the rate depends on the concentrations of both reactants. Varying the concentration of either the alkyl halide or the hydroxide ion will correspondingly affect the rate. However, when R is $C(CH_3)_3$ the rate law is

$$Rate = k[(CH_3)_3CCl]$$

indicating that the rate depends only on the concentration of the alkyl halide and is independent of the concentration of the hydroxide ion. Varying the concentration of the halide will affect the rate; varying that of the hydroxide ion will not. What do these rate laws indicate about the mechanisms of these reactions?

Solution   The fact that the two different reactions follow different rate laws (second order vs. first order overall) indicates that two different mechanisms are involved. In the first case the rate law is consistent with a single-step bimolecular mechanism:

$$CH_3 - \ddot{C}l: \; + \; :\ddot{O}H^- \longrightarrow CH_3 - \ddot{O}H \; + \; :\ddot{C}l:^-$$

Such a mechanism should exhibit a rate law that is first order in each reactant, since both are involved in the rate-limiting step (the *only* step, in this case). When R is $C(CH_3)_3$, the OH^- does not appear in the rate law, indicating that it must be involved *after* the rate-limiting step. A plausible mechanism would be

$$(CH_3)_3CCl \xrightarrow{\text{slow}} [I]$$

$$[I] \; + \; OH^- \longrightarrow (CH_3)_3COH \; + \; Cl^-$$

Of course, the rate law itself doesn't *prove* that this is the correct mechanism, because other mechanisms would also be consistent with the first-order rate law. Moreover, the rate law doesn't tell us anything about the nature of the intermediate, other than the fact that it is unstable.

Still, we *do* know that the tertiary chloride reacts by a different mechanism than does the primary chloride, a mechanism where the OH^- ion is not involved until after the rate-limiting step.

EXAMPLE 9-15   The reaction

$$2 NO + O_2 \longrightarrow 2 NO_2$$

proceeds according to the rate law

$$Rate = k[NO]^2[O_2]$$

(a) What is the order in each reactant, and the overall order of the reaction?
(b) Is this rate law consistent with a one-step mechanism?

Solution (a) Second order in NO, first order in O_2, third order overall. (b) Yes, because the individual orders match the stoichiometric coefficients of the net reaction. But remember, this alone does not *prove* that this reaction proceeds by a one-step mechanism—a three-body collision is a highly improbable event. If the mechanism *is* multistep, we do know that all three molecules must be involved in reaching the rate-limiting transition state.

EXAMPLE 9-16 A certain reaction follows the rate law

$$Rate = k[A][B]^2$$

When the reactant concentrations are [A] = 0.10 M and [B] = 0.40 M, the rate of the reaction is found to be 3.4×10^{-5} M s^{-1}. What is the value of k?

Solution Solve the rate law equation for k, then substitute the values of rate and concentration.

$$Rate = k[A][B]^2$$

$$k = \frac{rate}{[A][B]^2}$$

$$= \frac{3.4 \times 10^{-5} \text{M s}^{-1}}{(0.10 \text{ M})(0.40 \text{ M})^2}$$

$$= 2.1 \times 10^{-3} \text{M}^{-2} \text{ s}^{-1}$$

C. Effect of temperature

In Section 9-5 we noted that increasing the temperature of a reaction increases its rate by providing a greater fraction of reactant molecules with the required activation energy (see Figure 9-6). Examination of the general rate law [Eq. (9-8)] suggests that it is k that must increase with temperature, since concentrations are not appreciably affected by changes in temperature. The temperature dependence of k is given by the **Arrhenius equation**

ARRHENIUS EQUATION
$$\ln\left(\frac{k_2}{k_1}\right) = \frac{E_a}{R}\left(\frac{1}{T_1} - \frac{1}{T_2}\right) = \frac{E_a}{R}\left(\frac{T_2 - T_1}{T_2 T_1}\right) \qquad (9\text{-}12)$$

where the subscripts 1 and 2 refer to absolute temperatures T_1 and T_2, and E_a is the Arrhenius activation energy (Section 9-5).

EXAMPLE 9-17 A certain reaction has a rate constant of 3.5×10^{-3} s^{-1} at 25°C, and a rate constant of 7.3×10^{-2} s^{-1} at 50°C. (a) What is the order of the rate law? (b) Calculate the Arrhenius activation energy of the reaction.

Solution (a) From the dimension of k (s^{-1}), we know the rate law is first order overall. (b) Solve Eq. 9-12 for E_a, and be sure to convert to absolute temperatures (25°C = 298 K, 50°C = 323 K):

$$\ln\left(\frac{k_2}{k_1}\right) = \frac{E_a}{R}\left(\frac{1}{T_1} - \frac{1}{T_2}\right) = \frac{E_a}{R}\left(\frac{T_2 - T_1}{T_2 T_1}\right)$$

$$E_a = R\left(\frac{T_1 T_2}{T_2 - T_1}\right)\ln\left(\frac{k_2}{k_1}\right)$$

$$= (8.31 \text{ J K}^{-1} \text{ mol}^{-1})\left(\frac{(298 \text{ K})(323 \text{ K})}{323 \text{ K} - 298 \text{ K}}\right)\ln\left(\frac{7.3 \times 10^{-2} \text{s}^{-1}}{3.5 \times 10^{-3} \text{s}^{-1}}\right)$$

$$= 9.7 \times 10^4 \text{ J mol}^{-1} = 97 \text{ kJ mol}^{-1}$$

9-8. Transition-State Structure and Hammond's Postulate

So far in this chapter, we've found that the thermodynamic favorability of a reaction is determined by the sign and magnitude of $\Delta G°$, while the *rate* of approach to equilibrium is determined (among other things) by the height of the activation barrier ($\Delta G^{\ddagger}$ or E_a). Yet, there is no obvious connection between $\Delta G°$ and the activation barrier.

A. Nature of the transition state

Because transition states, unlike intermediates, have no finite lifetime (Section 9-6), it is impossible to isolate and study them the way we study the structures of reactants, products, and even intermediates. Nonetheless, a hypothesis has been advanced that allows us to make certain inferences about transition state structures. **Hammond's postulate** says that, for *any* step in a chemical reaction,

- The transition state will occur closer (in reaction coordinate, structure, and energy) to the reactant *or* product *of that step,* whichever has the higher free energy.

Let's explore the consequences of this idea. Figure 9-11 shows reaction profiles for two hypothetical one-step reactions, both of which are thermodynamically favorable ($\Delta G° < 0$).

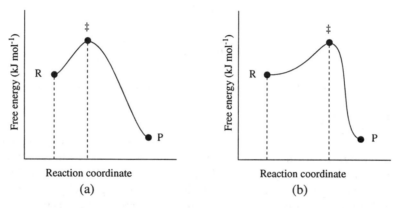

FIGURE 9-11. Reaction profile diagrams for reaction with $\Delta G°$ < 0. (a) Profile consistent with Hammond's postulate. (b) Profile inconsistent with Hammond's postulate. (R, reactant; P, product.)

Hammond's postulate tells us that for a *favorable* reaction, the transition state will occur closer to the *reactant* of that step because the reactant has a higher free energy than the product. This situation, depicted in Figure 9-11a, is described as having an "early" transition state (i.e., closer to the reactant). The postulate says that a favorable reaction will *not* have a "late" transition state, as in Figure 9-11b.

Just the converse is true for a reaction with $\Delta G° > 0$, as shown in Figure 9-12.

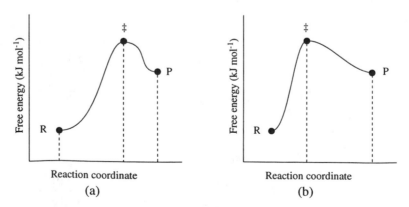

FIGURE 9-12. Reaction profile diagrams for reactions with $\Delta G°$ > 0. (a) Profile consistent with Hammond's postulate. (b) Profile inconsistent with Hammond's postulate. (R, reactant; P, product.)

Here, the transition state will occur closer to the *product* (which now has the higher free energy); this late transition state is shown in Figure 9-12a. Profiles such as in Figure 9-12b do not occur.

B. Transition state and $\Delta G°$

You may be wondering, "So what? What does it matter if the transition state is early or late?" Here's the answer: The transition state will resemble the structure (reactant or product) to which it is closest. Thus an early transition state (Figure 9-11a) will resemble (in reaction coordinate, structure, geometry, and energy) the *reactant* structure, while a *late* transition state (Figure 9-12a) will resemble the *product* of that step.

Consider the implications of this resemblance. In a reaction where $\Delta G° < 0$ (Figure 9-11a), factors that influence the stability of the reactant will similarly affect the stability of the transition state resembling it. In a reaction with $\Delta G° > 0$ (Figure 9-12a), factors that stabilize (or destabilize) the *product* will have a similar effect on the transition state.

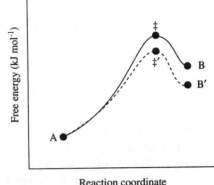

These qualitative predictions allow us to make a connection between the *rate* of a reaction step (as reflecting the height of the activation barrier) and the $\Delta G°$ for that step. For example, consider the reaction profile in Figure 9-13.

FIGURE 9-13. Reaction profile diagram for a reaction with $\Delta G° > 0$; the more stable product B′ forms more rapidly.

The solid line indicates the reaction A → B with $\Delta G° > 0$. The dashed line represents A (the same reactant) going to B′, a similar but different product that is more stable than B. Since the transition state resembles the product in both cases, and since B′ is lower in energy than B, the transition state leading to B′ will be lower in energy than the one leading to B. Therefore, the formation of B′ will be *faster* than the formation of B. Stated another way, for each "uphill" ($\Delta G° > 0$) step in a reaction, the more stable the product *of that step*, the more rapidly it will be formed.

C. Kinetic and thermodynamic control

The above discussion might seem to suggest that the more stable product is always formed more rapidly, but such is not necessarily the case. Remember that the **rate–product correlation** described above applies only to "uphill" *steps* in a reaction. For reaction steps with $\Delta G° < 0$ the transition state resembles the *reactant*, and we can't make such easy predictions. For example, Figure 9-14 shows two such situations. In (a) the *more* stable product (B′) is formed more rapidly, while in (b) the *less* stable product (B) is formed more rapidly.

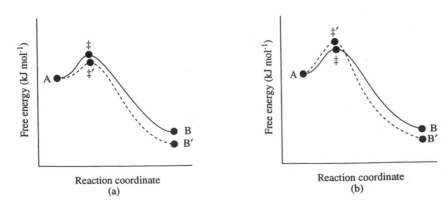

FIGURE 9-14. Reaction profile diagrams for reactions with $\Delta G° < 0$. (a) The more stable product B′ is formed faster. (b) The *less* stable product B is formed faster.

Because the transition states in both diagrams resemble reactant A, not products, the relative energies of B and B′ have little influence on transition state energies and structures. A reaction where the *less* stable product is formed more rapidly is said to be **kinetically controlled**, whereas a reaction in which the more stable product is formed in preference to the less stable one is said to be **thermodynamically controlled**.

EXAMPLE 9-18 For reasons that we will discuss in Chapter 10, a tertiary carbocation (R_3C^+) is more stable than a primary one (RCH_2^+). Assuming that R_3CCl and RCH_2Cl have comparable free energies, which of the two reactions below will occur more rapidly? Use an energy profile diagram to explain your answer.

$$R_3C\!-\!Cl \longrightarrow R_3C^+ + Cl^-$$

$$RCH_2\!-\!Cl \longrightarrow RCH_2^+ + Cl^-$$

Solution Since both of these reactions involve bond breaking that is not compensated by any bond making, both are expected to be "uphill," with $\Delta G° > 0$. Therefore, in each case the transition state will resemble the carbocation product. Because a tertiary carbocation is more stable, the transition state leading to it should be lower in energy than the one leading to the primary carbocation. As a result, the tertiary carbocation will be formed more rapidly. See Figure 9-15.

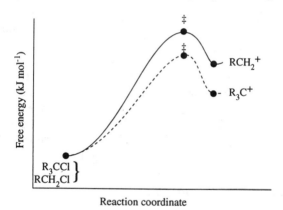

FIGURE 9-15. Reaction profile diagram showing the more stable tertiary carbocation forms faster than the less stable primary carbocation.

EXAMPLE 9-19 The reaction of 1,3-butadiene with HCl produces a mixture of 3-chloro-1-butene (**4**) and 1-chloro-2-butene (**5**, cis and trans isomers).

$$H_2C\!=\!CH\!-\!CH\!=\!CH_2$$

H^+, slow

[I]

Cl^-

$$H_3C\!-\!CH\!-\!CH\!=\!CH_2$$
$$\quad\;\;\;|$$
$$\quad\;\;\;Cl$$

4

$$H_3C\!-\!CH\!=\!CH\!-\!CH_2$$
$$\qquad\qquad\qquad\;\;|$$
$$\qquad\qquad\qquad\;\;Cl$$

5

Although **4** is known to be less stable than **5**, it is formed preferentially in the early stages of the reaction. **(a)** What does this indicate? **(b)** What will happen to the product mixture as a function of time?

Solution (a) We can draw a reaction profile for this system as shown in Figure 9-16.

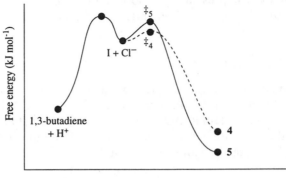

FIGURE 9-16. Reaction profile diagram for the formation of **4** and **5**: the less stable product (**4**) forms faster than the more stable product (**5**).

The preferential formation of the less stable product (**4**) is an indication of kinetic control, that is, it forms *faster* than **5**; the transition state leading to **4** is lower than the one leading to **5**. (b) Given enough time, **4** will equilibrate with the more stable **5** (via intermediate I). At equilibrium the ratio of **4** to **5** will depend only on their relative free energies, according to Eq. (9-7). At this point, the reaction is under thermodynamic control, that is, **5** will be the major product.

9-9. Putting It All Together: The Isomerization of Acetone Enol

In Section 9-1 we established that the isomerization of acetone enol (1-propen-2-ol, **1**) to acetone (**2**) was a favorable reaction from the standpoint of bond energies. Let's now discuss this reaction in more detail in order to see what steps are needed to investigate its mechanism. We'll assume for purposes of this discussion that the values of $\Delta H°$, $\Delta S°$, and $\Delta G°$ are not available from tables.

A. Stoichiometry

The first step in proposing a mechanism for any reaction is to write out the balanced equation for the net reaction. To do this, we must have studied the reaction in sufficient detail to know its complete stoichiometry: How many moles of each reactant are required, and how many moles of each product are produced. This, of course, requires that the *structure* of each product has been determined! It also requires that we do a **mass balance**, accounting for every mole of product initially present.

Most chemical reactions produce not only the desired product(s) but also one or more **side products**. The occurrence of these side products is usually undesirable from the standpoint of **yield** (the ratio of actual amount of product formed to the maximum amount theoretically possible). Nonetheless, the identity of such side products can often be mechanistically significant. For example, perhaps the side product is actually an *intermediate*, which is slowly converted to the desired product. Or perhaps the desired product itself slowly decomposes to the side product. Or, most commonly, perhaps the product and side product are formed by **competing** (parallel but independent) **reactions**.

It should be mentioned that organic chemists are notoriously lax about balancing reaction equations. Instead, they often write the structure of the primary reactant and product, and put all reagents (other reactants) over the reaction arrow. If you use this shorthand notation, at least be sure you know the stoichiometry of the reaction you're abbreviating. Thankfully, our enol isomerization has a very simple net reaction: One mole of enol cleanly gives one mole of acetone, with essentially no side products observed under most conditions.

B. Watch out for impurities

In carrying out the reaction, it is most important to avoid variables and contaminants that might interfere with the reaction. Thus it was discovered that, although enol **1** rapidly isomerizes to acetone under a variety of conditions, the reaction is extremely slow when the reactant and solvent are highly purified, and the reaction vessel extremely clean. It was eventually found that even traces of acid or base present in the solvent or on the reaction vessel were sufficient to accelerate the reaction by orders of magnitude. This fact will be incorporated into our mechanistic picture.

C. Calculate K, $\Delta G°$, $\Delta H°$, and $\Delta S°$

Once we have determined the identity and yield of each product, we calculate the equilibrium constant, if possible, by measuring the concentrations of all relevant species present at equilibrium. From this, we can calculate $\Delta G°$, using Eq. (9-7). Often, the amounts of unconsumed reactants are so small that they are below the detection limit of our instruments. In such a case, we can only estimate limits to the values of K and $\Delta G°$.

EXAMPLE 9-20 At equilibrium (25°C) the ratio of enol to acetone is less than 0.01%. What limits can you place on the values of K and $\Delta G°$?

Solution From Eq. (9-6) we know that the K value is given by

$$K = \frac{[\text{acetone}]}{[\text{enol}]} > \frac{99.99}{0.01} = 9999$$

From Eq. (9-7') we can estimate $\Delta G°$:

$$\Delta G° = -RT \ln K$$
$$< -(8.31 \text{ J K}^{-1} \text{ mol}^{-1})(298 \text{ K}) \ln 9999$$
$$< -22,800 \text{ J mol}^{-1} = -22.8 \text{ kJ mol}^{-1}$$

Thus, we know that K is at least 9999, and that $\Delta G°$ is at most -22.8 kJ mol^{-1}, and probably even more negative.

Now, how do we obtain $\Delta H°$ and $\Delta S°$? Although the detailed equation [it closely resembles Eq. (9-12)] is beyond the scope of this book, it is possible to calculate $\Delta H°$ by measuring the temperature dependence of the value of K. Once we know (or have estimates of) $\Delta G°$ and $\Delta H°$, we can use Eq. (9-4) to calculate $\Delta S°$.

D. Determine the rate law

Next, we'll measure the rate of the reaction and its dependence on concentration (to get the rate law) and temperature (to get the activation energy). We've already noted that the "clean" isomerization is very slow, but that the reaction becomes very fast in the presence of acid or base. Since the acid or base can be added in carefully monitored quantities to the "clean" reaction, the concentration dependence on them can also be determined. It should be noted that the acid or base is not consumed during the reaction, so each is serving as a true catalyst (Section 9-5) and not as reactants.

Let us presume that the rate laws for the three versions of the reaction look like this:

Uncatalyzed (u):	*Rate* $= k_u[\text{enol}]$	**(9-13u)**
Acid-catalyzed (a):	*Rate* $= k_a[\text{enol}][\text{acid}]$	**(9-13a)**
Base-catalyzed (b):	*Rate* $= k_b[\text{enol}][\text{base}]$	**(9-13b)**

In each case the reaction is first order in enol and, if present, first order in catalyst. Not surprisingly, we also find that the magnitude of rate constants k_a and k_b, though dependent on the particular structure of the acid and base, are both much larger than k_u. This indicates that the catalyzed activation barriers are much lower than the uncatalyzed barrier.

E. Proposed mechanisms

Having collected all the appropriate thermodynamic and kinetic parameters, we're ready to suggest some mechanisms. The simplest mechanism consistent with Eq. (9-13u) is a process where all the bond breaking and bond making occurs in the same step. Such a one-step mechanism is said to be **concerted** and can be drawn like this:

Mechanism I

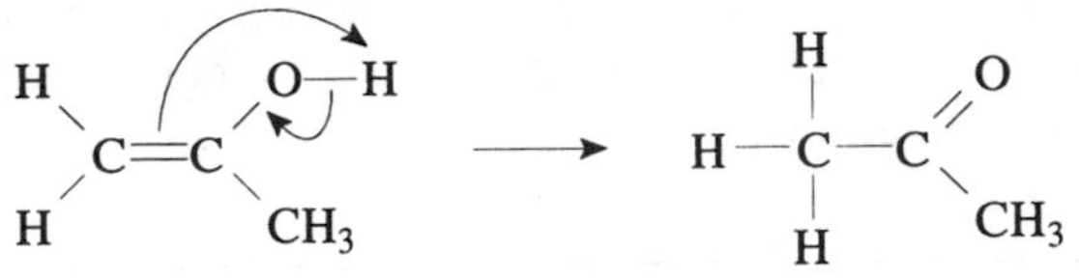

(Remember from Section 4-2 that arrows indicate the movement of *electrons*, not nuclei.) There are several problems with this mechanism, not the least of which is: How does the OH hydrogen know how to get back to the carbon?

Another possible mechanism, also consistent with Eq. (9-13u), is a two-step process:

Mechanism II

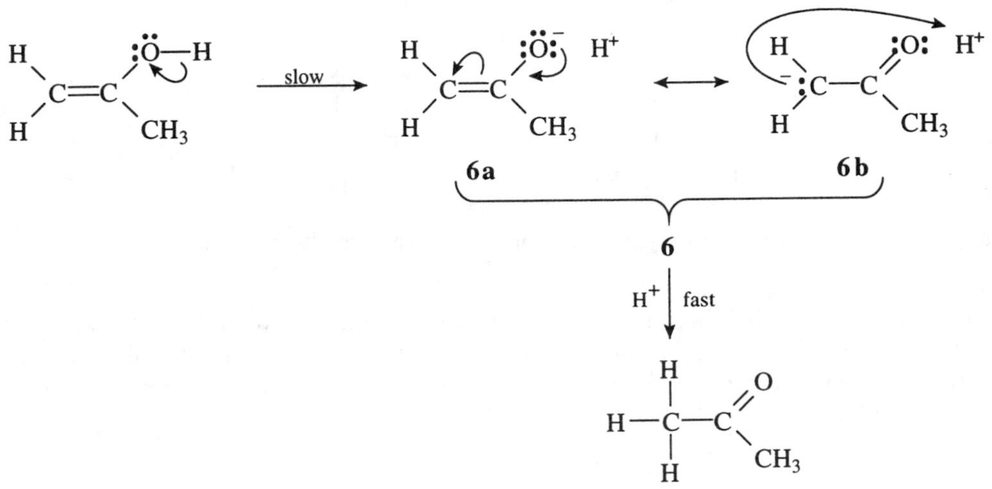

Here, the O–H bond breaks heterolytically (Sections 4-1 and -2), with the more electronegative oxygen receiving the electron pair. The intermediate (**6**) formed in the rate-limiting step is called an **enolate ion**, and it subsequently recaptures the H⁺ ion to form acetone. Note that the enolate ion exists as a resonance hybrid (Sections 4-4 and -5) of two forms (**6a** and **6b**), analogous to the two resonance forms of the allyl anion (Example 4-6).

There is a way we might be able to distinguish between these two mechanisms. The first one requires an **intramolecular** (i.e., within the same molecule) hydrogen transfer. But with mechanism II there is the likelihood that two (or more) nearby enol molecules could be undergoing the reaction at the same moment, and might engage in an **intermolecular** (i.e., *between* two molecules) hydrogen ion transfer:

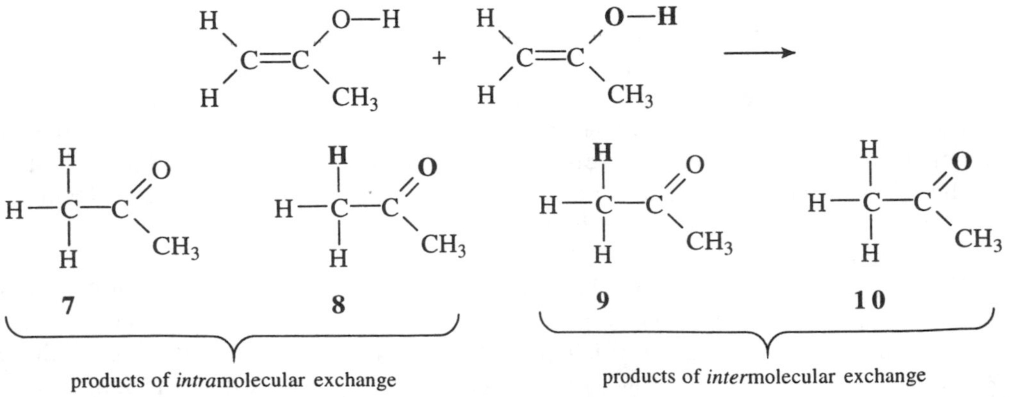

One experimental way to test for this **hydrogen scrambling** is to prepare a sample of the enol using different isotopes (Section 1-1) of hydrogen and oxygen at the OH group. The isotopically labeled atoms are shown above in boldface. Then, a mixture of doubly labeled and unlabeled enol is allowed to undergo isomerization, and the resulting acetone is analyzed (most easily by mass spectrometry, Section 7-2) to determine if all of the labeled atoms remain together (no scrambling, Mechanism I), or whether they are scrambled into otherwise unlabeled molecules (Mechanism II). Of course, for this labeling experiment to be valid, it must be demonstrated that neither the reactant nor the product undergoes isotopic scrambling by any other process than isomerization.

EXAMPLE 9-21 (a) What would be the outcome of a labeling experiment like the one described above, if the isomerization involved the concerted *bimolecular* hydrogen transfer below?

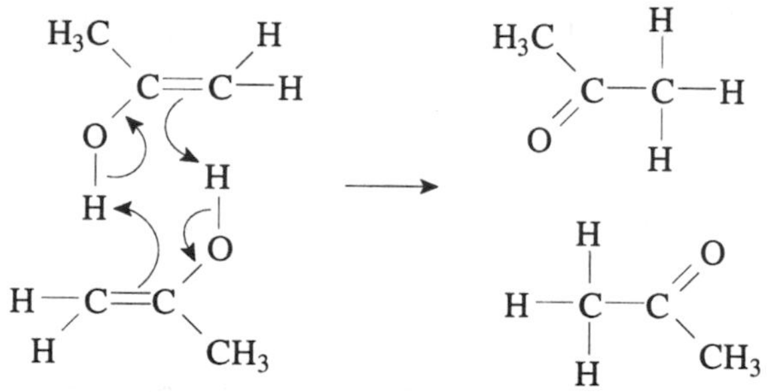

(b) What evidence for or against this mechanism do we already have?

Solution

(a) Such a mechanism involving one labeled enol and one unlabeled enol *would* show scrambling of labeled hydrogen, giving **9** and **10**. (Remember that two *unlabeled* molecules could react together, giving two molecules of **7**, while reaction of two *labeled* enols would give two molecules of **8**.)

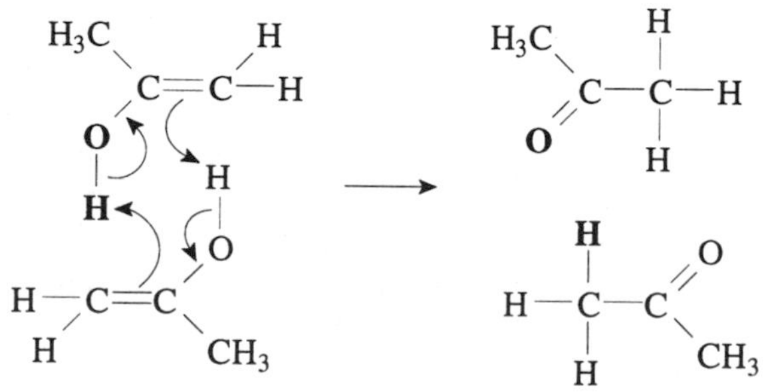

(b) This mechanism would follow a rate law that is *second* order in enol. Such a rate law is not observed.

It's possible in principle that both Mechanisms I and II are operating simultaneously. But, in general, it is highly unlikely that there are two parallel mechanisms that have sufficiently similar kinetic and thermodynamic parameters to give the same product under the same conditions. Nonetheless, the isotope scrambling data would provide a direct indication of whether one or both mechanisms were operative. No scrambling would indicate Mechanism I. If the label is scrambled at the same rate that acetone is produced, only Mechanism II is indicated. On the other hand, if the label is scrambled slower than acetone is produced, two mechanisms are indicated.

We can account for the observed base catalysis by recognizing that the base B: can facilitate the heterolytic cleavage of the O–H bond in Mechanism II. Although we'll postpone a detailed discussion of acids and bases until Chapter 10, it is useful at this point to recall that an *acid* is a *proton donor*, while a *base* is a *proton acceptor*. (Remember that a proton and an H⁺ are exactly the same thing.) The formation of the H–B bond would help compensate for the breakage of the O–H bond, lowering the activation barrier of the rate-limiting step.

Mechanism IIb

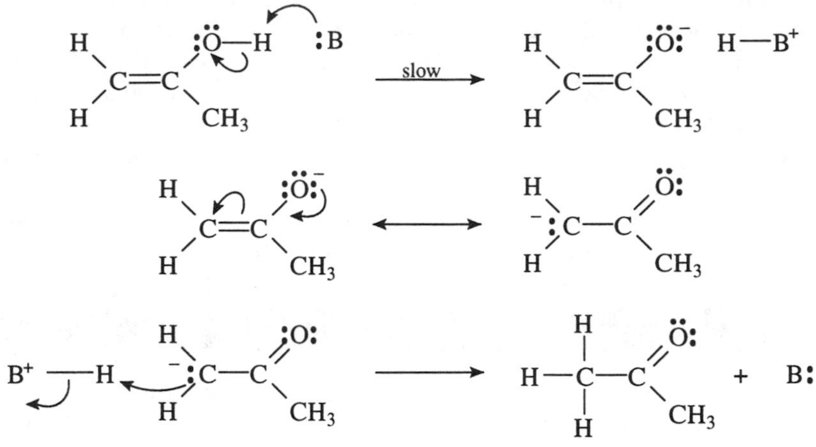

The base is released unchanged in the second step. (This is still a two-step reaction. The middle "step" is not a step at all, but rather a reminder that the enolate ion has two principal resonance forms.) Notice how isotope scrambling would again be predicted to occur in the base-catalyzed mechanism, because H–B$^+$ (unlabeled) could mix with **H–B$^+$** before returning the borrowed proton.

The base-catalyzed Mechanism IIB, like mechanism II, involves O–H bond breaking before C–H bond making. Do you suppose it is possible to have C–H bond making precede O–H bond breaking? The answer is yes, and that is exactly what happens in the acid-catalyzed reaction. We'll write our acid as H–A.

Mechanism III

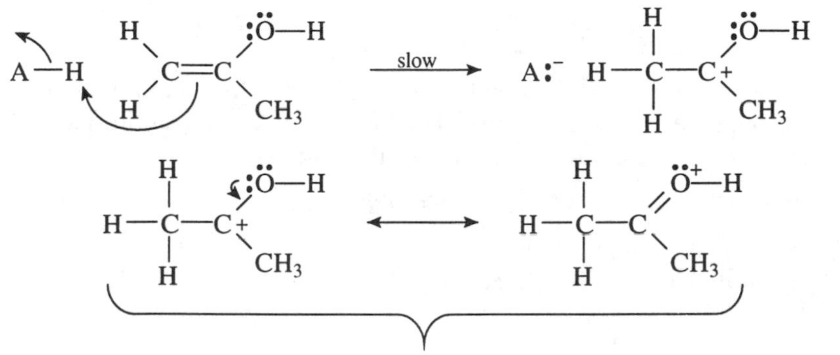

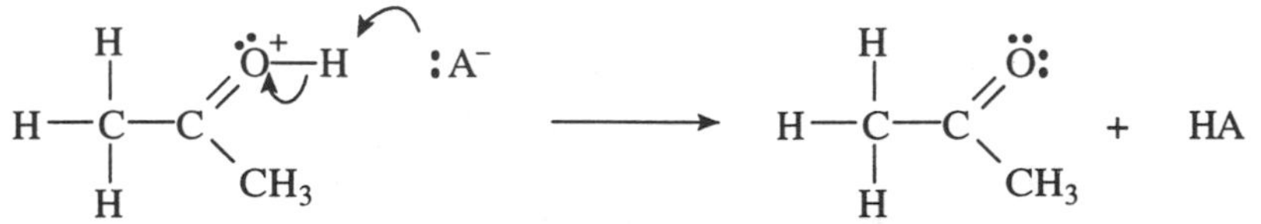

This two-step mechanism involves intermediate **11** (simply called protonated acetone), which has two principal resonance forms. Notice how the only difference between the base-catalyzed mechanism (IIB) and the acid-catalyzed one (III) is the sequence of proton donation and proton retrieval.

It turns out that Mechanisms IIB and III are believed to be the ones operative in the isomerization of enols in the presence of bases or acids. The uncompensated bond breaking in Mechanism II is too costly energetically to be important. That is why the "clean" reaction is so slow. In Chapter 10 we will discuss factors that control concerted reactions such as that in Mechanism I.

Now you have been introduced to the steps required to suggest a mechanism for a reaction: Ascertaining the net reaction, measuring the thermodynamic and kinetic parameters, and determining the structure of any side products or intermediates. In some cases it will be most useful to do isotope scrambling experiments, or to examine the stereochemical consequences of the reaction. Only when all the data are in is it time to suggest a mechanism.

One final word about mechanisms. The goal of science is to explain and predict the behavior of nature. If our proposed mechanism is consistent with all known facts and correctly predicts the outcome of all experiments, it is a valid mechanistic hypothesis. But no mechanism can ever be "proved" in the sense that we are sure it is "true." All we can do is generate a picture in our minds that reflects our interpretation of the data we have collected. If, sometime in the future, a verifiable experimental result occurs that is at odds with our mechanism, we must alter our picture to accommodate the new result. This is the basis of scientific evolution.

SUMMARY

1. The enthalpy (heat content) change (ΔH) for a reaction is related to the total bond strength of the reactant(s) compared to that of the product(s). It can be calculated either from standard heats of formation (ΔH_f°) with Eq. (9-2), or estimated from average bond dissociation energies (ΔH_d) with Eq. (9-1).

$$\Delta H^\circ = \sum \Delta H_{f(products)}^\circ - \sum \Delta H_{f(reactants)}^\circ \quad \textbf{(9-2)}$$

$$\Delta H = \sum \Delta H_{d(reactants)} - \sum \Delta H_{d(products)} \quad \textbf{(9-1)}$$

A *negative* ΔH° corresponds to a process that is favorable from the standpoint of enthalpy (i.e., heat energy is liberated).

2. The entropy change (ΔS°) of a reaction is related to the changes in disorder attending the reaction. A favorable process from the standpoint of entropy is one with a *positive* ΔS°, that is, where disorder *increases*. The value of ΔS° can be calculated from tabulated values of standard entropies (S°), according to Eq. (9-3).

$$\Delta S^\circ = \sum S_{products}^\circ - \sum S_{reactants}^\circ \quad \textbf{(9-3)}$$

3. The ultimate thermodynamic criterion for overall favorability (spontaneity) of a reaction is that the free energy change (ΔG°) is *negative*. The free energy change comprises the enthalpy change and the entropy change according to the relationship

$$\Delta G^\circ = \Delta H^\circ - T\Delta S^\circ \quad \textbf{(9-4)}$$

where T is the absolute temperature.

4. The equilibrium constant K is related to ΔG° by Eq. (9-7) or (9-7').

$$K = e^{-\Delta G^\circ / RT} \quad \textbf{(9-7)}$$

$$\Delta G^\circ = -RT \ln K = -2.3RT \log K \quad \textbf{(9-7')}$$

A negative ΔG° corresponds to a K value greater than one.

5. The mechanism of a chemical reaction is a detailed description of the exact molecular path by which reactants are converted to products. A complete mechanism is consistent with all thermodynamic and kinetic information known about the reaction.

6. The mechanism of a reaction can be graphically represented with a reaction (or energy) profile diagram, which shows the free energy of reactant(s), product(s), intermediate(s), transition state(s), and the path connecting them.

7. The activation barrier of a reaction step is the energetic peak that a reactant molecule must surmount to be converted to product.

8. At a given temperature the free energy of a collection of molecules is distributed in such a way that the average free energy is ΔG_f°, but that most molecules have either more or less energy. A small fraction of molecules possess enough energy to surmount the activation barrier, and this fraction increases with increasing temperature.

9. An intermediate is a substance that is formed in one step of a reaction, then consumed in a subsequent step.

10. The number of steps in a multistep reaction is equal to the number of transition states.

11. The rate law for a chemical reaction is an equation that describes the dependence of the reaction rate on reactant concentrations.
12. The kinetic order of a reactant, its exponent in the rate law, is related to the number of molecules of that reactant involved in reaching the rate-limiting transition state of the reaction.
13. The temperature dependence of the rate constant k is given by the Arrhenius equation

$$\ln\left(\frac{k_2}{k_1}\right) = \left(\frac{E_a}{R}\right)\left(\frac{1}{T_1} - \frac{1}{T_2}\right) \qquad (9\text{-}12)$$

where E_a is the activation energy.
14. Hammond's postulate states that for a reaction step with a *negative* $\Delta G°$, the transition state of that step will resemble the reactant of that step. Conversely, for a reaction step with a *positive* $\Delta G°$, the transition state will resemble the product of that step.
15. A kinetically controlled reaction is one where the preferred product is the one formed most rapidly, *not* necessarily the most stable one. A thermodynamically controlled reaction is one where the most stable product (the one with the lowest free energy) *is* preferred.
16. A catalyst is a substance that accelerates a reaction but is not consumed or permanently changed by the reaction. The action of a catalyst is to provide a change in the mechanism of the reaction and a lowering of the activation barrier.

RAISE YOUR GRADES

Can you define...?

- physical organic chemistry
- enol, enolate ion
- structural and stereo-isomerizations
- bond-dissociation energy
- endothermic, exothermic, thermoneutral
- enthalpy and $\Delta H°$
- standard state
- standard heat of formation ($\Delta H_f°$)
- entropy and $\Delta S°$
- standard entropy values ($S°$)
- Gibbs free energy and $\Delta G°$
- free energy of formation ($\Delta G_f°$)
- equilibrium, equilibrium constant
- stoichiometric coefficients
- mass-action expression
- activation barrier
- free energy of activation, activation energy
- transition state (activated complex)
- reaction coordinate
- reaction (or energy) profile diagram
- principle of microscopic reversibility
- reaction mechanism
- intermediate
- catalyst
- chemical kinetics
- rate law
- (kinetic) order
- rate constant
- rate-limiting step
- unimolecular, bimolecular
- Arrhenius equation
- Hammond's postulate
- kinetic control
- thermodynamic control
- net reaction
- mass balance
- side product
- yield
- competing reaction
- concerted reaction
- isotope scrambling
- intra- vs. intermolecular

Can you explain...?

- how to calculate $\Delta H°$ and how to determine if a reaction is favorable from the standpoint of enthalpy
- how to calculate $\Delta S°$ and how to determine if a reaction is favorable from the standpoint of entropy
- how to calculate $\Delta G°$ and how to determine if a reaction is favorable
- the three laws of thermodynamics
- how to draw and label a reaction profile diagram
- all the steps and information needed to formulate a mechanistic hypothesis
- how energy is distributed among a collection of molecules, and how this distribution is affected by changes in temperature

☑ the relationship between the number of steps in a mechanism and the numbers of transition states and intermediates
☑ the key differences between a transition state and an intermediate
☑ how to interpret the rate law of a reaction, especially the kinetic orders
☑ how to calculate the activation energy from the temperature dependence of the rate constant
☑ How Hammond's postulate allows us to make guesses about the structure of transition states
☑ how a catalyst works

SOLVED PROBLEMS

PROBLEM 9-1 (a) Using the bond dissociation energies in Table 9-1, estimate ΔH for the conversion of dimethyl ether to ethanol in the gas phase.

(b) Is this a favorable reaction from the standpoint of enthalpy?
(c) What type of reaction is this?
(d) What is the minimum number of bonds that must be broken and made to carry out this transformation?

Solution

(a) The total bond strengths for each compound (in kJ mol^{-1}) are:

CH_3OCH_3		CH_3CH_2OH	
6 C(sp^3)–H	6(410)	5 C(sp^3)–H	5(410)
2 C(sp^3)–OC(sp^3)	2(335)	1 C(sp^3)–OH	383
Total	3130	1 C(sp^3)–C(sp^3)	368
		1 H–OC(sp^3)	427
		Total	3228

Using Eq. (9-1) we can estimate ΔH:

$$\Delta H = \sum \Delta H_{d(\text{reactants})} - \sum \Delta H_{d(\text{products})}$$
$$= 3130 - 3228 = -98 \text{ kJ mol}^{-1}$$

(b) Yes, because ΔH is negative.
(c) A structural isomerization.
(d) Bonds broken: One C(sp^3)–OC(sp^3) and one C(sp^3)–H; bonds made: One C(sp^3)–C(sp^3) and one C(sp^3)O–H. This is shown diagrammatically below.

Note that this is *not* a mechanism, but simply an accounting of bonds that need to be broken and made. Using only the strengths of these bonds (C–O and C–H broken, C–C and O–H made) we can make an even cruder estimate of ΔH:

$$\Delta H = (335 + 410) - (368 + 427) = -50 \text{ kJ mol}^{-1}$$

PROBLEM 9-2   Consider the thermodynamic data below (valid at 25°C):

	CH_3OCH_3 (g)	CH_3CH_2OH (g)
ΔH_f° (kJ mol^{-1})	−185	−235
S° (J K^{-1} mol^{-1})	266	282

(a) Calculate ΔH° for the reaction discussed in Problem 9-1. Explain any differences between this value and the ΔH value calculated in Problem 9-1.
(b) Calculate ΔS° for the same reaction. What does this value indicate?
(c) Calculate ΔG° for the same reaction. Is this reaction thermodynamically favorable?
(d) What is the value of K for this reaction? What fraction of the dimethyl ether remains at equilibrium?

Solution
(a) Use Eq. (9-2) to find ΔH°:

$$\Delta H^\circ = \sum \Delta H^\circ_{f(products)} - \sum \Delta H^\circ_{f(reactants)}$$
$$= (-235) - (-185) = -50 \text{ kJ mol}^{-1}$$

The ΔH° value calculated here is more accurate than the ΔH value calculated in Problem 9-1. This is because the latter value comes from average bond energies, while the former value comes from heats of formation for the specific compounds involved. Again, the closer agreement with the crude value of ΔH is fortuitous.
(b) Use Eq. (9-3) to find ΔS°:

$$\Delta S^\circ = \sum S^\circ_{products} - \sum S^\circ_{reactants}$$
$$= 282 - 266 = 16 \text{ K J}^{-1} \text{ mol}^{-1}$$

Because ΔS° is positive, it means that the product is more disordered than the reactant, and therefore this reaction *is* favorable from the standpoint of entropy.
(c) Use Eq. (9-4) to find ΔG°; be sure to adjust the units of ΔS°:

$$\Delta G^\circ = \Delta H^\circ - T\Delta S^\circ$$
$$= -50 \text{ kJ mol}^{-1} - (298 \text{ K})(0.016 \text{ kJ K}^{-1} \text{ mol}^{-1})$$
$$= -50 \text{ kJ mol}^{-1} - 4.8 \text{ kJ mol}^{-1} = -55 \text{ kJ mol}^{-1}$$

Because ΔG° is *negative*, this *is* a thermodynamically favorable reaction.
(d) Use Eq. (9-7) to obtain K from ΔG°:

$$K = e^{-\Delta G^\circ / RT} = \exp(-\Delta G^\circ / RT)$$
$$= \exp \frac{-(-55 \text{ kJ mol}^{-1})}{(0.00831 \text{ kJ K}^{-1} \text{ mol}^{-1})(298 \text{ K})}$$
$$= \exp(22)$$
$$= 4.4 \times 10^9$$

Alternatively, we can use Eq. (9-7′):

$$\Delta G° = -2.3RT \log K$$

$$\log K = \frac{-\Delta G°}{2.3RT}$$

$$= \frac{-(-55 \text{ kJ mol}^{-1})}{2.3(0.00831 \text{ kJ K}^{-1} \text{ mol}^{-1})(298 \text{ K})}$$

$$= 9.66$$

$$K = 4.5 \times 10^9$$

From Eq. (9-6) we know that the mass action–equilibrium expression for this reaction is

$$K = \frac{[\text{ethanol}]}{[\text{dimethyl ether}]} = 4.4 \times 10^9$$

Therefore,

$$\frac{[\text{dimethyl ether}]}{[\text{ethanol}]} = \frac{1}{K} = \frac{1}{4.4 \times 10^9} = 2.3 \times 10^{-10}$$

Thus, only about one molecule of dimethyl ether remains for each 4.4 billion ethanol molecules, or 2.3 molecules of the ether for each 10 billion of ethanol.

PROBLEM 9-3 (a) Using the bond dissociation energies in Table 9-1, estimate ΔH for the conversion of ethylene oxide to acetaldehyde in the gas phase.

(b) Is this a favorable reaction from the standpoint of enthalpy? (c) What type of reaction is this? (d) What is the minimum number of bonds that have to be broken and made to carry out this transformation? (e) What structural feature in ethylene oxide is neglected in this calculation?

Solution
(a) The total bond strengths for each compound (in kJ mol^{-1}) are:

ethylene oxide		acetaldehyde	
4 C(sp^3)–H	4(410)	3 C(sp^3)–H	3(410)
1 C(sp^3)–C(sp^3)	368	1 C(sp^2)–H	431
2 C(sp^3)–OC(sp^3)	2(335)	1 C(sp^3)–C(sp^2)	372
Total	2678	1 C=O	732
		Total	2765

Using Eq. (9-1) we can estimate ΔH:

$$\Delta H = \sum \Delta H_{d(\text{reactants})} - \sum \Delta H_{d(\text{products})}$$

$$= 2678 - 2765$$

$$= -87 \text{ kJ mol}^{-1}$$

(b) Yes, because ΔH is negative.
(c) Like the reaction in Problem 9-1, this is another example of a structural isomerization.
(d) Bonds broken: One C(sp^3)–OC(sp^3) and one C(sp^3)–H; bonds made: one C(sp^3)–H and one C=O. This is shown diagrammatically as follows.

(e) Ring strain (Section 7-3), which *increases* the free energy of ethylene oxide, is neglected in this calculation. Thus the true $\Delta H°$ is undoubtedly more negative than this value, because the ring strain is relieved during the isomerization.

PROBLEM 9-4 Consider the thermodynamic data below (valid at 25°C):

	ethylene oxide (g)	acetaldehyde (g)
$\Delta H_f°$ (kJ mol^{-1})	-51	-166
$S°$ (J K^{-1} mol^{-1})	243	266

(a) Calculate $\Delta H°$ for the reaction discussed in Problem 9-3. Does this calculation take into account the factor mentioned in Problem 9-3e?
(b) Calculate $\Delta S°$ for the same reaction. What does this value indicate?
(c) Calculate $\Delta G°$ for the same reaction. Is this reaction thermodynamically favorable?
(d) What is the value of K for this reaction? What fraction of the ethylene oxide has been converted to acetaldehyde when equilibrium is reached?

Solution
(a) Use Eq. (9-2) to find $\Delta H°$:

$$\Delta H° = \sum \Delta H_{f(\text{products})}° - \sum \Delta H_{f(\text{reactants})}°$$
$$= (-166) - (-51)$$
$$= -115 \text{ kJ mol}^{-1}$$

Yes. The standard heat of formation of ethylene oxide does take into account its ring strain, which is why the value of $\Delta H°$ calculated here is much more negative than the value estimated from bond energies.
(b) Use Eq. (9-3) to find $\Delta S°$:

$$\Delta S° = \sum S_{\text{products}}° - \sum S_{\text{reactants}}°$$
$$= 266 - 243 = 23 \text{ J K}^{-1} \text{ mol}^{-1}$$

Because $\Delta S°$ is positive it means that the product is more disordered than the reactant, and therefore this reaction *is* favorable from the standpoint of entropy.
(c) Use Eq. (9-4) to find $\Delta G°$; be sure to adjust the units of $\Delta S°$:

$$\Delta G° = \Delta H° - T\Delta S°$$
$$= -115 \text{ kJ mol}^{-1} - (298 \text{ K})(0.023 \text{ kJ K}^{-1} \text{ mol}^{-1})$$
$$= -122 \text{ kJ mol}^{-1}$$

Because $\Delta G°$ is *negative*, this *is* a thermodynamically favorable reaction.

(d) Using Eq. (9-7):

$$K = e^{-\Delta G^\circ / RT} = \exp\left(\frac{-\Delta G^\circ}{RT}\right)$$

$$= \exp\left[\frac{-(-122 \text{ kJ mol}^{-1})}{(0.00831 \text{ kJ K}^{-1} \text{ mol}^{-1})(298 \text{ K})}\right]$$

$$= \exp 4922$$

$$= 2.5 \times 10^{21}$$

Using Eq. (9-7'):

$$\Delta G^\circ = -2.3 RT \log K$$

$$\log K = \frac{\Delta G^\circ}{2.3 RT}$$

$$= -\left[\frac{-122 \text{ kJ mol}^{-1}}{2.3(0.00831 \text{ kJ K}^{-1} \text{ mol}^{-1})(298 \text{ K})}\right]$$

$$= 21.42$$

$$K = 2.6 \times 10^{21}$$

From Eq. (9-6) we know that the mass action–equilibrium expression for this reaction is

$$K = \frac{[\text{acetaldehyde}]}{[\text{ethylene oxide}]} = 2.5 \times 10^{21}$$

At equilibrium there are 2.5×10^{21} molecules of acetaldehyde for one of ethylene oxide. The fraction of ethylene oxide unconverted is the reciprocal of this, $1/(2.5 \times 10^{21})$, or 4×10^{-22}. Therefore, only about four ethylene oxide molecules out of 10^{22} remain at equilibrium.

PROBLEM 9-5 **(a)** The isomerization of dimethyl ether to ethanol (Problems 9-1 and 9-2) is thermodynamically favorable at 298 K. At what temperature, if any, would the reaction become unfavorable? **(b)** From the fact that it is thermodynamically favorable, what can you say about the rate of this isomerization?

Solution **(a)** Because ΔH° is negative and ΔS° is positive (Problem 9-2), this reaction will be favorable at all temperatures. See Example 9-5a. **(b)** The fact that the isomerization is thermodynamically favorable tells us *nothing* about how fast it occurs.

PROBLEM 9-6 The isomerization of ethylene oxide to acetaldehyde (Problems 9-3 and 9-4) is found to be subject to acid catalysis. Suppose that the mechanism of the acid-catalyzed reaction can be shown to involve four steps: (1) protonation at oxygen, (2) heterolytic cleavage of a C–O bond, (3) shift of a **hydride** ($H:^-$) and concomitant formation of an oxygen-protonated C=O, and (4) deprotonation of the oxygen.
(a) Draw structures for the reactant, product, and all intermediates, using arrows to show how each is converted to the next.
(b) Which step is likely to be rate limiting, and why?
(c) Draw the reaction profile diagram for this mechanism, labeling both axes and all critical points.
(d) Specifically, how does the acid catalyze the reaction?

Solution

(b) The second step (opening of the ring) is the only one involving breakage of a bond that is not compensated by some other bond making, so it is likely to be the rate-limiting step.

(c) See Figure 9-17.

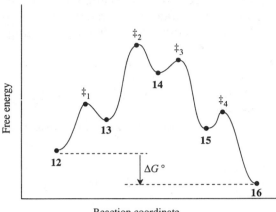

FIGURE 9-17. The reaction profile diagram for Problem 9-6c.

(d) The acid supplies the proton that attacks the oxygen in the first step. This protonation weakens the C–O bond and makes it easier to cleave (heterolytically) in the second step.

PROBLEM 9-7 Explain why the racemization of an optically active compound is expected to be spontaneous.

Solution The two enantiomers of any chiral compound will have identical free energies of formation and standard entropies (Section 8-5). Therefore, $\Delta H°$ for their interconversion is necessarily zero. However, a racemic mixture will have a random (R or S) configuration at each chiral center. This is a more disordered arrangement than having all chiral centers with one configuration. Hence, $\Delta S°$ for the racemization is positive, and $\Delta G°$ is negative: A favorable reaction.

PROBLEM 9-8 For the thermodynamically favorable reaction A + B + C → D the following separate mechanistic observations were made:

(1) When the concentration of A was doubled, the rate of formation of D doubled.
(2) When the concentration of B was halved, the rate of formation of D fell to one-fourth its original value.
(3) Doubling the concentration of C had no effect on the rate.

(a) What is the order in each reactant and the rate law for this reaction? What is the overall order?
(b) What can you determine about the point when C becomes involved in the reaction?

Solution
(a) The rate is directly proportional to the concentration of A, so it is first order in A. But the rate is proportional to the *square* of the concentration of B (one-half squared is one-fourth), so it is second order in B. The order in C is zero. So, the rate law is given by

$$Rate = k[\text{A}][\text{B}]^2$$

which is third order overall.
(b) Reactant C must be involved *after* the rate-limiting step of the reaction.

PROBLEM 9-9 Isobutene (2-methylpropene, **17**) reacts with HCl to form *tert*-butyl chloride (**18**). The reaction is first order in both **17** and HCl.

$$\begin{array}{c} \text{H}_3\text{C} \\ \diagdown \\ \text{C}{=}\text{CH}_2 \quad + \quad \text{HCl} \quad \longrightarrow \quad \text{H}_3\text{C}{-}\overset{\displaystyle \text{CH}_3}{\underset{\displaystyle \text{Cl}}{\overset{|}{\underset{|}{\text{C}}}}}{-}\text{CH}_3 \\ \diagup \\ \text{H}_3\text{C} \end{array}$$

$$\textbf{17}\textbf{18}$$

(a) Write the rate law for this reaction.

(b) What is the overall order of the reaction?

(c) Propose a one-step (concerted) mechanism for the reaction, showing with structures and arrows how the reactants are converted to the product.

(d) Draw a reaction profile diagram (labeling the axes and all critical points) for your mechanism in part (c).

(e) Propose a two-step mechanism for the same reaction involving a carbocation intermediate. Draw each structure and use arrows to show how they interconvert.

(f) Draw a reaction profile diagram (labeling the axes and all critical points) for your mechanism in part (e).

(g) How might you experimentally differentiate between the two mechanisms?

Solution

(a) $Rate = k[\mathbf{17}][HCl]$

(b) First order in each reactant, second order overall.

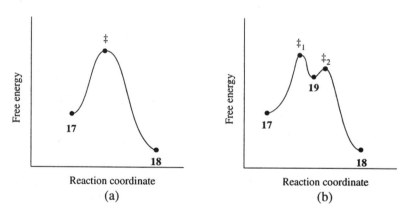

(c)

(d) See Figure 9-18a.

FIGURE 9-18. The reaction profile diagrams for (a) Problem 9-9d and (b) Problem 9-9f.

(e)

17   **19**   **18**

(f) See Figure 9-18b. The first step is likely to be rate limiting because charged species are formed from uncharged ones, and two bonds are broken while only one is made.

(g) By using a mixture of doubly isotopically labeled **HCl** plus unlabeled HCl, we could determine whether both labeled atoms stay together in product **18**, or whether they were scrambled with the labeled H in one molecule and the labeled Cl in another. The occurrence of scrambling would be evidence in support of the two-step mechanism. (However, the *absence* of scrambling does not preclude the two-step mechanism. Why?)

PROBLEM 9-10   The rate of a certain reaction doubles when the temperature is raised from 25°C to 35°C. What is the activation energy of this reaction?

Solution   Use Eq. (9-12), converting both temperatures to Kelvin (298 K and 308 K, respectively).

$$\ln\left(\frac{k_2}{k_1}\right) = \frac{E_a}{R}\left(\frac{T_2 - T_1}{T_1 T_2}\right)$$

$$\ln\left(\frac{2k}{k}\right) = \frac{E_a}{8.31 \text{ J K}^{-1} \text{ mol}^{-1}}\left[\frac{308 - 298}{(298)(308)}\right]$$

$$E_a = \frac{(\ln 2)(8.31 \text{ J K}^{-1} \text{ mol}^{-1})(308 \text{ K})(298 \text{ K})}{308 \text{ K} - 298 \text{ K}}$$

$$= 5.3 \times 10^4 \text{ J mol}^{-1} = 53 \text{ kJ mol}^{-1}$$

PROBLEM 9-11 Assume for the moment that the addition of HCl to **17** (Problem 9-9) proceeds by the two-step mechanism shown in parts (e) and (f). Using Hammond's postulate and appropriate reaction profile diagrams, explain why the addition does *not* produce significant amounts of 1-chloro-2-methylpropane (**20**), an isomer of **18**.

$$\textbf{17} \; + \; \text{HCl} \quad \xrightarrow{\;/\!\!/\;} \quad \underset{\substack{| \\ \text{H} \quad \text{Cl}}}{\overset{\substack{\text{CH}_3 \\ |}}{\text{H}_3\text{C}\!-\!\text{C}\!-\!\text{CH}_2}}$$

20

(*Reminder:* A tertiary carbocation is more stable than a primary one.)

Solution The carbocation intermediate that would lead to **20** has structure **21**.

$$\underset{\substack{| \\ \text{H}}}{\overset{\substack{\text{CH}_3 \\ |}}{\text{H}_3\text{C}\!-\!\text{C}\!-\!\text{CH}_2^+}}$$

21

Being a primary carbocation, it is less stable than **19** (a tertiary carbocation). Formation of either carbocation from **17** + HCl is rate limiting, a step with $\Delta G° > 0$. Therefore, Hammond's postulate predicts that the transition state leading to the carbocation will resemble that carbocation (the product of that step). Thus the transition state leading to **21** is expected to be higher in energy than the one leading to **19**. It follows that carbocation **19** (and product **18**) will form faster than carbocation **21** (and product **20**). See Figure 9-19.

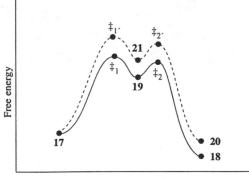

FIGURE 9-19. The reaction profile diagram for Problem 9-11.

PROBLEM 9-12 After reviewing the mechanism for the isomerization of acetone enol (Section 9-9), suggest a mechanism for the base-catalyzed racemization of (+)-**3** [(+)-2-methylcyclohexanone; Section 9-2].

(+)-**3**

Solution   In order to account for the racemization of (+)-**3**, we must propose a mechanism that involves a symmetrical (i.e., achiral) intermediate. We know that the base-catalyzed isomerization of acetone enol follows the mechanistic steps below:

Mechanism IIB

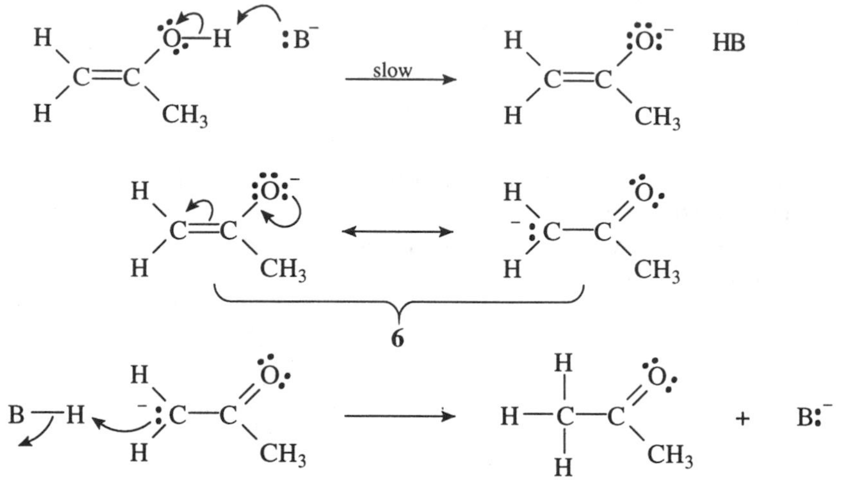

Of course, like any reaction (in principle), both of these steps are reversible, although the overall reverse reaction is thermodynamically unfavorable. Nonetheless, it should be possible to reverse the second step, using a base to **deprotonate** acetone to give enolate **6** (the equilibrium arrows of different length indicate the preferred side of the equilibrium):

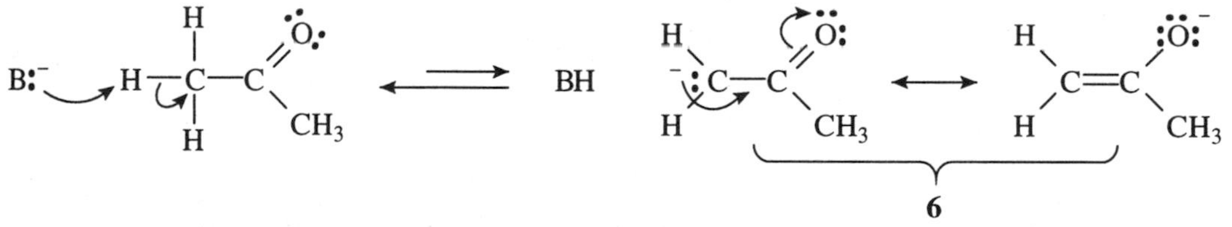

If we were to do the same thing to (+)-**3**, deprotonating at C-2, we would be left with enolate **22**:

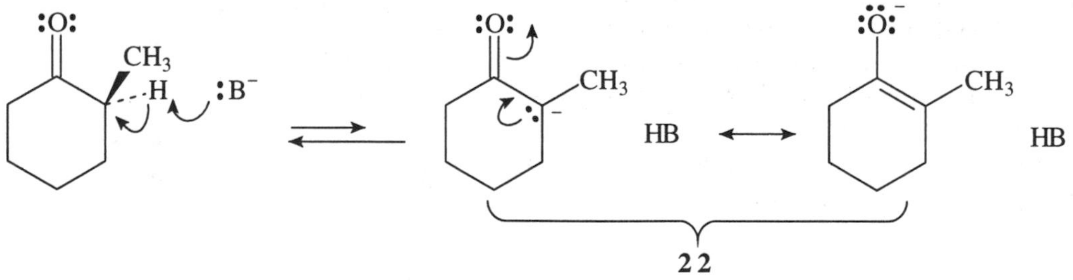

Once deprotonated, C-2 undergoes a change in hybridization from sp^3 to sp^2 (in **22**) to permit resonance of the nonbonding pair with the C=O (Section 4-5), and all chirality is lost. So, when **22** is reprotonated to re-form **3**, the proton can be delivered equally well from either side of the ring, giving both (+)-**3** and (−)-**3** in equal amounts.

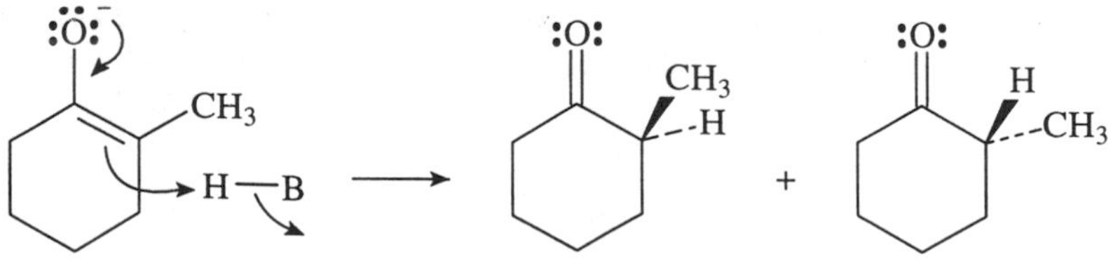

(+)-**3**   (−)-**3**

10 TYPES OF ORGANIC REACTIONS

THIS CHAPTER IS ABOUT

☑ **Classifying Organic Reactions**
☑ **Polar Reactions**
☑ **Acid–Base Reactions**
☑ **Relative Strengths of Acids and Bases**
☑ **The Effect of Structure on Acidity**
☑ **Free-Radical Reactions**
☑ **Oxidation–Reduction (Redox) Reactions**
☑ **Pericyclic Reactions**
☑ **Predicting Reaction Mechanisms**

10-1. Classifying Organic Reactions

Each of the functional groups (Chapter 6) of organic molecules has its own unique pattern of reactivity, which can be described in detail. But before we go into details, it will be helpful to characterize the general types of organic reactions. In this chapter we'll see how virtually all organic reactions can be classified on the basis of their mechanisms into one of three basic types: *Polar reactions*, *free-radical reactions*, and *pericyclic reactions*. Oxidation–reduction (redox) reactions can sometimes belong to more than one of these simple categories. At any rate, each of these categories has its own characteristics, which we'll now begin to describe. The more familiar we become with these characteristics, the easier it will be for us to examine a given set of reactants and conditions and predict not only the products of the reaction, but also the mechanisms by which they arise.

10-2. Polar Reactions

By far the vast majority of organic reactions occur by polar mechanisms. Included are all reactions that involve heterolytic bond cleavage and/or heterogenic bond formation (Section 4-2), as well as reactions that proceed by way of ionic intermediates such as carbocations and carbanions (Section 4-1). The isomerization of acetone enol, discussed at some length in Section 9-9, is an example of a polar mechanism.

How do you recognize a polar mechanism? Certain features of the reactions, their reactants or products, should serve as red flags.

The first molecular structure feature that should alert you to the possibility of a polar mechanism is that *polar covalent bonds are being broken or made*. The functional groups discussed in Chapter 6 exhibit significant polarity, and this polarity is "felt" by the atoms near the functional group. These groups, their principal resonance forms, and their net polarity are reviewed in Table 10-1. Because typical carbon–hydrogen and carbon–carbon single bonds have very little or no polarity (since there is little or no difference in electronegativity between the atoms), they are rarely involved in polar reactions. However, carbon–carbon *multiple* bonds, as well as carbon–hydrogen bonds that are activated by a nearby polar functional group, *can* take part in polar reactions.

The next red flag that suggests a polar mechanism is the *presence of ions among the reactants, products, or intermediates*. In this regard, remember that H⁺ (a proton) is an ion; thus, virtually all acid-catalyzed (as well as base-catalyzed) reactions proceed by polar mechanisms. (More about this in the next section.)

Finally, there is the effect of solvent. Many polar reactions have steps in which the transition state is more polar than the reactant(s). This is because polar bonds are being stretched in preparation for cleavage, and polarity (as measured by dipole moment, Section 3-1) is directly related to the distance separating the charges. Such polar transition states and intermediates are stabilized by polar solvent molecules through dipole–dipole interactions (Section 7-7). The result is that *polar reactions often proceed faster in polar solvents* (such as water, alcohols, carboxylic acids, and dimethyl sulfoxide) than they do in nonpolar solvents (such as hydrocarbons and ethers). Such **solvent effects** can often be used to diagnose the type of mechanism occurring in a given reaction.

TABLE 10-1. Polarity of the Common Functional Groups

Functional group class	Structure and polarity
Terminal alkyne	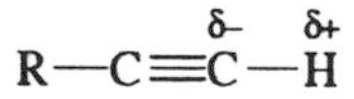 $R{-}C{\equiv}C{-}H$ ($\delta-$ on C, $\delta+$ on H)
Alkyl halide (X = F, Cl, Br)	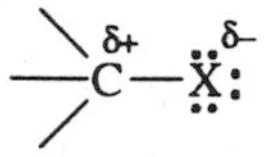 $\overset{\delta+}{C}{-}\overset{\delta-}{X}{:}$
Alcohol	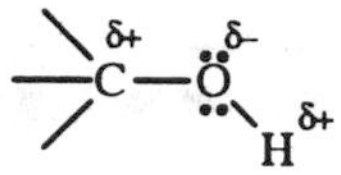 $\overset{\delta+}{C}{-}\overset{\delta-}{O}{-}\overset{\delta+}{H}$
Ether	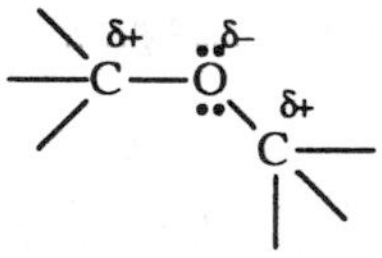 $\overset{\delta+}{C}{-}\overset{\delta-}{O}{-}\overset{\delta+}{C}$
Amine	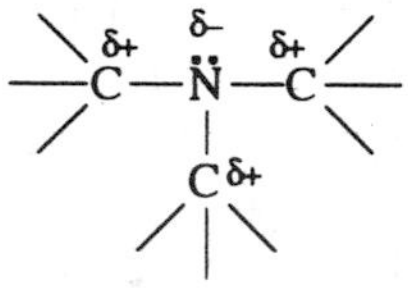 $\overset{\delta+}{C}{-}\overset{\delta-}{N}{-}\overset{\delta+}{C}$; $\overset{\delta+}{C}$
Ketone, aldehyde	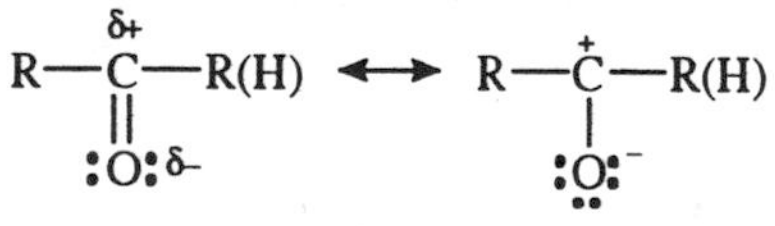 $R{-}\overset{\delta+}{C}({=}\overset{\delta-}{\ddot{O}}){-}R(H) \longleftrightarrow R{-}\overset{+}{C}({-}\ddot{O}{:}^-){-}R(H)$
Carboxylic acid (X = OH) and derivatives (X = OR′, NR₂, halide)	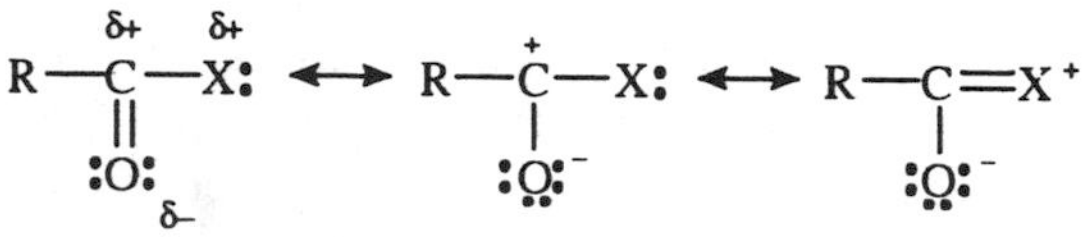 $R{-}\overset{\delta+}{C}({=}\overset{\delta-}{\ddot{O}}){-}\overset{\delta+}{X}{:} \longleftrightarrow R{-}\overset{+}{C}({-}\ddot{O}{:}^-){-}X{:} \longleftrightarrow R{-}C({-}\ddot{O}{:}^-){=}X^+$
Nitro	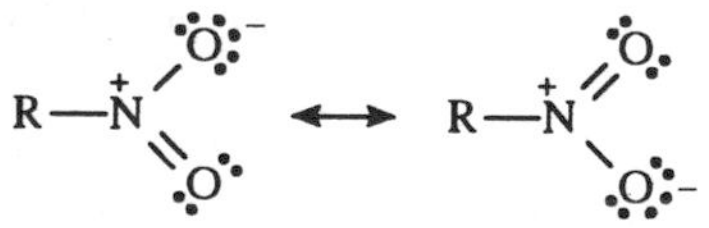 $R{-}\overset{+}{N}(\nearrow\ddot{O}{:}^-)({=}\ddot{O}) \longleftrightarrow R{-}\overset{+}{N}({=}O)(\searrow\ddot{O}{:}^-)$
Nitrile	 $R{-}\overset{\delta+}{C}{\equiv}\overset{\delta-}{N} \longleftrightarrow R{-}\overset{+}{C}{=}\ddot{N}{:}^-$
Organometallic	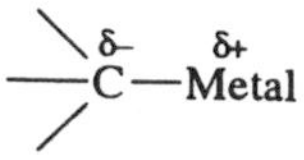 $\overset{\delta-}{C}{-}\overset{\delta+}{Metal}$

10-3. Acid–Base Reactions

Many organic reactions involve acids and bases of one sort or another. In this section we'll learn to identify several types of acids and bases and the kinds of products they form.

A. Arrhenius concept

The original generalized concept of acids and bases was formulated by Arrhenius. An Arrhenius acid is a compound that, *when dissolved in water,* increases the concentration of H^+ (or H_3O^+). Similarly, an Arrhenius base is a compound that, *when dissolved in water,* increases the concentration of OH^- (hydroxide ion). Typical Arrhenius acids include the common mineral acids such as HCl, HBr, HI, H_2SO_4, $HClO_4$, and HNO_3, while Arrhenius bases include the common metal hydroxides such as NaOH and KOH. The products from the reaction between an Arrhenius acid and base are water ($H^+ + OH^- \rightarrow H_2O$), and the salt of the remaining ions (e.g., NaCl).

B. Brønsted–Lowry concept

The Arrhenius concept of acids and bases is quite useful, provided that your reactions take place in an aqueous medium. Unfortunately, most organic reactions cannot be carried out in water as the solvent because most organic compounds are not appreciably soluble in water (Section 7-7). When nonaqueous media are involved, the most useful acid–base concept is the one enunciated by Brønsted and Lowry: A Brønsted–Lowry acid is a *proton donor,* a Brønsted–Lowry base is a *proton acceptor.* Thus the Brønsted–Lowry concept focuses attention on the *proton transfer* between the acid (HA) and the base (B:):

$$B\!: \; + \; H{-}A \; \rightleftharpoons \; (B{-}H)^+ \; + \; :A^-$$

There are two important things to remember about a Brønsted–Lowry acid–base reaction. First, virtually all proton transfers are reversible, with the equilibrium favoring the *weaker* acid and base. Second, the products of the reaction are another acid (BH⁺) and base (:A⁻). The acid BH⁺ is called the **conjugate acid** of B:, while :A⁻ is the **conjugate base** of HA.

EXAMPLE 10-1 Identify the Brønsted–Lowry acids and bases in the reaction below (which is very similar to the one we discussed in Section 9-9). Indicate all conjugate relationships.

Solution

In the forward reaction, A⁻ is the conjugate base of acid HA, and BH is the conjugate acid of base B⁻. In the reverse reaction, HA is the conjugate acid of base A⁻, and B⁻ is the conjugate base of acid BH.

Proton transfers are among the fastest reactions known, indicating that they generally have low activation barriers. Another way of saying the same thing is that proton-transfer reactions tend to come to equilibrium very rapidly. The position of equilibrium (i.e., which side of the equation is favored) is determined by the relative strengths of the acids and bases.

C. G. N. Lewis concept

There is one other type of acid–base concept that is also quite useful in organic chemistry. G. N. Lewis focused on the *bonds* (i.e., electron pairs) that are being made and broken in the reaction. According to the Lewis concept an acid is an *electron-pair acceptor* and a base is an *electron-pair donor*. The product of a Lewis acid–base reaction is a "complex," a molecule (or ion) with the newly formed bond. One of the simplest Lewis acid–base reactions (which is also part of an Arrhenius acid–base reaction) is this one:

$$H^+ \ + \ :OH^- \longrightarrow HOH$$

Notice how H^+ serves as the *acid* because it accepts (begins to share) an electron pair donated by the base (OH^-).

EXAMPLE 10-2 Identify the Lewis acid and base in the reaction below (which comes from Problem 9-9):

$$(H_3C)_3C^+ \ + \ :\ddot{C}l:^- \longrightarrow (H_3C)_3CCl$$

Solution

$$(H_3C)_3C^+ \ + \ :\ddot{C}l:^- \longrightarrow (H_3C)_3CCl$$

$$\text{acid} \qquad\qquad \text{base}$$

By now, you can begin to appreciate that a Lewis acid must possess an *empty orbital* in its valence shell in order to accommodate the newly shared pair. Similarly, a Lewis base must have one or more available electron pairs to share. Usually these are nonbonding pairs, but sometimes σ- or π-bonding pairs can be donated.

EXAMPLE 10-3 The formation of the carbocation in Example 10-2 can occur via protonation of isobutene (Problem 9-9):

$$\begin{array}{c} H_3C \\ \\ H_3C \end{array}\!\!\!\!\text{C}=\text{CH}_2 \ + \ H^+ \longrightarrow (H_3C)_3C^+$$

Can this reaction be regarded as a Lewis acid–base process? If so, identify the acid and base and the type of electron pair donated.

Solution Yes! Isobutene is the electron-pair donor (donating a π-bonding electron pair), while H^+ is the electron pair acceptor.

Notice that these three concepts (Arrhenius, Brønsted–Lowry, Lewis) do not contradict or displace each other. Rather, they are complementary. The Brønsted–Lowry concept extends the Arrhenius concept, especially with regard to bases and solvents other than water. The Lewis concept extends both the Arrhenius and Brønsted–Lowry concepts to reactions not involving proton transfer.

You might recall the terms *electrophile* and *nucleophile* from Section 4-2. From their definitions, you can see that an electrophile is exactly the same as a Lewis acid, and a nucleophile is synonymous with a Lewis base. Many polar organic reactions involve electrophiles and nucleophiles and can therefore be regarded as Lewis acid–base reactions.

10-4. Relative Strengths of Acids and Bases

The strength of an Arrhenius acid (HA) is determined by the extent of its ionization and dissociation in water according to the reaction

$$HA + H_2O \rightleftharpoons H_3O^+ + A^-$$

If the equilibrium lies far to the right, the acid is essentially fully dissociated in water. Such an acid is said to be a **strong acid** (in water), and examples include the mineral acids listed in the previous section. **Weak acids** are those that are *soluble*, but only a small fraction (typically 1% or less) of the dissolved acid undergoes dissociation.

We can quantify the strength of an acid by measuring the equilibrium constant (K_a) for the dissociation reaction.

ACID DISSOCIATION CONSTANT
$$K_a = \frac{[H_3O][A^-]}{[HA]} \qquad (10\text{-}1)$$

The concentration of water, essentially a constant in dilute aqueous solution, is factored into the value of K_a. As with any equilibrium constant (Section 9-4), a large value of K_a ($>>1$) corresponds to a strong acid, while a small (but always positive!) value ($0 < K_a << 1$) indicates a weak acid.

EXAMPLE 10-4   (a) Acetic acid has a K_a value (in water at 25°C) of 1.7×10^{-5}, while HCN has a value of 4.9×10^{-10}. Which is the stronger acid? (b) If an acid is fully (100%) dissociated, what would its K_a value be?

Solution
(a) Because 1.7×10^{-5} is greater than 4.9×10^{-10}, acetic acid is considerably stronger than HCN. However, both of these acids are considered weak because both have K_a values less than one.
(b) If the acid is 100% dissociated, the concentration of *dissolved but undissociated* acid ([HA]) is zero. By Eq. (10-1), this would lead to an infinite value for K_a.

Because the spectrum of acidity ranges from greater than 10^{10} (very strong) to less than 10^{-50} (extremely weak), it is common practice to express K_a values in logarithmic form according to the relation:

DEFINITION OF pK_a
$$pK_a = -\log K_a \qquad (10\text{-}2)$$

This provides a less unwieldy number with which to compare acidities.

EXAMPLE 10-5   Calculate the pK_a values for acetic acid and HCN (Example 10-4).

Solution   Use Eq. (10-2):

For acetic acid: $pK_a = -\log(1.7 \times 10^{-5}) = 4.77$
For HCN: $pK_a = -\log(4.9 \times 10^{-10}) = 9.31$

Notice from this example [as well as from Eq. (10-2)] that the *smaller* the value of K_a, the *larger* the value of the corresponding pK_a. Moreover, the *larger* the pK_a, the *weaker* the acid!

The quantitative measurement of acid strengths *in water* allows us to set up an *acidity scale*, which ranges in strength from the acidity of H_3O^+ (pK_a –1.7) to that of water (pK_a 16). Acids stronger than

H_3O^+ are essentially fully dissociated in water, and therefore their strengths cannot be differentiated *in water*. This is often referred to as the leveling effect of water. Similarly, acids that are weaker than water are so weak that they don't contribute as much to the acidity of the solution as does the **autodissociation** of water itself.

$$H_2O + H_2O \rightleftharpoons H_3O^+ + OH^-$$

Note that when an Arrhenius acid dissociates in water, the water is serving as a Brønsted–Lowry base. If we want to measure the acidities outside the pK_a range of -1.7 to 16, it is necessary to change the solvent from water to another substance (we'll denote as HSol) that can also serve as a Brønsted–Lowry base:

$$HA + HSol \rightleftharpoons H_2Sol^+ + A^-$$

$$K_a = \frac{[H_2Sol^+][A^-]}{[HA]} \qquad (10\text{-}3)$$

Again, the stronger the acid HA, the further the equilibrium will lie to the right, and the larger will be the value of K_a. Another way of saying this is,

- At equilibrium, the weaker side (that is, the Brønsted–Lowry acid and base with the lower free energy) will be favored.

Thus if, for a certain acid, the equilibrium favors the *left* side of the proton-transfer equation, then HA is a *weaker* acid than H_2Sol^+, and HSol is a weaker base than A^-.

$$HA + HSol \overset{K<1}{\rightleftharpoons} H_2Sol^+ + A^-$$

weaker	weaker	stronger	stronger
acid	base	acid	base

If the right side is favored, the converse is true:

$$HA + HSol \overset{K>1}{\rightleftharpoons} H_2Sol^+ + A^-$$

stronger	stronger	weaker	weaker
acid	base	acid	base

From these relationships we can also infer that

- The stronger an acid is, the weaker its conjugate base is (as a base).

EXAMPLE 10-6 What is the conjugate base of acetic acid? of HCN? Which of these two conjugate bases is stronger (as a base)?

Solution We know (from Chapter 6) that the structure of acetic acid is CH_3CO_2H; so its conjugate base is the acetate ion $CH_3CO_2^-$. The conjugate base of HCN is CN^-. Because HCN is a weaker acid than acetic acid (Example 10-3), CN^- is a stronger base than acetate ion.

From the above discussion, a little thought should convince you that

- The strongest acid that can exist *undissociated* in significant concentration in the solvent HSol is the conjugate acid of that solvent (H_2Sol^+).

You can reason it out this way: If you were to take an acid HA that was *stronger* than H_2Sol^+, then the equilibrium would favor the *right* side, i.e., HA would donate its proton to HSol. Thus very little *undissociated* HA would remain.

EXAMPLE 10-7 Complete this statement: The strongest *base* that can exist in significant concentration in solvent HSol is _____.

Solution The conjugate *base* of solvent HSol is Sol⁻. In a reaction with a base (B:), the solvent acts as an *acid*:

$$B\text{:} \; + \; HSol \; \rightleftharpoons \; BH^+ \; + \; Sol^-$$

If B: is a stronger base than Sol⁻, then it will deprotonate the solvent HSol to form Sol⁻. Thus, little unprotonated B: will remain. The answer is therefore: The strongest base that can exist in significant concentration in solvent HSol is <u>Sol⁻</u>.

Now, suppose we wish to determine the relative acid strengths of HCl and HClO₄, both of which are fully dissociated in water; then we would need to use a less basic solvent. For example, we might choose to use acetic acid as solvent. Being itself a weak acid, acetic acid is less basic than water.

EXAMPLE 10-8 (a) Write the Brønsted acid–base reaction between HCl (the acid) and acetic acid (the base). (b) Suppose that HClO₄ is still fully dissociated in acetic acid, but HCl is only partially dissociated. Which acid is stronger, HCl or HClO₄?

Solution

(a) $$HCl \; + \; CH_3CO_2H \; \rightleftharpoons \; CH_3CO_2H_2^+ \; + \; Cl^-$$

acid ◄─────────── conjugates ───────────► base
base ◄─── conjugates ───► acid

(b) HClO₄ would thus be demonstrated to be stronger than HCl in acetic acid (and hence presumably in water also).

In the case of organic acids much weaker than water, we need to choose a solvent that is more basic than water. A good candidate is ammonia, NH₃, which is liquid below –33°C. For example, suppose we wish to measure the acidity of a terminal acetylenic hydrogen according to the reaction:

$$RC\equiv CH \; + \; \text{:}B^- \; \rightleftharpoons \; RC\equiv C\text{:}^- \; + \; BH$$

When OH⁻ is used as the base, with water as solvent, no appreciable reaction occurs. The left side of the equilibrium is so highly favored that no significant amount of the *acetylide ion* (RC≡C:⁻) can be detected. By contrast, when NH₂⁻ (amide ion) is used as the base, in liquid ammonia as solvent, deprotonation of the acetylene is essentially complete. These results establish that the acetylide ion is a stronger *base* than OH⁻, but a weaker base than NH₂⁻. It also establishes that the terminal acetylene is a stronger acid than NH₃, but a weaker acid than water.

As you can see, acid and base strengths are always relative, involving comparisons of proton-donating and proton-accepting capabilities. But, by extending our range of solvents we can tabulate a broad spectrum of relative acidities, as shown in Table 10-2. It is not so important that you remember these exact pK_a values, but you will find it useful to memorize the relative acidity order of these various functional groups.

From the data in Table 10-2 we can easily determine how strong a base must be to deprotonate a given acid. For example, a terminal alkyne can be deprotonated by any base appearing below the acetylide ion. But those bases appearing above the acetylide are too weak to accomplish the deprotonation.

TABLE 10-2. Acid Strengths of Common Organic Brønsted Acids[a]

Class	Structure	pK_a	Conjugate base
Mineral acids	HClO₄	−10	ClO₄⁻
	HCl	−7	Cl⁻
Sulfonic acids	RSO₃H	−6.5	RSO₃⁻
	H₃O⁺	−1.7	H₂O
	HF	3	F⁻
Carboxylic acids	RCO₂H	5	RCO₂⁻
	H₂S	7	HS⁻
Phenol	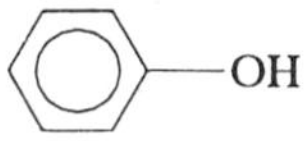 ⟨benzene⟩—OH	10	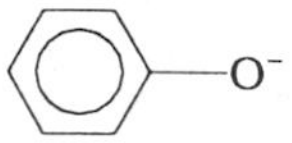 ⟨benzene⟩—O⁻
Water	H₂O	16	OH⁻
Alcohol	ROH	17	RO⁻
Hydrogen alpha to a carbonyl	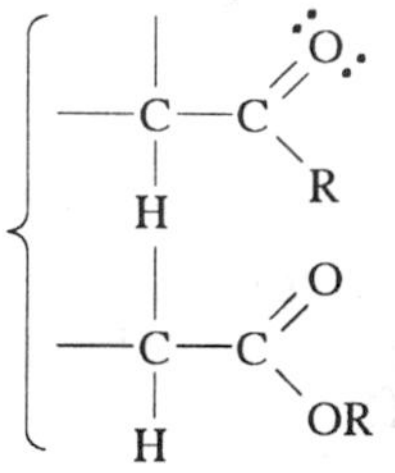 (α to ketone structure)	20	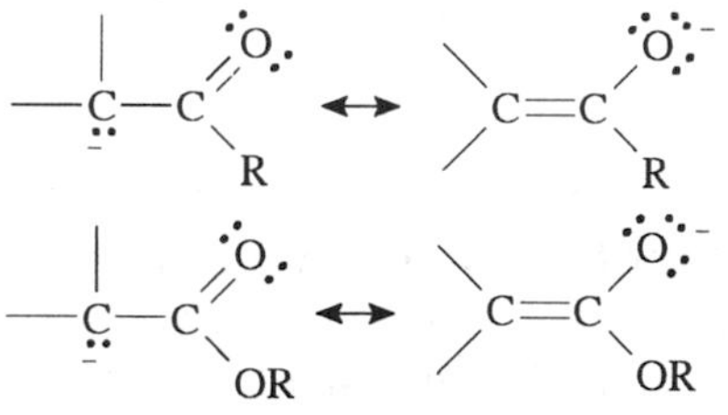 (enolate resonance structures)
	(α to ester structure)	25	(ester enolate resonance structures)
Terminal alkyne	RC≡CH	25	R–C≡C:⁻
Ammonia	NH₃	34	H₂N:⁻
Amine	R₂NH	38	R₂N:⁻
Terminal alkene	[(H)R]₂C=CH₂	44	[(H)R]₂C=C̈H⁻
Alkane	H₃CCH₃	50	H₃CC̈H₂⁻

[a]Data taken from J. March, *Advanced Organic Chemistry*, 4th Ed., John Wiley & Sons, New York, New York, 1992.

EXAMPLE 10-9 Is an alkoxide ion (RO⁻) a strong enough base to completely deprotonate a hydrogen alpha to an ester carbonyl? A phenolic hydrogen?

Solution An alkoxide ion is the conjugate base of an alcohol, which has a pK_a around 17. It can therefore deprotonate any acid with a pK_a significantly less than 17. Thus, a phenolic hydrogen (pK_a 10) would be readily deprotonated, but the hydrogen alpha to an ester (pK_a 25) would not.

EXAMPLE 10-10 Indicate the expected order of acidity (1 = most acidic) of the three types of hydrogens in propanoic acid (below). Also, estimate the pK_a of each set of hydrogens.

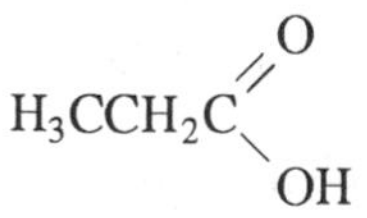

$$H_3CCH_2C\overset{O}{\underset{OH}{}}$$

Solution

$$(pK_a\ 50)\ \text{③}H_3CCH_2C\overset{O}{\underset{\text{②}\ OH\ \text{①}}{}}\ (pK_a\ 5)$$
$$(pK_a\ 25)$$

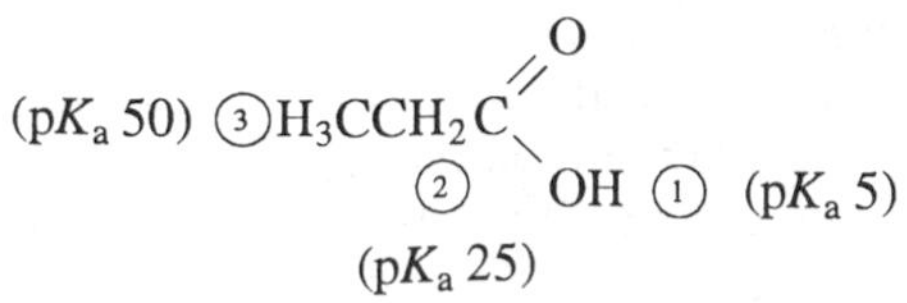

10-5. The Effect of Structure on Acidity

As we look down the list of acid structures in Table 10-2, a logical question is: Why this order? Why is $HClO_4$ a stronger acid than HCl? Why is a carboxylic acid stronger than a phenol? Since one of our prime goals is to be able to correlate structural differences with patterns in reactivity, let's now discuss some of the structural factors in a molecule that control its intrinsic acidity, that is, its proton-donating ability.

A. Effect of neighboring atom

The most important factor controlling acid strength is the atom (element) to which the hydrogen is bonded. If we compare the acidity of H_2O (pK_a 16) with that of H_2S (pK_a 7), we can see that a hydrogen bonded to oxygen is much less acidic than one bonded to sulfur. From this and similar comparisons, we infer that:

- When comparing elements *within a* (vertical) *group* (e.g., O, S, Se, Te), the corresponding hydride (compound with hydrogen) becomes *more* acidic as you move *down* the group.

The reason for this order is that the bond to the hydrogen becomes successively weaker as the size of the atom to which it's bonded becomes larger. But remember that this size–bond-strength effect is valid only for comparisons within a *group*.

If the comparison involves atoms *within a* (horizontal) *period* (e.g., C, N, O, F), atomic size *decreases* from left to right (Section 1-7). And yet,

- Acidity of the corresponding hydride *increases* from left to right across a period.

This is because comparisons within a period are correlated not with size differences, but with differences in electronegativity (Section 1-7), which increases from left to right across a period. As the atom to which the hydrogen is bonded becomes more electronegative, the electrons in the bond become more polarized toward the other atom, making the hydrogen (proton) more positive and more easily donated. This is why, in Table 10-2, the order of acidity is $CH_3CH_3 < R_2NH < H_2O < HF$.

B. Resonance effect

When comparing one oxygen acid with another (e.g., a carboxylic acid with an alcohol), the hydrogen is connected to the same element (oxygen) in both cases, and therefore group and period comparisons are no longer the relevant criteria. Now other, more subtle, factors come into play, most notably resonance and inductive stabilization of the conjugate base. For example, while the negative charge on an alkoxide ion (RO^-, the conjugate base of an alcohol) is localized on the oxygen, the negative charge on a carboxylate ion is distributed to both oxygens by resonance (Chapter 4), and this greatly stabilizes the carboxylate ion relative to the alkoxide ion:

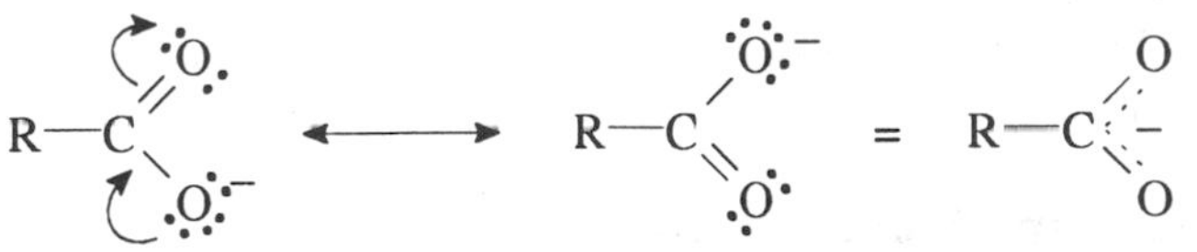

Because the conjugate base of a carboxylic acid is more stable (lower free energy), the dissociation equilibrium is shifted to the right, and the acidity of the carboxylic acid is thus increased relative to the alcohol.

In the same context, a phenol (which is also an oxygen acid) is more acidic than an alcohol because of resonance charge delocalization of the phenoxide ion (Example 4-14):

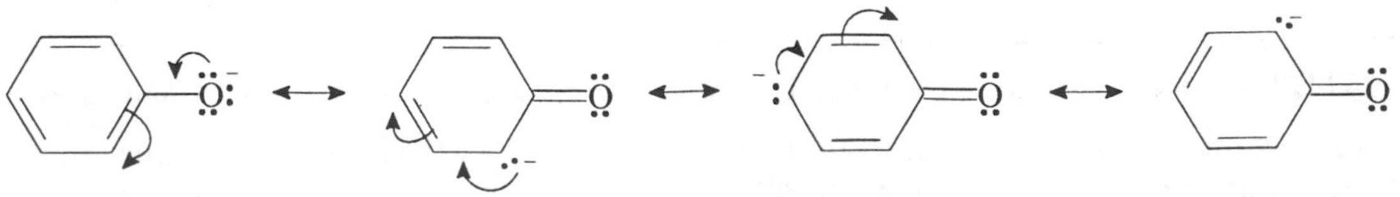

C. Inductive effect

Because of their ability to accept electrons from a negatively charged atom through resonance, functional groups such as the carbonyl and the phenyl ring are said to have an **electron-withdrawing resonance effect** (denoted –R). But certain atoms and groups can also exert an electron-withdrawing effect, even though they are not in a position to engage in resonance interactions. Such groups have a large effective electronegativity and thereby polarize the σ-bonding electrons toward themselves and away from the charge-bearing atom. These groups are said to have an **electron-withdrawing inductive effect** (denoted –I). For example, fluoroacetic acid (pK_a 2.7) is stronger than acetic acid itself (pK_a 4.8) because the highly electronegative fluorine atom, though four bonds away from the hydrogen, polarizes the intervening bonds by induction. This polarization has two consequences: The bond to the hydrogen is weakened, and the conjugate base (once the hydrogen is removed) is stabilized.

By the same token, acetic acid is somewhat weaker than formic acid (pK_a 3.8), indicating that a methyl group has an **electron-donating inductive effect** (+I) compared to hydrogen.

A methyl group is also believed to exert a weak +R effect, donating electron density to the carbonyl group by what is called **hyperconjugation** (or no-bond resonance):

We can generalize by saying that

- Electron-withdrawing groups (–R or –I) tend to be acid strengthening (and conjugate base weakening), while electron-donating groups (+R or +I) are acid weakening (and conjugate base strengthening).

D. Effects of various functional groups

Table 10-3 lists many of the functional groups we discussed in Chapter 6 and classifies them as either +R, –R, +I, or –I. Note that some groups can have several different effects depending on their position in the molecule and the electronic demands placed on them. For example, a phenyl group is electron withdrawing by induction (for reasons we'll discuss), but it can be either electron withdrawing or electron donating by resonance. Compare the direction of the inductive effect of each group in Table 10-3 with the polarity of the group shown in Table 10-1.

EXAMPLE 10-11   Explain how a halogen atom such as F can exhibit both +R and –I effects.

Solution   Because of its high electronegativity a fluorine atom will always exert a strong inductive attraction on the σ electrons in the molecule:

$$\underset{\underset{\delta-}{F}}{\longleftarrow}\underset{\overset{+}{\underset{\delta+}{C}}}{C}\diagdown$$

However, because it possesses three nonbonding pairs, it can also donate an electron pair by resonance, provided it is attached directly to an atom with a parallel *p* orbital:

$$:\ddot{F}\!-\!C\!=\!C\diagdown \quad\longleftrightarrow\quad :\overset{+}{\ddot{F}}\!=\!C\!-\!\overset{\cdot\cdot}{C}:^{-}$$

TABLE 10-3. Electronic Effects of the Common Functional Groups[a]

Group	+I	−I	+R	−R
−R (alkyl)	*		*	
−$\ddot{\text{F}}$:, −$\ddot{\text{C}}$l:, −$\ddot{\text{B}}$r:		*	*	
−$\ddot{\text{O}}$H, −$\ddot{\text{O}}$R		*	*	
−$\ddot{\text{O}}$:$^{-}$	*		*	
$-\text{C}\diagup\!\!\!\!\overset{\ddot{\text{O}}\cdot}{\diagdown}_{\text{X}}$ (X = H, R, OR′, halogen, NR$_2$)		*		*
$-\text{C}\diagup\!\!\!\!\overset{\ddot{\text{O}}\cdot}{\diagdown}_{\ddot{\text{O}}^{-}}$	*			
−$\ddot{\text{N}}$H$_2$, −$\ddot{\text{N}}$R$_2$		*	*	
−$\overset{+}{\text{N}}$R$_3$		*		
−NO$_2$		*		*
−CH=CH$_2$		*	*[b]	*[b]
−phenyl		*	*[b]	*[b]
−C≡CH		*		*
−SO$_2$X		*		*
−CN		*		*
−$\ddot{\text{S}}$H, −$\ddot{\text{S}}$R		*	*	

[a]+, Electron donating; −, electron withdrawing; I, inductive effect, R, resonance effect; all effects are relative to a comparably situated hydrogen.
[b]Direction of resonance effect determined by electronic demands placed on the group.

While perusing the effects in Table 10-3, you may have wondered why groups such as −CH=CH$_2$, −phenyl, and −C≡CH have electron-*withdrawing* inductive effects, while typical saturated alkyl groups have electron-*donating* inductive effects. This is because an *sp*- or *sp*^{2}-hybridized carbon atom is effectively *more* electronegative than hydrogen, while an *sp*^{3} carbon atom is *less* electronegative. You might recall that the energy of a hybrid orbital is lowered (becoming more stable) as its fraction of *s* character increases (see Section 3-3). Because such an orbital is lower in energy, it attracts electrons toward itself, making it effectively more electronegative than a hybrid with less *s* character. This is also the reason why a terminal acetylenic hydrogen is more acidic than an alkene (vinyl) hydrogen, which in turn is more acidic than an alkane hydrogen (Table 10-2).

EXAMPLE 10-12   Which molecule would you expect to be more acidic, ethane or cyclopropane? Explain. (You might want to review Problem 3-9.)

Solution   Table 10-2 shows ethane to be the least acidic compound listed, with a pK_a of 50. Unfortunately, cyclopropane isn't listed in the table. Nonetheless, we do know (from Problem 3-9) that the C–H bonds in cyclopropane involve $sp^{2.46}$ carbon hybrids compared to the sp^3 hybrids used in ethane. Thus the cyclopropane C–H bonds involve carbon hybrids that are richer in *s* character, and therefore lower in energy, than those in cyclopropane. As a result, cyclopropane should be the more acidic of the two compounds. In fact, cyclopropane has a pK_a of 46.

E. Effect of solvent

Before leaving the topic of structure–acidity relationships, we should point out one very important disclaimer. Over the past few years it has become increasingly clear that the solvent often plays a dramatic role in determining the relative acidities and basicities of compounds. Thus some of the effects that were previously believed to be intrinsic (that is, only a function of the molecule's structure) are now believed to be a result of solvent–solute interactions. The order of acidities given in Table 10-2 and the effects listed in Table 10-3 are very useful for rationalizing a wide variety of properties and reactions of organic molecules. However, there can be some significant changes in relative acidity when reactions are carried out in the (solvent-free) gas phase.

10-6. Free-Radical Reactions

A free radical, you will recall from Section 4-1, is an *uncharged* species with one (or more) *unpaired* electrons. A carbon radical is *trivalent* (three groups attached) with one unpaired electron. Most carbon radicals are high-energy species with short lifetimes that occur as intermediates in free-radical reactions. But a few free radicals are stable enough to be **persistent**, that is, they have sufficiently long lifetimes to be readily studied or even isolated. Free radicals are usually formed from the homolytic cleavage (Section 4-2) of a covalent bond.

A. Nature of free-radical reactions

Reactions that involve carbon free-radical intermediates are less common than polar reactions. This is because free-radical reactions can have very complicated mechanisms, and they often result in a variety of products. As a result, free-radical reactions, as a class, have been less useful and less studied than polar reactions. Still, there are some unique and interesting things about free-radical reactions, and these are worth discussing.

Free-radical reactions often involve relatively nonpolar bonds to terminal (Section 3-1) monovalent atoms such as hydrogens attached to carbon. Recall that most hydrogens bonded to carbon (unless they are activated by a nearby functional group) are quite *un*reactive under polar conditions. (In fact, one of the hottest new areas of organic research is to synthesize catalysts that will activate otherwise inactive C–H bonds.) But these same "nonpolar" hydrogens can become quite a bit more reactive under free-radical conditions.

Because C–H bonds are quite strong (Section 9-1), they will not undergo spontaneous homolytic cleavage to form radicals except at *very* high temperatures. Still, the homolytic cleavage of a C–H bond can be brought about at low temperature by reaction with a previously formed radical (R·). In this reaction, which is called a **hydrogen atom abstraction**, the R· radical becomes bonded to the hydrogen, leaving behind a new carbon radical:

(Recall that we use fishhook arrows to indicate the motion of one electron, Section 4-2.) But the questions remains, where does the initiating radical R· come from?

B. Initiation

A common characteristic of most radical reactions is the need for an **initiation step** in which the initiating radical R• is formed by the homolytic cleavage of a specific bond. This cleavage can be brought about in either of two ways. If the bond is sufficiently weak, it can be cleaved thermally when the temperature is raised only modestly (less than 100°C). Compounds that undergo such facile cleavage are called **free-radical initiators**, and they include the compounds shown below (Δ indicates a thermal process):

$$(CH_3)_3C-O-O-C(CH_3)_3 \quad \xrightarrow{\Delta} \quad 2\ (CH_3)_3C-O\cdot$$

$$(CH_3)_2\underset{\underset{CN}{|}}{C}-N=N-\underset{\underset{CN}{|}}{C}(CH_3)_2 \quad \xrightarrow{\Delta} \quad 2\ (CH_3)_2\underset{\underset{CN}{|}}{C}\cdot\ +\ N_2\uparrow$$

$$I-I \quad \xrightarrow{\Delta} \quad 2\ I\cdot$$

Note that free-radical initiators are characterized by at least one weak bond.

Alternatively, free-radical reactions can be initiated photochemically. A photon of high-energy (usually ultraviolet) light is absorbed by a specific bond, giving the electronically excited state of that bond that then cleaves homolytically. While unactivated C–H bonds are not very **photolabile** (prone toward photochemical cleavage), certain bonds such as the one in Cl_2 are readily **photodissociated**:

$$:\ddot{C}l-\ddot{C}l: \quad \xrightarrow{h\nu} \quad \left[:\ddot{C}l-\ddot{C}l:\right]^* \quad \longrightarrow \quad 2\ :\ddot{C}l\cdot$$
excited state

In this context, $h\nu$ is the energy of the photon whose frequency is ν, and the asterisk ($*$) indicates an electronically excited state. Note also that the chlorine atoms are themselves free radicals.

C. Propagation

Here is another unique feature of many free-radical reactions. Often, only a small amount of initiator or a short period of irradiation is all that is required to start the reaction. Once started, the reaction is self-sustaining and requires no further initiation. These are the characteristics of a free-radical **chain mechanism**. The feature that makes a chain mechanism unique is that, once initiated, the reaction consists of a series of **propagation steps** that involve repeated formation and consumption of free-radical intermediates called **chain carriers**. The reaction continues as long as the chain carriers are able to propagate.

The classic example of a free-radical chain mechanism is the photo-initiated chlorination of a hydrocarbon such as methane. The *net* reaction for this process is

$$CH_4\ +\ Cl_2 \quad \longrightarrow \quad CH_3Cl\ +\ HCl$$

From the standpoint of thermodynamics this is a favorable reaction with a $\Delta G°$ (gas phase) of -103 kJ mol^{-1}. The reaction is initiated by the photodissociation of a chlorine molecule into two chlorine atoms. The two alternating propagation steps are these:

$$H-\underset{\underset{H}{|}}{\overset{\overset{H}{|}}{C}}-H \quad \cdot\ddot{C}l: \quad \longrightarrow \quad H-\underset{\underset{H}{|}}{\overset{\overset{H}{|}}{C}}\cdot\ +\ H-\ddot{C}l:$$

$$H-\underset{\underset{H}{|}}{\overset{\overset{H}{|}}{C}}\cdot \quad :\ddot{C}l-\ddot{C}l: \quad \longrightarrow \quad H-\underset{\underset{H}{|}}{\overset{\overset{H}{|}}{C}}-\ddot{C}l:\ +\ \cdot\ddot{C}l:$$

In the first propagation step, a chlorine atom abstracts a hydrogen from methane to form HCl (one of the ultimate products) and a methyl radical. In the second step the methyl radical abstracts a chlorine atom from Cl_2, forming CH_3Cl (the other ultimate product) and another chlorine atom. This chlorine atom then repeats the first step, which leads to a repetition of the second step, and so on. All it takes (in principle) is *one* chlorine atom to begin the chain, which then continues until one or both of the reactants is completely consumed. Above all, note that

- The algebraic sum of the propagation steps is the same as the net reaction.

$$CH_4 + Cl\cdot \longrightarrow \cdot CH_3 + HCl$$
$$\cdot CH_3 + Cl_2 \longrightarrow CH_3Cl + \cdot Cl$$
$$\overline{CH_4 + \cancel{Cl}\cdot + \cdot\cancel{CH_3} + Cl_2 \longrightarrow \cdot\cancel{CH_3} + HCl + CH_3Cl + \cdot\cancel{Cl}}$$

Here the chain carriers are the chlorine atom (radical) and the methyl radical. In this case the initiating radical ($Cl\cdot$) was also a chain carrier, but this is not usually the case.

D. Termination

As you might have guessed, there is a way that the free-radical chain can be broken, and that is when a chain carrier is somehow intercepted and prevented from propagating. Although the chain can be started with but a single chlorine atom, in practice there will be some number of initiating radicals formed during the initiation step—not a large number when measured in grams or moles, but a respectable number nonetheless. And these initiating radicals will either be converted into chain carriers or (as is the case with the chlorine atoms) they will serve as chain carriers themselves. However, if any two of these radicals happen to encounter each other, rather than their intended reactant molecule, they will undergo **coupling** (homogenic bond formation, Section 4-2), taking two chain carriers out of commission. There are three so-called **termination steps** possible in the chlorination of methane:

$$2\ :\!\ddot{C}l\cdot \longrightarrow Cl_2$$
$$2\ \cdot CH_3 \longrightarrow CH_3CH_3$$
$$:\!\ddot{C}l\cdot + \cdot CH_3 \longrightarrow CH_3Cl$$

Clearly, the chain reaction can continue only as long as the chain carriers avoid termination.

E. Inhibition

Because there are relatively few chain carriers present at any instant (compared to the number of reactant molecules), free-radical chain reactions are very sensitive to even small amounts of substances that can "trap" free radicals. One such substance is molecular oxygen, which itself is a **biradical** (a species with *two* unpaired electrons, Problem 4-8). Free-radical chain reactions are notoriously sensitive to the presence of oxygen, and it must be scrupulously avoided if a chain reaction is to be carried out or studied. On the other hand, it is often useful to *prevent* free-radical processes when they are undesirable side reactions. This can be done by adding a small amounts of **free-radical inhibitors** (also known as **free-radical scavengers**)—compounds that are capable of intercepting chain-carrying radicals and converting them to less reactive species that are not able to propagate the chain. Common antioxidants such as BHT and BHA serve just this purpose in a wide variety of applications.

BHT   BHA $+$ = *t*-butyl

EXAMPLE 10-13 Both BHT and BHA intercept chain-carrying radicals by donating the phenolic hydrogen atom. Explain why the resulting phenoxy radical is very stable and incapable of supporting further chain reaction.

Solution The phenoxy radical is stabilized by a resonance interaction with the phenyl ring, which delocalizes the unpaired electron.

This is an exceptionally stable free radical, not only because the unpaired electron is delocalized, but also because the sterically bulky tertiary butyl groups inhibit the phenoxy radical from undergoing any coupling or abstraction reactions.

EXAMPLE 10-14 Propose a mechanism for the net reaction

given the following facts: (1) The reaction requires a free-radical initiator or photochemical activation, but once initiated it is self-sustaining; and (2) the reaction is suppressed in the presence of BHT.

Solution The characteristics of the reaction (involves a nonpolar C–H bond, requires initiation, responds to inhibitors) suggest that it goes by a free-radical chain mechanism. We'll assume the initiation step produces some initiating radical R·, which abstracts a hydrogen from $\geq$C–H, leaving a chain-carrying carbon radical:

The propagation steps are

The chain carriers are the carbon radical and the peroxy radical, and the algebraic sum of the propagation steps gives the net reaction. The termination steps are any encounters between two chain-carrying radicals. This reaction, called **autoxidation**, is usually an undesirable process leading to the degradation of organic compounds in foods, medicines, polymers, etc. But autoxidation can be suppressed by the presence of compounds such as BHA and BHT.

F. Multiple substitution during free-radical reactions

Although the free-radical chlorination of methane might seem to be a good way to synthesize chloromethane, the fact is that there are several complications. In addition to chloromethane, there are other side products formed including dichloromethane, trichloromethane (chloroform), and even tetrachloromethane (carbon tetrachloride). The relative amounts of these products depend on such factors as the initial ratio of methane to chlorine and the length of time the reaction is allowed to proceed.

The side products begin to arise when a chain-carrying chlorine atom abstracts a hydrogen from *chloromethane* instead of from methane. This leads to a second set of propagation steps in competition with the original ones:

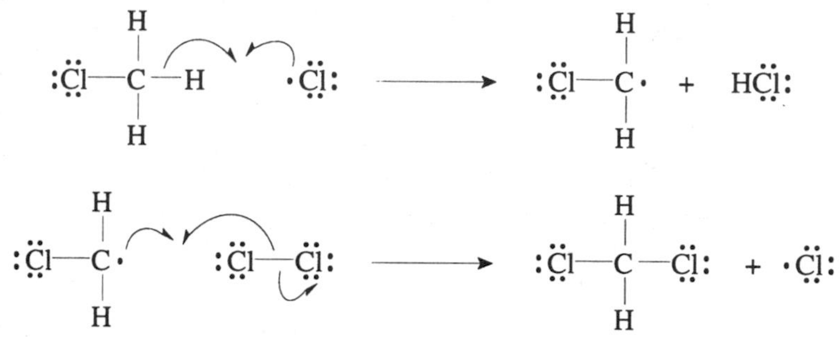

This diversion of an original chain carrier into a new set of propagation steps becomes more probable as the concentration of methane decreases and the concentration of chloromethane (the primary product) increases. And, of course, as the concentration of *dichloromethane* increases, so does the likelihood of its suffering hydrogen abstraction to start a set of propagation steps leading to chloroform, and so on.

EXAMPLE 10-15 How many *dichlorinated* products are possible in the photo-initiated chlorination of ethane? Can you think of any reasons why one of these should be favored over the other(s)?

Solution The primary product of chlorination will be chloroethane. From symmetry considerations (Section 8-2) we can quickly see that the second chlorine can replace either one of the two equivalent methylene hydrogens (to give 1,1-dichloroethane) or one of the three equivalent methyl hydrogens (giving 1,2-dichloroethane):

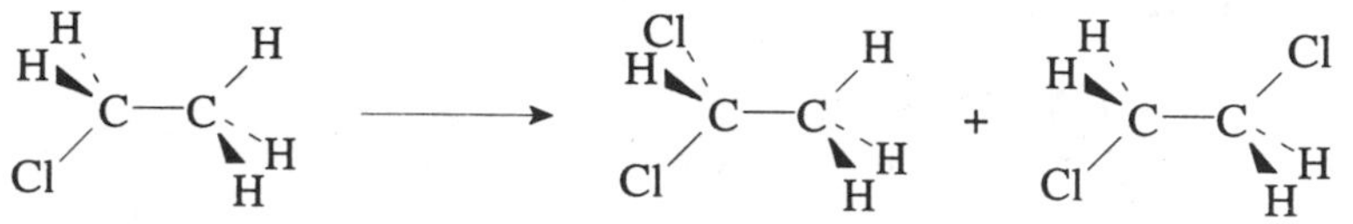

chloroethane   1,1-dichloroethane   1,2-dichloroethane

Initially, there seem to be only two reasons why one of these might be favored. The first is a steric effect (Section 7-3). Perhaps the fairly large chlorine on the methylene carbon would inhibit attack by the chain-carrying chlorine atom on a methylene hydrogen. This would favor formation of the 1,2 isomer. Or there may be a statistical preference for attack at the methyl hydrogens (of which there are three) compared to the methylene hydrogens (of which there are only two). However, the original chlorine may also have a (de)stabilizing effect on the radical intermediate.

EXAMPLE 10-16 (a) Write the structures of all possible *structural* isomers (you may neglect stereoisomers) that would result from the *mono*chlorination of methylcyclohexane.

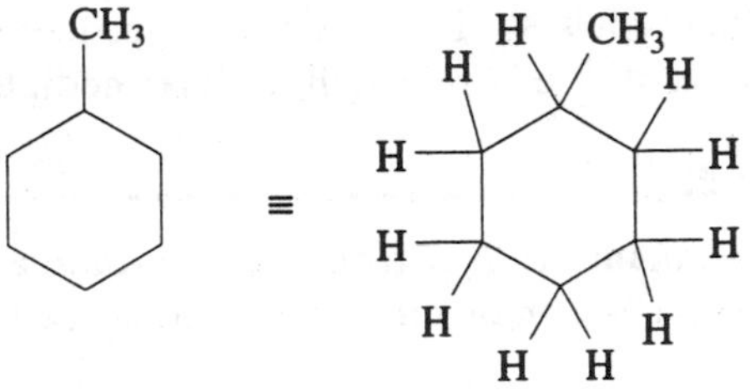

For each structure, give the number of equivalent hydrogens whose abstraction would lead to that product. **(b)** Which product(s) would be favored by statistical factors? Which by steric factors?

Solution

(a) By symmetry, there are five different monochlorinated isomers. The number of equivalent hydrogens whose abstraction would lead to each product is shown in parentheses below each structure.

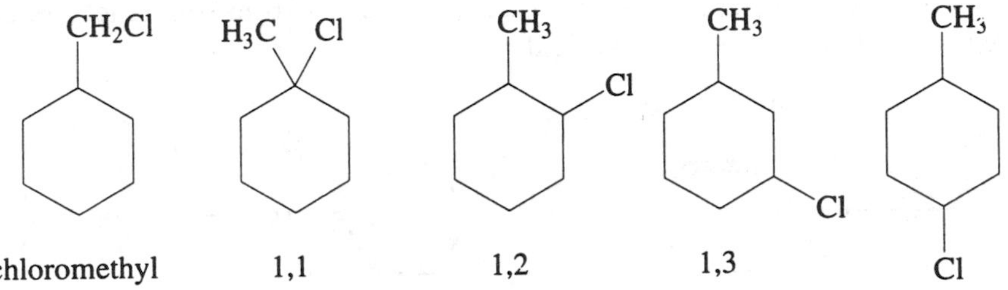

chloromethyl	1,1	1,2	1,3	1,4
(3)	(1)	(4)	(4)	(2)

(b) Statistically, the 1,2 and 1,3 isomers would be favored, because there are four ways to make each one. Steric effects would seem to favor formation of the chloromethyl isomer and the 1,3 and 1,4 isomers, because interference *by* the methyl is avoided in formation of each of these.

G. Selectivity in free-radical reactions

After working Example 10-16, it may come as a shock to find that the 1,1 isomer is the favored product from the free-radical chlorination of methylcyclohexane! Therefore, there must be another factor even more important than statistical and steric effects. There is, and it is an electronic effect relating to the stabilities of the radical precursor of each product. These radical precursors are shown below:

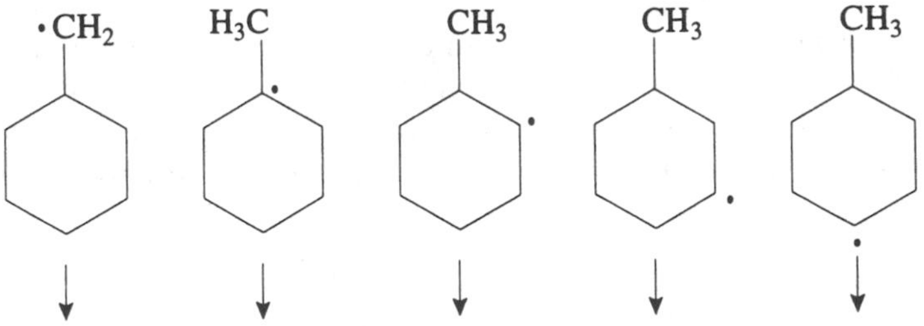

chloromethyl	1,1	1,2	1,3	1,4

You will notice that the first of these radicals is a *primary* one (*one* carbon attached to the carbon bearing the unpaired electron), the second one is *tertiary* (*three* carbons), while the rest are *secondary*. As was also true for carbocations (Problem 9-11), the stability order for carbon free radicals is: Tertiary more stable than secondary and secondary more stable than primary (all other things being equal). That is, alkyl groups attached to the carbon bearing the unpaired electron tend to stabilize the free radical by means of their electron-donating inductive and resonance (hyperconjugation) effects (Section 10-5). The transition state for hydrogen atom abstraction is so "late" that the most stable radical is formed the most rapidly, a manifestation of Hammond's postulate (Section 9-8). Thus the major product is the one that is derived from the most stable (i.e., most substituted) free-radical intermediate.

EXAMPLE 10-17 The predominant product from the free-radical bromination of cyclohexene with *N*-bromosuccinimide (NBS) is 3-bromocyclohexene, the so-called allylic isomer. Explain why this product is preferred.

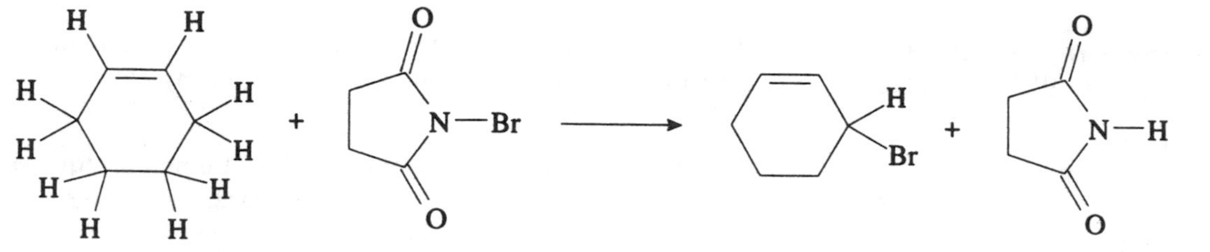

Solution There are three possible monobrominated products from this reaction. Each is shown below, together with its precursor free radical.

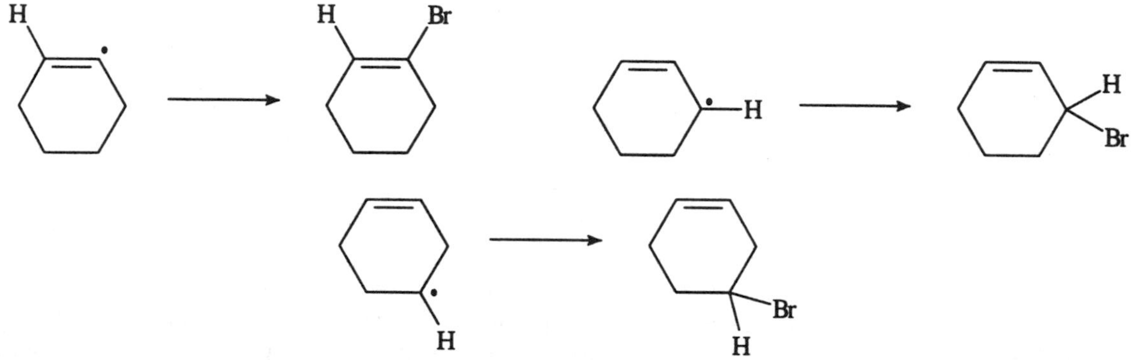

The fact that the 3-bromo isomer is favored indicates that the free radical leading to it is the most stable of the three precursor radicals. A moment's reflection should suggest why. This radical is stabilized by resonance with the double bond and is an example of an allylic radical (Example 4-11):

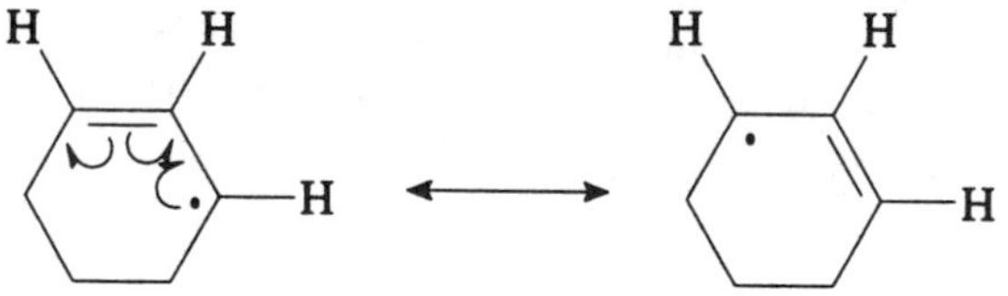

That this resonance interaction really occurs can be directly demonstrated by an isotope labeling experiment (Section 9-9). If we were to substitute the vinyl hydrogens with deuterium (D = ^{2}H), we could trace the fate of the original unpaired electron-bearing carbon. We would find that the 3-bromocyclohexene product is actually a mixture of the 1,2-dideuterio isomer and the 2,3-dideuterio isomer in equal amounts:

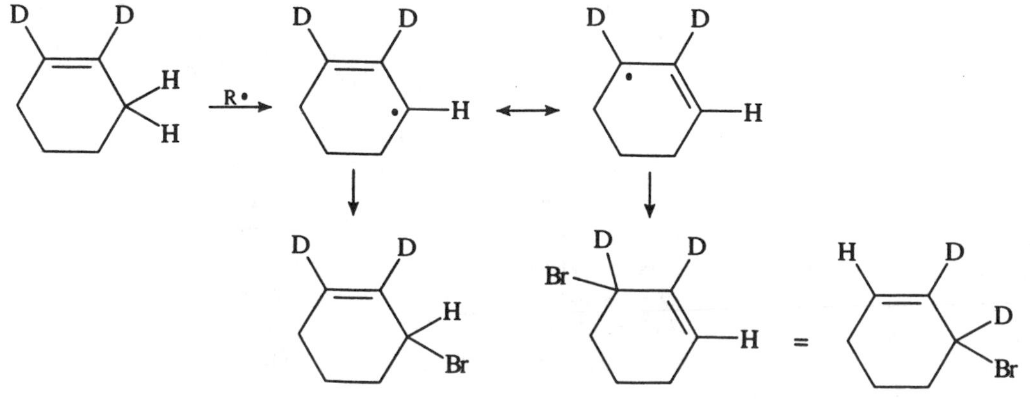

(1,2-)   (2,3-)

Finally, because they generally involve nonpolar reactants, products, intermediates, and transition states, free-radical reactions usually show little or no effect when solvent polarity is changed. Beware, though, that very few solvents are completely inert toward free radicals. Indeed, often the solvent itself becomes a reactant by providing an abundant source of abstractable hydrogens!

10-7. Oxidation–Reduction (Redox) Reactions

A. Oxidation state (number)

An atom or molecule is said to be **oxidized** when it *loses* one or more electrons. It is **reduced** when it *gains* one or more electrons. In most redox reactions electrons are transferred among atoms and molecules, so they're not really lost or gained permanently, but simply change position. For this reason, we can say that every oxidation is accompanied by a reduction, and vice versa. We can determine whether a given atom has been oxidized or reduced (or neither) during a reaction by noting any changes in its **oxidation number** (or **oxidation state**). The oxidation number (ON) of a single (unbonded) atom is equal to its formal charge (Section 4-1). When an atom is *oxidized*, its ON *increases* (becomes more positive or less negative), whereas *reduction* is accompanied by a *decrease* in ON (becoming more negative or less positive).

EXAMPLE 10-18 Identify which of the reactions below involve oxidation and which involve reduction. Provide the ON for each species.

$$Li\cdot \longrightarrow Li^+ + e^-$$

$$2e^- + \cdot\ddot{O}\cdot \longrightarrow :\ddot{\underset{..}{O}}:^{2-}$$

$$Fe^{2+} \longrightarrow Fe^{3+} + e^-$$

Solution

$$\left.\begin{array}{ccc} Li\cdot & \longrightarrow & Li^+ + e^- \\ \text{ON:} \quad 0 & & +1 \end{array}\right\}\text{oxidation}$$

$$\left.\begin{array}{ccc} 2e^- + \cdot\ddot{O}\cdot & \longrightarrow & :\ddot{\underset{..}{O}}:^{2-} \\ \text{ON:} \quad\quad 0 & & -2 \end{array}\right\}\text{reduction}$$

$$\left.\begin{array}{ccc} Fe^{2+} & \longrightarrow & Fe^{3+} + e^- \\ \text{ON:} \quad +2 & & +3 \end{array}\right\}\text{oxidation}$$

The ON of an atom *in a molecule* is the formal charge it *would* have if all the bonds in the molecule were ionic, assigning all bonding electrons to the more electronegative atom. (If two atoms have the same electronegativity, the electrons in the bond connecting them are equally divided between them.) Thus because carbon is (slightly) more electronegative than hydrogen, methane is viewed (for the purposes of calculating ON values) as

$$\begin{array}{c} H \\ | \\ H-C-H \\ | \\ H \end{array} = \quad H^+ \quad :\ddot{\underset{..}{C}}:^{4-} \quad H^+ \quad\quad \begin{array}{c} H^+ \\ \\ \\ H^+ \end{array}$$

In methane, the carbon has an ON of –4, while each hydrogen has an ON of +1. Notice that the overall charge on a molecule or ion (zero in the case of methane) is the sum of the ON values of each of its atoms.

EXAMPLE 10-19 Give the ON of each atom in the structures below by first redrawing each structure as if its bonds were ionic.

$$CH_3CH_3 \qquad CO_2 \qquad H_2CO \qquad HCOBr$$

Solution

Thus the ON of carbon can range from +4 (as in CO_2) to -4 (as in CH_4).

EXAMPLE 10-20 Determine the ON of each atom in the molecules below. Identify the redox reactions, showing which atoms (if any) are oxidized and which are reduced.

(a) $\qquad\qquad CH_3Br + OH^- \longrightarrow CH_3OH + Br^-$

(b) $\qquad\qquad H_2C=CH_2 + H_2 \longrightarrow CH_3CH_3$

(c) $\qquad\qquad H_2C=CH_2 + Br_2 \longrightarrow CH_2CH_2$
$\qquad\qquad\qquad\qquad\qquad\qquad\qquad\quad | \quad |$
$\qquad\qquad\qquad\qquad\qquad\qquad\qquad\ Br\ \ Br$

(d) $\qquad\qquad CH_4 + Cl_2 \longrightarrow CH_3Cl + HCl$

Solution

(a)

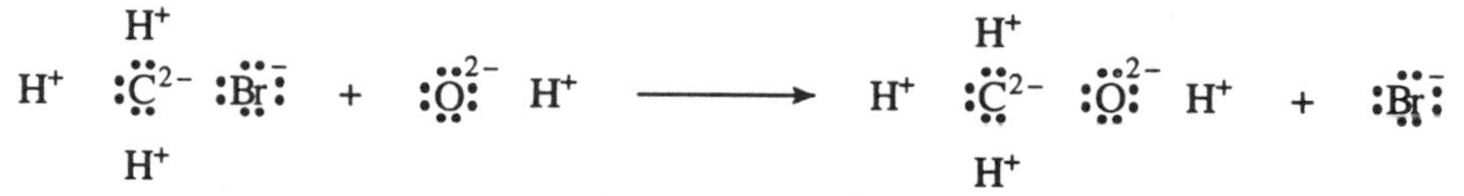

No atom changes ON, so this is not a redox reaction.

(b)

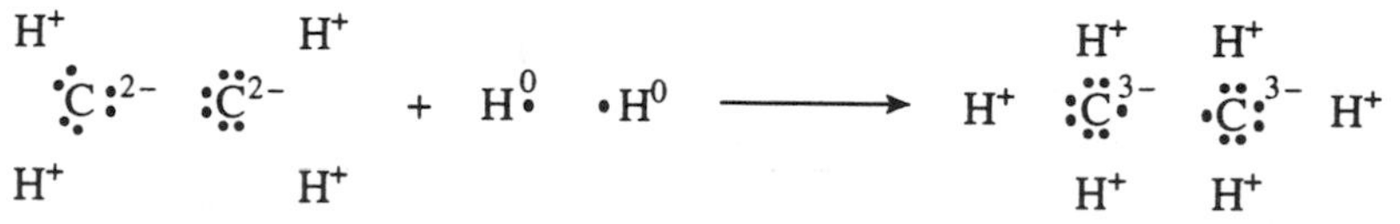

Each carbon is reduced by one electron, each hydrogen in H_2 is oxidized by one electron.

(c)

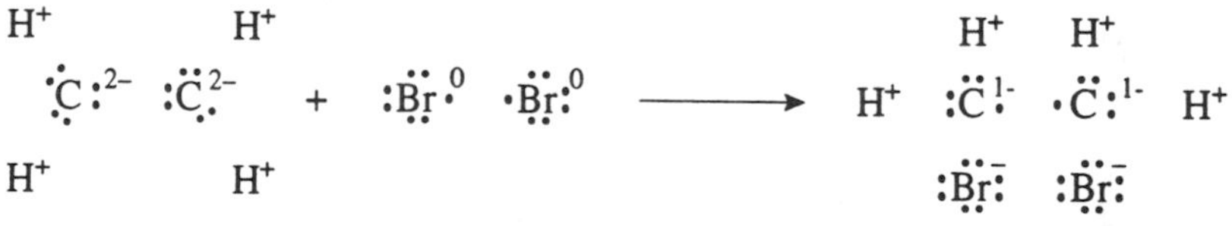

Each carbon is oxidized by one electron, each bromine in Br_2 is reduced by one electron.

(d)

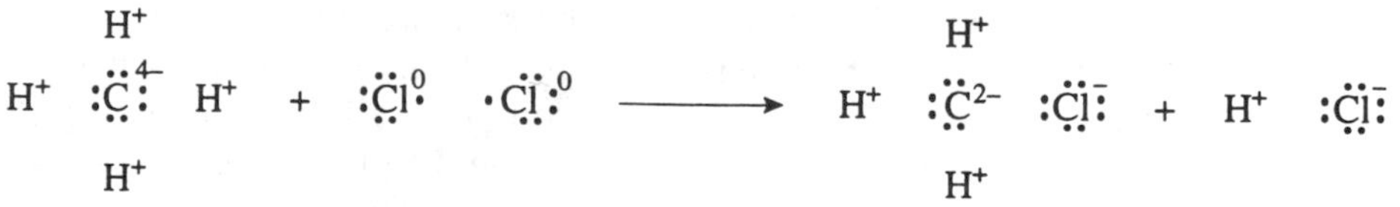

The carbon is oxidized by two electrons, while each chlorine in Cl_2 is reduced by one electron.

In all of these reactions except (a) there is both an oxidation and a reduction. You can't have one without the other!

B. Single-electron transfer

By far the vast majority of organic redox reactions are similar to the ones shown in Example 10-20, where the oxidations and reductions are indicated by changes in ON. It is quite rare for an organic molecule to simply lose or gain a single electron (a so-called **single-electron transfer**, or **SET**) in the same way that inorganic atoms and complexes do. Nonetheless, there are certain conditions under which such one-electron redox processes do occur. For example, during the ionization by electron impact that precedes mass spectroscopic analysis (Section 7-2), a molecule loses the most easily ionized electron to form the derived **radical cation**, so-called because it has both an unpaired electron and a positive charge. Thus ethylene loses one of its π-bonding electrons to give the ethylene radical cation:

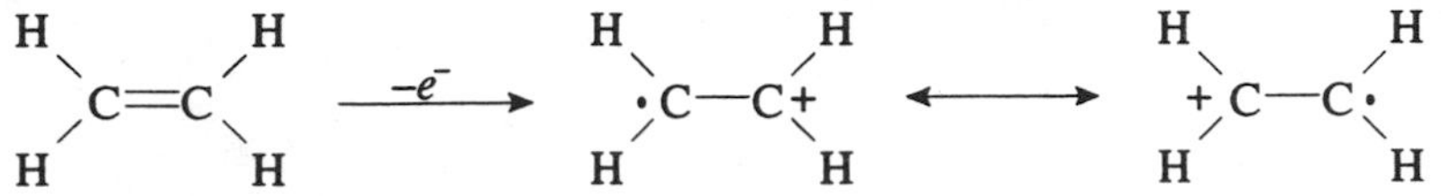

This process does involve oxidation at carbon, though the positive charge is equally shared by both carbons because of resonance.

Certain organic molecules can undergo one-electron reduction by accepting an electron into a low-energy vacant antibonding orbital (Section 2-7). For example, the aromatic molecule naphthalene (Example 4-13) reacts with metallic potassium to form the naphthalene **radical anion**. The electron goes into the vacant π^* (antibonding) orbital with lowest energy:

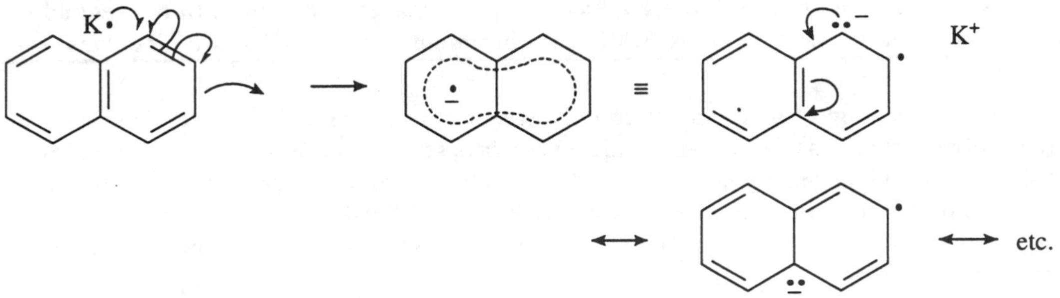

Here, the potassium atom is oxidized and naphthalene is reduced, though once again the charge and the unpaired electron are delocalized throughout the naphthalene molecule.

C. Complex free-radical redox reactions

Most organic redox reactions (except for those carried out electrochemically or by electron impact) occur either by polar or free-radical mechanisms, mostly the former. However, some such redox reactions proceed by very complicated mechanisms. For example, the combustion of a hydrocarbon with oxygen in a flame is one of the most complicated reactions known. Similarly, the relatively simple-appearing oxidation of a secondary alcohol to a ketone, using CrO_3, which has the net reaction

$$3 \ R_2C\overset{\displaystyle OH}{\underset{\displaystyle H}{\big<}} + \ 2 \ CrO_3 \longrightarrow 3 \ R_2C{=}O + 2 \ Cr(OH)_3$$

actually proceeds by two parallel series of steps, of which only one series is shown below:

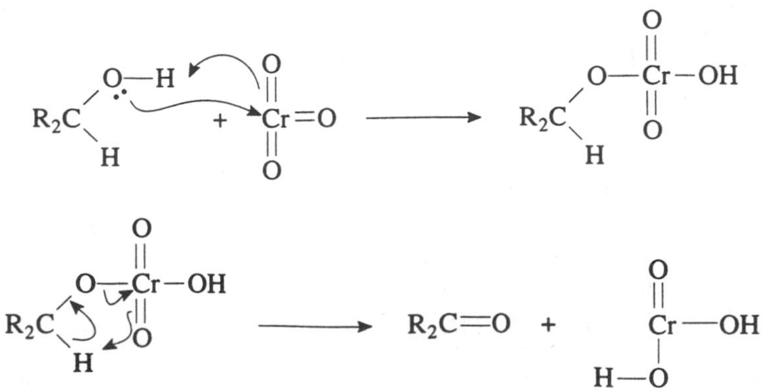

EXAMPLE 10-21 In the two steps above, assign ON values to each carbon and chromium atom. In which step does the oxidation/reduction actually occur?

Solution

The first step does not change any ON values, so it is not a redox reaction. In the second step the carbon is oxidized by two electrons, while the chromium is reduced by two electrons.

As you encounter more and more redox reactions in organic chemistry, you will begin to recognize certain classes of compounds that always serve as **oxidizing agents** (compounds that bring about oxidations by being reduced themselves), while others serve as **reducing agents** (compounds that bring about reductions by being oxidized themselves). But remember that some redox reactions, such as combustion and the CrO_3 oxidation, proceed by rather complicated mechanisms.

EXAMPLE 10-22 After reviewing Examples 10-20 and 10-21, state whether each of the reagents below is an oxidizing agent, a reducing agent, or neither:

$$OH^-, H_2, Br_2, Cl_2, CrO_3$$

Solution H_2 is a reducing agent; Br_2, Cl_2, and CrO_3 are all oxidizing agents; OH^- is neither.

10-8. Pericyclic Reactions

There are reactions of a certain type that appear so simple and straightforward that they used to be described as "no-mechanism" reactions. Of course, all reactions have mechanisms, but these reactions all seemed to go by one-step **concerted** (Section 9-9) mechanisms with no intermediates or other complications. Because each of these reactions could be viewed as involving a reorganization of electrons by way of a cyclic transition state, they have come to be known as **pericyclic reactions.**

As chemists began to study the mechanisms of pericyclic reactions in detail, they became aware of some very interesting peculiarities they presented. For example, many of these reactions were found to have an essentially exclusive stereochemical outcome (Section 8-9). That is, although two (or more) stereoisomers of a product seemed possible, only one was formed. In other cases, one of two very similar-appearing reactions would be found to go readily under thermal conditions (but not photochemically), while the other would go photochemically (but not thermally).

Eventually, these studies led to the **principle of orbital symmetry conservation**, which has been described in qualitative terms by the **Woodward–Hoffmann rules.** These rules allow us to predict the conditions under which a pericyclic reaction will occur, and what the stereochemical outcome will be. This work was sufficiently important to have resulted in the 1981 Nobel Prize in chemistry.

Although the principle of orbital symmetry conservation is applicable in some form to all reactions, it is especially useful in the context of pericyclic reactions. We'll discuss two subcategories of pericyclic reactions in detail, *electrocyclic reactions* (also called *valence isomerizations*) and *cycloadditions*. As we describe these reactions, remember that they occur by concerted mechanisms, with all bond making and bond breaking in a single step, and they proceed by a cyclic transition state.

A. Electrocyclic reactions

Electrocyclic ring closures are isomerizations in which a molecule with a conjugated π system is converted to a cyclic molecule with a new σ bond between the two termini (the atoms at the ends of the original system), with a change in position of the remaining π bonds. The reverse of this reaction is called an **electrocyclic ring opening**. An example of an electrocyclic ring closure is the thermal cyclization of *cis*-1,3,5-hexatriene (**1**) to 1,3-cyclohexadiene (**2**). This reaction can be written in the following way to emphasize the cyclic nature of the isomerization:

This reaction is thermodynamically favorable mainly because, in effect, a carbon–carbon π bond is being converted to a (stronger) carbon–carbon σ bond (Table 9-1). Yet there is a strange stereochemical preference that becomes apparent only when we start with an appropriately labeled derivative of **1**. Thus the dideuterio derivative below (**1-D$_2$**) gives only the cis isomer of **2-D$_2$**, and none of the trans isomer:

Even when the deuterium atoms are traded for larger methyl groups, the more sterically congested cis isomer is still formed in preference to the more stable trans isomer, an example of kinetic control (Section 9-8).

This **stereospecificity**, in which one particular stereoisomer of the reactant goes to one specific stereoisomer of the product, requires that the termini (the end CH$_2$ groups) of triene **1** must rotate in *opposite* directions (one clockwise, the other counterclockwise) as the new ring is formed. This sort of rotation is called **disrotation**.

By contrast, we can see that the trans isomer could arise only if the termini rotate in the *same* direction (both clockwise, or both counterclockwise). This sort of rotation is called **conrotation**.

Thus the electrocyclization of **1** is stereospecifically disrotatory, with no conrotatory component.

EXAMPLE 10-23 What symmetry element(s) (Section 8-2), if any, does the disrotation of **1** to **2** preserve? That is, are there any symmetry elements that are valid at all points along the reaction coordinate? How about for the conrotation?

Solution Triene **1** and diene **2** both have a plane of symmetry, which is preserved at all points along the disrotation reaction coordinate. This may be seen by drawing a representation of the transition state:

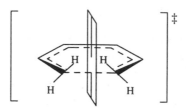

Compounds **1** and **2** also both possess a C_2 axis in the plane of the ring (passing between C-1 and C-6 and between C-3 and C-4), which would be preserved during conrotation:

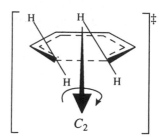

Now here's the real kicker. Like many pericyclic reactions, the isomerization of **1** to **2** can also be carried out photochemically. A photon of the appropriate frequency is absorbed by **1**, promoting one of the electrons in the molecule's highest energy bonding orbital to the next higher (antibonding) orbital, thereby giving an electronically excited state of **1** (denoted **1***). If the photoreaction is carried out at a sufficiently low temperature to suppress the thermal reaction, we find that the reaction goes by a stereospecifically conrotatory path! Thus **1-D$_2$** yields the trans isomer:

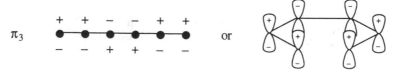

The question, then, is this: Why does the cyclization of **1** to **2** go by a disrotatory path under thermal conditions (which involve the electronic **ground state**), while it follows a conrotatory path under photochemical conditions (when the excited state is involved)?

There are several equivalent ways to rationalize the observed stereospecificity, but the easiest way is to examine the symmetry of the **highest occupied molecular orbital (HOMO)**. This orbital is the highest energy orbital with electrons in it. In the case of triene **1**, the HOMO is π_3 (Example 4-12), which looks like this:

$$\pi_3 \qquad \begin{array}{cccccc} + & + & - & - & + & + \\ \bullet & \bullet & \bullet & \bullet & \bullet & \bullet \\ - & - & + & + & - & - \end{array} \qquad \text{or}$$

In order for the termini to become bonded, the two orbital lobes that overlap to make the new σ bond must have the *same* phase (sign of the wave function, Section 2-7). In the case of π_3 this can happen only by disrotation (only one of the two possible disrotatory paths is shown:

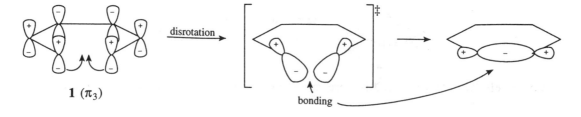

1 (π_3)

If π_3 were to close in a conrotatory fashion, an *antibonding* interaction would develop between the termini:

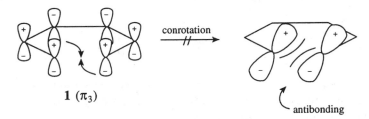

1 (π_3)

We thus describe the disrotatory ring closure as **thermally allowed** (by symmetry), and the conrotatory process as **thermally forbidden**.

In the photochemical isomerization of **1** to **2**, triene **1** absorbs a photon to form the first excited state (**1***), which has configuration $\pi_1^2 \pi_2^2 \pi_3^1 \pi_4^1$.

$$
\begin{array}{ccc}
\pi_6 & \text{———} & \text{———} \\[4pt]
\pi_5 & \text{———} & \text{———} \\[4pt]
\pi_4 & \text{———} & \uparrow\!\downarrow\text{ (↓)} \\[4pt]
\pi_3 & \uparrow\!\downarrow & \uparrow \\[4pt]
\pi_2 & \uparrow\!\downarrow & \uparrow\!\downarrow \\[4pt]
\pi_1 & \uparrow\!\downarrow & \uparrow\!\downarrow \\[4pt]
 & \mathbf{1} & \mathbf{1*}
\end{array}
$$

$\xrightarrow{h\nu}$

Now π_4 is the HOMO, and it looks like this (Example 4-12):

$$\pi_4 \quad \underset{-\;\;+\;\;+\;\;-\;\;-\;\;+}{\overset{+\;\;-\;\;-\;\;+\;\;+\;\;-}{\bullet\!-\!\bullet\!-\!\bullet\!-\!\bullet\!-\!\bullet\!-\!\bullet}} \quad \text{or}$$

The phase relationship between the termini is exactly reversed compared to π_3. Now *conrotation* leads to the bonding relationship (again, only one of two possible conrotatory paths is shown):

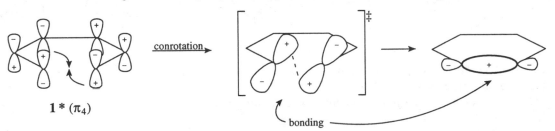

1* (π_4)

bonding

Thus the *conrotatory* cyclization of **1** is photochemically allowed and thermally forbidden. When a reaction is symmetry *forbidden*, it means that there is a large symmetry-imposed component to the activation barrier, making the reaction essentially impossible *by a concerted mechanism*. For a symmetry-*allowed* reaction, there is no symmetry component to the activation barrier, though other factors (steric, electronic, and so on) may still give rise to a substantial barrier.

From the principle of microscopic reversibility (Section 9-5), we know that if a disrotatory path is allowed for the *forward* thermal reaction, it will also be allowed for the *reverse* (i.e., ring-opening) thermal reaction. Thus if we were to cause diene **2** to undergo thermal ring opening back to **1** (a thermodynamically unfavorable process requiring the input of energy), it would proceed by both of the two possible disrotatory paths:

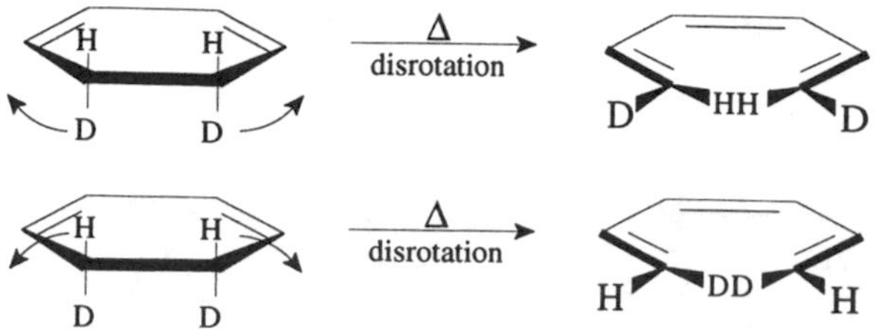

To summarize, when trying to deduce the allowed stereochemical result of an electrocyclic reaction, it is easiest to examine the HOMO of the open (noncyclic) molecule, regardless of whether it's the reactant or the product.

EXAMPLE 10-24 Predict the symmetry-allowed stereochemical outcome for the thermal and photochemical ring opening of labeled cyclobutene **3** to 1,3-butadiene derivatives **4a–c**; i.e., which product(s) is (are) formed under each set of conditions?

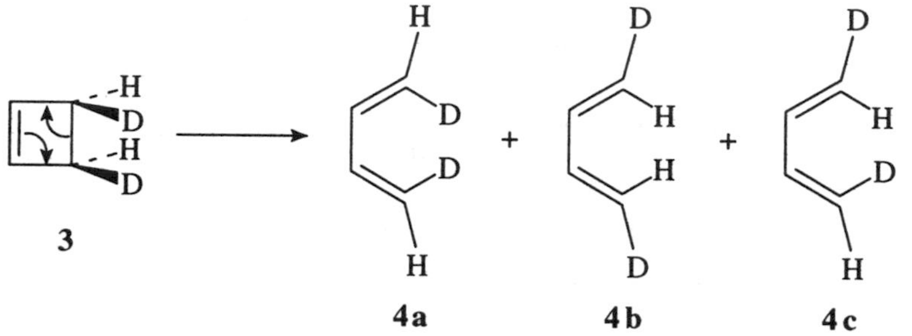

Solution For both the thermal and photochemical processes we'll examine the HOMO of the open form, the 1,3-butadiene product. For the thermal reaction the HOMO is π_2 (Section 4-6), which looks like this:

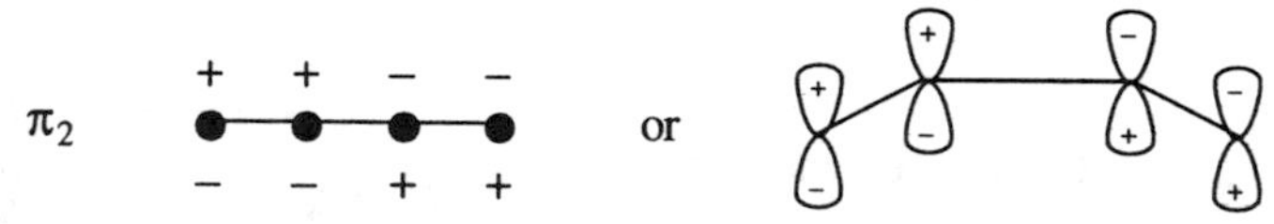

Clearly, only *conrotation* will result in a bonding interaction between the termini:

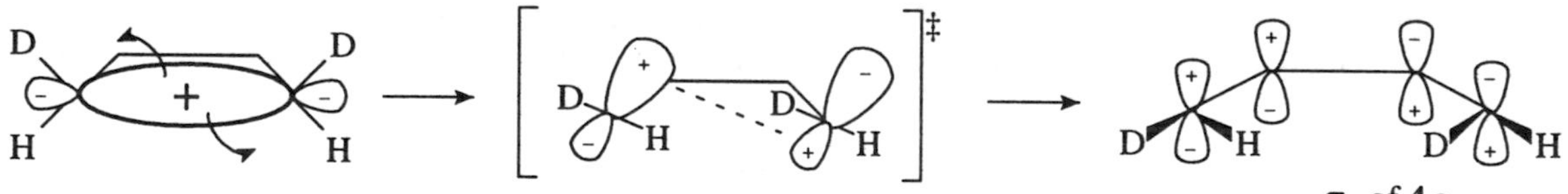

Therefore, the *thermal* ring opening of **3** will give only **4c**, since either conrotation (both clockwise or both counterclockwise) gives **4c**.

On the other hand, the *photochemical* ring opening involves π_3 of butadiene as the HOMO; it looks like this:

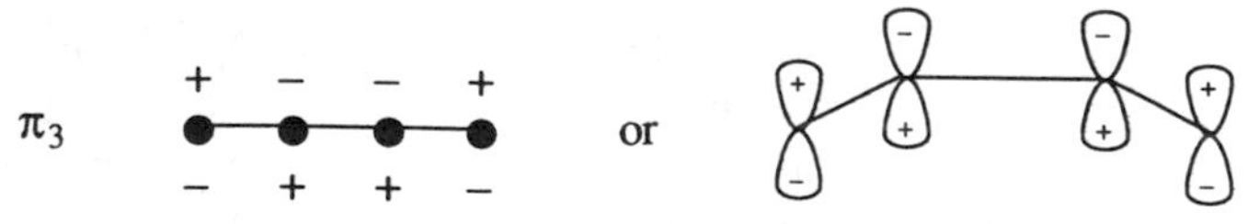

In this case, only *disrotation* provides the needed bonding relationship:

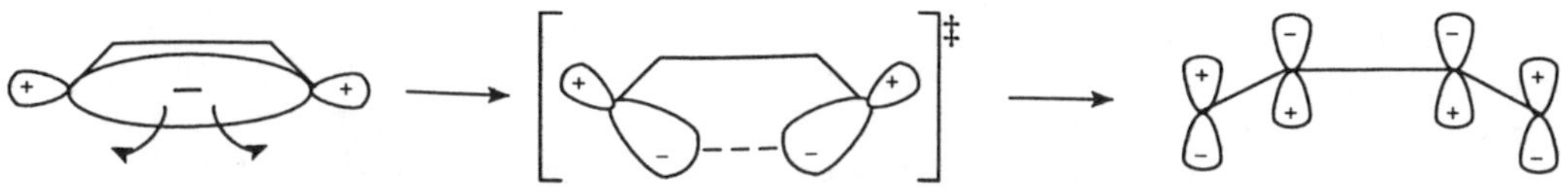

Thus the photochemical ring opening can occur by either disrotatory path, giving **4a** and **4b**:

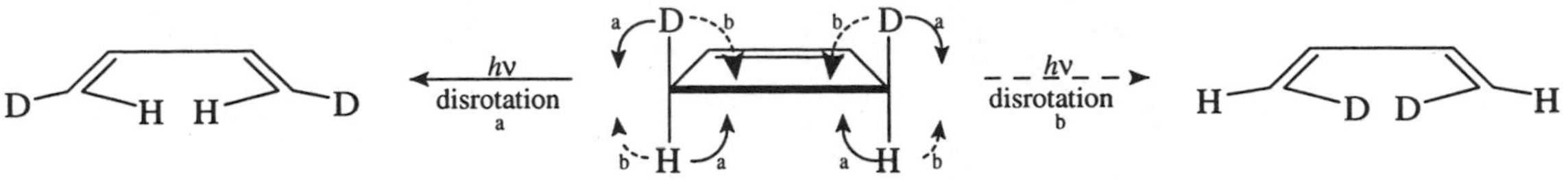

B. Cycloadditions

A second very important class of pericyclic reactions involves **cycloadditions**, in which two unsaturated molecules join by converting two π bonds into two new σ bonds between their termini. The classic example of a cycloaddition is the **Diels–Alder reaction**, where a diene such as 1,3-butadiene (or one of its derivatives) reacts with an alkene to form a cyclohexene derivative:

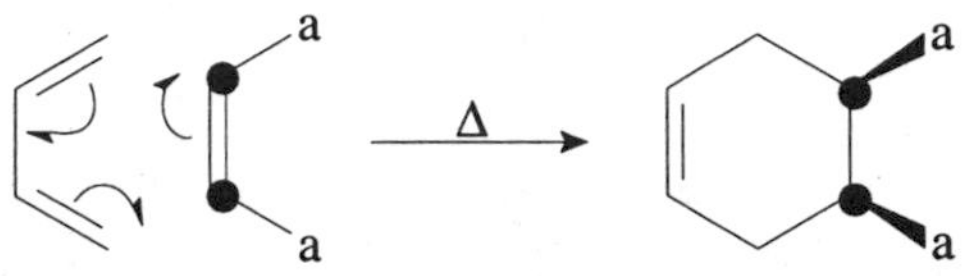

In this drawing the alkene carbons are shown as dots so we can keep track of them.

The Diels–Alder reaction is one of the most important and versatile synthetic reactions in all of organic chemistry. It usually occurs under relatively mild (thermal) conditions, and it preserves the stereochemistry of the alkene fragment. That is, the groups (e.g., the a's above) that were cis in the alkene remain cis in the cyclohexene product. We describe the Diels–Alder reaction as a 4π + 2π cycloaddition, which means that one fragment (the diene) supplies four π electrons to the reaction, while the other (the alkene) supplies two π electrons. A total of six electrons are involved in the reaction.

The alkene partner is often called the **dienophile** ("diene lover"). The most reactive dienophiles are those with one or more electron-withdrawing substituent groups (Section 10-5). This seems to indicate that the transition state of the Diels–Alder reaction involves a shifting of electron density *from* the diene *toward* the dienophile. Particularly good dienophiles include the ones shown below:

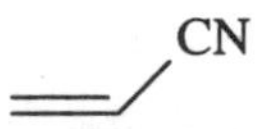

acrylonitrile

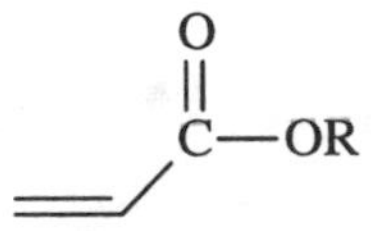

acrylate esters

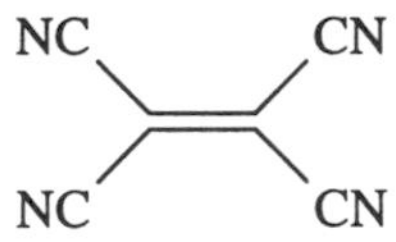

tetracyanoethylene

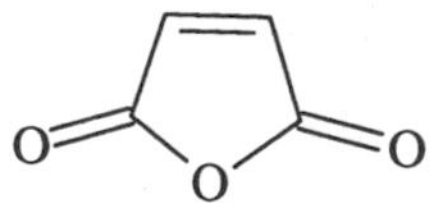

maleic anhydride

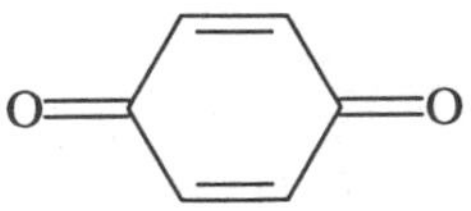

benzoquinone

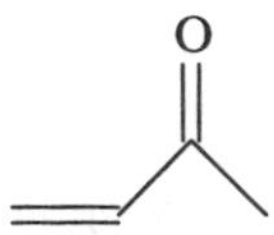

methyl vinyl ketone

The list of dienophiles can be extended to include molecules with other types of reactive π bonds, such as

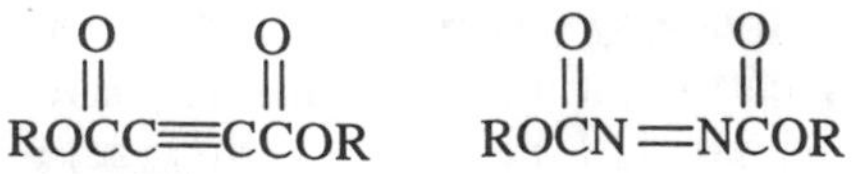

ROCC≡CCOR   ROCN=NCOR

EXAMPLE 10-25 Predict the structure of the product of each Diels–Alder reaction below:

(a)

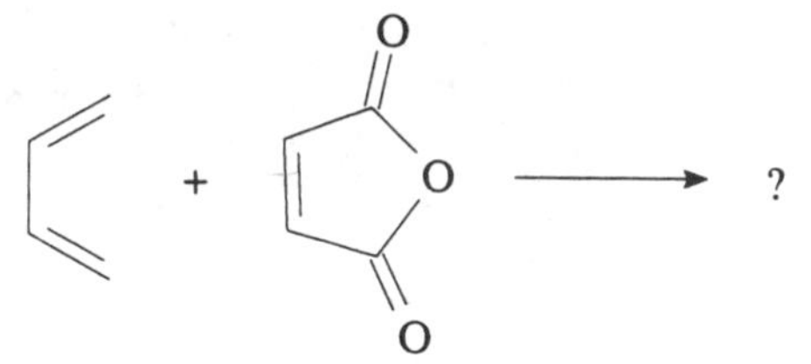

(b)

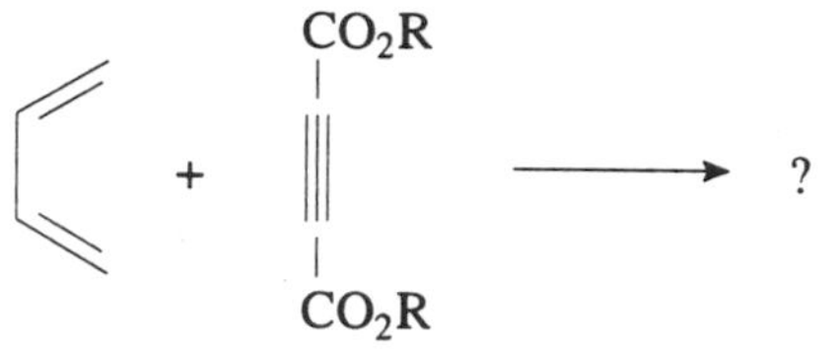

(c)

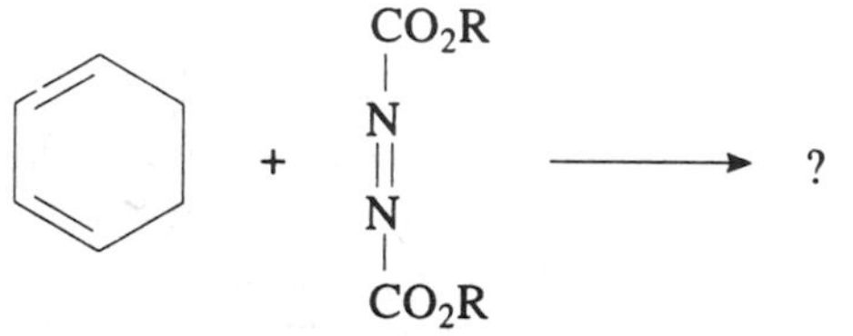

Solution

(a)

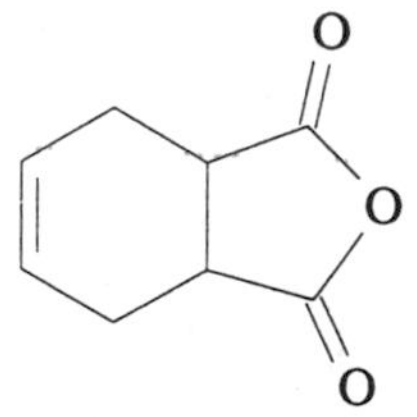

(b)

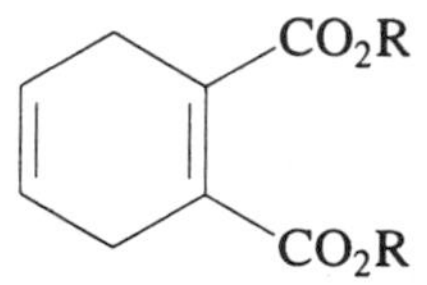

(c)

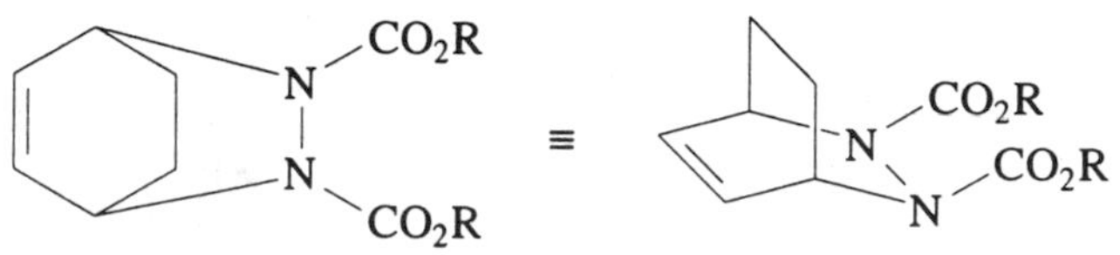

A reaction that bears a close similarity to the Diels–Alder reaction is the cycloaddition of two alkenes to form a cyclobutene derivative, with preservation of the stereochemistry of both alkene fragments.

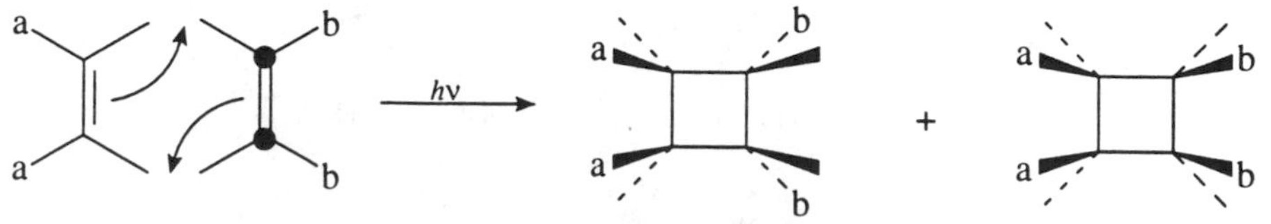

Yet this reaction, which we can describe as a $2\pi + 2\pi$ cycloaddition, does *not* proceed thermally. Instead, it occurs only when one (but not both!) of the reactants is photochemically converted to its excited state. Can symmetry considerations help us explain why the $4\pi + 2\pi$ cycloaddition is thermally allowed, while the $2\pi + 2\pi$ reaction is thermally forbidden and photochemically allowed? You bet!

A successful cycloaddition reaction requires that, in the transition state, there be a developing bonding interaction between the two reactants at both their termini. This can be accomplished if the symmetry (orbital phases) of the HOMO of the diene matches the symmetry of the **lowest unoccupied molecular orbital (LUMO)** of the dienophile. (The HOMO and LUMO are sometimes referred to as **frontier orbitals**.) Let's see how this works.

Suppose we compare the phases of the diene HOMO (π_2) with the LUMO of the dienophile

(π^*, Section 3-5). Note how the two sets of termini have the appropriate symmetry for bonding interactions:

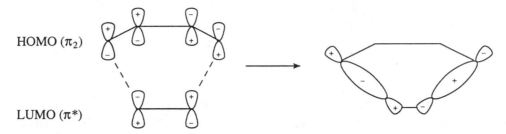

By contrast, the *thermal* $2\pi + 2\pi$ cycloaddition would involve the HOMO of one alkene (π) and the LUMO (π^*) of the other. Clearly, the termini do not possess the required symmetry, and therefore this reaction is symmetry forbidden.

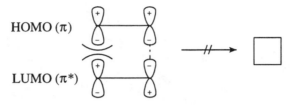

On the other hand, if one of the alkene partners is photochemically converted to its excited state, the HOMO becomes π^*:

Now, the HOMO symmetry of the excited alkene (π^*) *does* match that of the LUMO of the ground state partner (also π^*), so the $2\pi + 2\pi$ cycloaddition *is* photochemically allowed:

EXAMPLE 10-26 Predict the outcome of the *photochemical* reaction between 1,3-butadiene (which absorbs the photon) and ethylene, assuming that the temperature is kept low enough to suppress the Diels–Alder reaction.

Solution Because the $4\pi + 2\pi$ reaction is thermally allowed (and therefore photochemically forbidden), whereas the $2\pi + 2\pi$ reaction is just the opposite, this reaction will result in cyclobutane formation. There are several $2\pi + 2\pi$ reactions that are possible:

(See also Problem 10-9.)

EXAMPLE 10-27 Account for the fact that the cycloaddition below is found to occur thermally, albeit at somewhat elevated temperature. (*Hint*: Note the stereochemical outcome.)

Solution The most important thing to recognize is that the stereochemistry of the alkene on the left was *not* preserved. The "a" groups, originally cis in the reactant, are trans in the product. There are two possible explanations for this result. First, it is possible that the reaction is going by a stepwise (nonconcerted) mechanism. If this is true, the symmetry restrictions no longer apply because bond making and bond breaking occur in different steps. For example, perhaps only one bond is created in the first step, leaving a biradical intermediate (**5**), which undergoes ring closure in the second step:

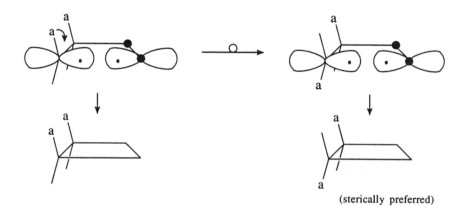

5

Because **5** would have a finite lifetime, there is the likelihood that it would undergo bond rotations prior to ring closing. These rotations can scramble the stereochemistry of each alkene fragment:

(sterically preferred)

The other possibility is that the reaction *does* go by a concerted mechanism, but that it follows a different approach geometry (symmetry) than the "face-to-face" approach shown above for the 2π + 2π cycloaddition. For example, a "crossed" approach of the HOMO (π) and LUMO (π*) *would* have the appropriate phase relationships for bonding between the two sets of termini.

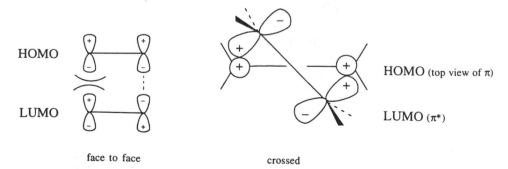

face to face   crossed

In the crossed geometry shown, we say that the HOMO has reacted **suprafacially** (from lobes on the *same* side of the molecular plane), while the LUMO has reacted **antarafacially** (from lobes on opposite sides). This is why the stereochemistry of the antarafacial fragment is switched (cis groups become trans, and vice versa), while the stereochemistry of the suprafacial fragment is preserved.

For completeness' sake, we'll close this section by stating that, for the most part, pericyclic reactions have relatively nonpolar transition states, so that changes in solvent polarity have little effect on reaction rates. Nonetheless, certain Diels–Alder reactions can be facilitated by Lewis acids or ionic catalysts!

10-9. Predicting Reaction Mechanisms

In the preceding chapter we discussed the kinds of experimental information we must have to formulate a reaction mechanism. In this chapter we have described the basic categories of organic reactions and their key characteristics. As you encounter organic reactions, you'll often be interested in at least *suggesting* a plausible mechanism, even though you may not have all of the thermodynamic and kinetic data you'd like. If this is the case, here's a procedure to help you get started. First, write the *balanced* net reaction. Identify all the bonds that are being made, broken, or relocated in the reaction. Then answer these questions:

1. Are these bonds polar or nonpolar?
2. Can the reaction be viewed as a Lewis acid–base reaction, or is there any indication of acid or base catalysis? (Polar reaction)
3. Are any of the reactants or products ionic? (Polar reaction)
4. Does an increase in solvent polarity increase the reaction rate? (Polar reaction)
5. Does the reaction require an initiator or photo-initiation? (Free-radical reaction)
6. Is there any evidence for free-radical intermediates or a chain mechanism? (Free-radical reaction)
7. Is the reaction affected by free radical inhibitors? (Free-radical reaction)
8. Do the oxidation numbers of any atoms change during the reaction? (Redox reaction)
9. Does the reaction involve the interconversion of π and σ bonds; if so, can the reaction be pictured as occurring through a cyclic reorganization of electrons, and does it seem to be concerted? (Pericyclic reaction)

Once you have gone through this checklist, you should have at least a start in proposing a mechanism for the reaction. Of course, your greatest asset in suggesting a mechanism is a backlog of experience in studying the mechanisms of a wide variety of organic reactions. You will get this by studying Volume II!

SUMMARY

1. Organic reactions can be divided into three broad categories; these categories and their characteristics are:
    (a) Polar reactions
        (1) polar bonds are broken heterolytically and made heterogenically
        (2) acid or base catalysis is often involved
        (3) reactions can be viewed as Lewis acid–base (electrophile–nucleophile) processes
        (4) ions are often involved as reactants, products, or intermediates
        (5) solvent polarity influences the rate of the reaction
    (b) Free-radical reactions
        (1) reactions involve free-radical intermediates formed by homolytic cleavage and abstraction processes of nonpolar bonds.
        (2) reactions often require an initiation step and show other characteristics of free-radical chain reactions
        (3) reactions are sensitive to small amounts of free-radical inhibitors
    (c) Pericyclic reactions
        (1) reactions involve a *concerted* electronic reorganization (interconversion of π and σ bonds) through a cyclic transition state
2. The oxidation number (ON) of an atom in a molecule is the formal charge it *would* have *if* all bonds in the molecule were ionic.
3. Oxidation involves the loss of electrons (or the increase in oxidation number), while reduction

involves the gain in electrons (or a decrease in oxidation number).

4. Organic redox (oxidation–reduction) reactions involve the transfer of electron density in such a way that the oxidation number of two or more atoms changes. Such reactions can proceed by polar, free-radical, or even pericyclic mechanisms.

5. There are three common acid–base concepts; their names and characteristics are summarized below:

Concept	Acid	Base	Reaction product
Arrhenius	produces H⁺ in water	produces OH⁻ in water	H_2O + salt
Brønsted–Lowry	proton donor	proton acceptor	conjugate acid + base
Lewis	electron-pair acceptor	electron-pair donor	complex

6. The strength of a Brønsted acid, an indication of its proton-donating ability, can be quantitatively expressed by the equilibrium constant K_a (or pK_a) for the reaction of the acid with some standard basic solvent such as water.

7. The stronger an acid is, the weaker its conjugate base is (as a base).

8. The strongest acid that can exist in significant concentration in generic solvent HSol is H_2Sol^+. The strongest base is Sol^-.

9. Acidity of hydrides (compounds with hydrogen) *increases* from left to right across periods, and from top to bottom down the groups, of the Periodic Table. In addition, resonance and inductive electron-withdrawing effects tend to strengthen acids and weaken their conjugate bases.

10. A free-radical reaction consists of the following steps:
    (a) Initiation step(s), in which a thermal or photochemical homolytic cleavage forms initiating radicals, which in turn produce chain-carrying radicals
    (b) Propagation steps, in which the chain carriers react with reactant molecules to produce products and other chain carriers; the algebraic sum of the propagation steps is the net reaction
    (c) Termination steps, where two chain carriers undergo coupling

11. In general, the predominant product from a free-radical reaction is the one formed from the most stable free-radical intermediate. The order of stability of carbon free radicals is tertiary (or resonance-stabilized) > secondary > primary.

12. The principle of orbital symmetry conservation can be stated as follows: for a concerted reaction to be symmetry allowed, bonding (i.e., like-phase) interactions must develop between the appropriate termini of the HOMO (and LUMO in the case of cycloadditions) during the transition state of the reaction. Applying this principle allows us to predict the stereochemical outcome of a variety of pericyclic reactions.

RAISE YOUR GRADES
Can you define...?

- ☑ polar reaction
- ☑ solvent effects
- ☑ Arrhenius acid and base
- ☑ Brønsted–Lowry acid and base
- ☑ Lewis acid and base
- ☑ conjugate acid and base
- ☑ K_a and pK_a
- ☑ autodissociation
- ☑ resonance effects (+R, –R)
- ☑ inductive effects (+I, –I)
- ☑ free-radical reaction
- ☑ chain mechanism
- ☑ atom-abstraction reaction
- ☑ free-radical initiator, initiation step
- ☑ propagation steps

- ☑ pericyclic reaction
- ☑ electrocyclic reaction
- ☑ cycloaddition
- ☑ stereospecific reaction
- ☑ conrotatory, disrotatory
- ☑ symmetry allowed
- ☑ symmetry forbidden
- ☑ excited state
- ☑ HOMO, LUMO
- ☑ frontier orbital
- ☑ Diels–Alder reaction
- ☑ dienophile
- ☑ suprafacial, antarafacial
- ☑ oxidation, reduction
- ☑ redox reaction

☑ termination steps
☑ free-radical inhibitor
☑ free-radical coupling
☑ chain carrier
☑ biradical
☑ autoxidation

☑ oxidation number
☑ oxidizing agent
☑ reducing agent
☑ radical cation
☑ radical anion
☑ SET

Can you explain...?

☑ the differences between polar, free radical, and pericyclic reactions
☑ how to determine when a reaction involves oxidation and reduction
☑ how to predict the relative acidity of a compound
☑ how acidity is quantitatively measured and expressed
☑ which structural features of a molecule are important in controlling its acidity
☑ which groups are +R, –R, +I, and –I, and why
☑ how a chain reaction mechanism works
☑ how to predict the favored product in a free radical chlorination
☑ how to predict when a given pericyclic reaction will be allowed or forbidden, and
   what the stereochemical outcome will be

SOLVED PROBLEMS

PROBLEM 10-1   Complete these statements:
(a) At equilibrium in an acid–base reaction, the stronger/weaker (circle one) acid and base is favored.
(b) The strongest acid that can exist (in significant concentration) in CH_3CO_2H is _____.
(c) The strongest base that can exist (in significant concentration) in CH_3CO_2H is _____.
(d) When measuring the acidity of a compound much less acidic than water, it is necessary to use a base whose conjugate acid is stronger/weaker (circle one) than H_2O.
(e) The stronger an acid is, the stronger/weaker (circle one) its conjugate base is (as a base).

Solution   **(a)** weaker, **(b)** $CH_3CO_2H_2^+$, **(c)** $CH_3CO_2^-$, **(d)** weaker, **(e)** weaker.

PROBLEM 10-2   **(a)** Indicate the most acidic hydrogen in each compound shown below, then arrange the compounds from most acidic to least.

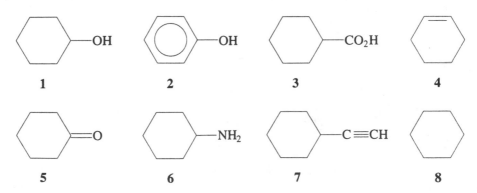

(b) Arrange phenol, *p*-cresol (*p*-methylphenol), and *p*-nitrophenol from most acidic to least, and explain your reasoning.
(c) Suggest a reason why hydrogens alpha to a ketone (or aldehyde) carbonyl (pK_a 20) are more acidic than hydrogens alpha to an ester carbonyl (pK_a 25).
(d) Explain why an allylic (methylene) hydrogen in 1,3-cyclopentadiene (pK_a 16) is far more acidic than the allylic hydrogens in cyclopentene.

Solution

(a)

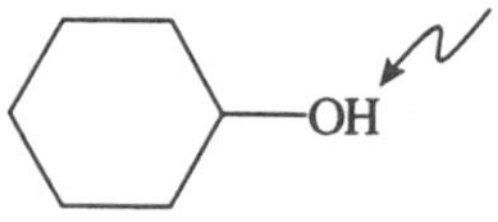

1 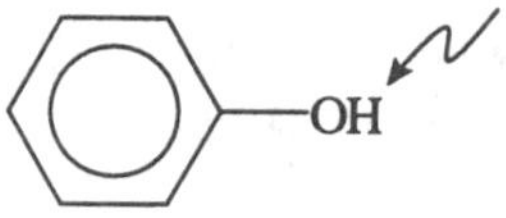   2 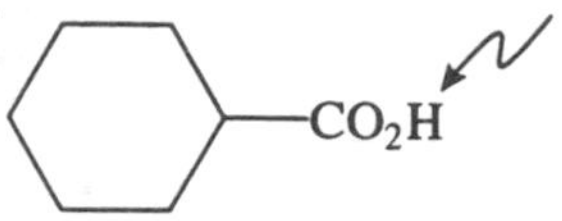   3    4

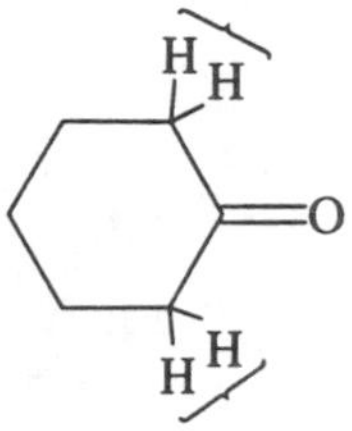

5 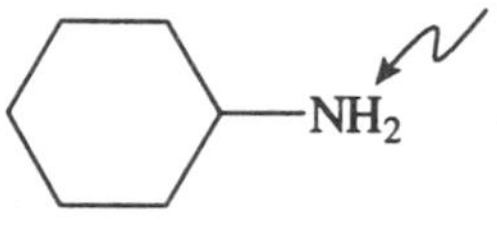   6 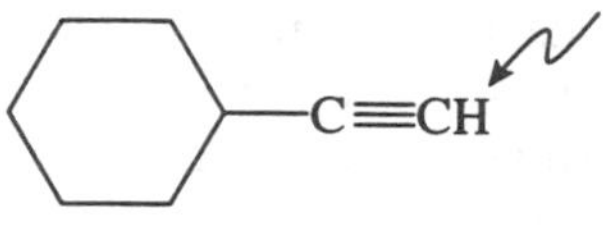   7 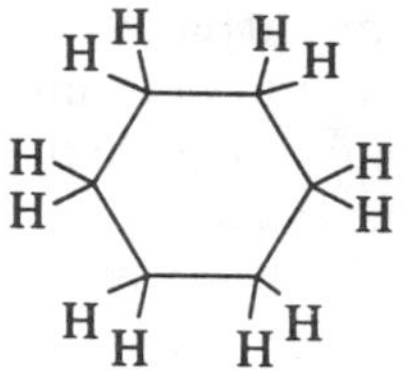   (all equivalent)

8

Acidity order: **3 > 2 > 1 > 5 > 7 > 6 > 4 > 8**

(b) A nitro group is electron withdrawing by both resonance and induction, and therefore has an acid-strengthening effect (compared to hydrogen) on the phenolic hydrogen. A methyl group, on the other hand, is electron donating by both resonance (hyperconjugation) and induction, so it has an acid-weakening effect on the phenolic hydrogen. Thus the acidity order is: *p*-Nitrophenol (pK_a 7.16) > phenol (pK_a 9.99) > *p*-cresol (pK_a 10.26).

(c) The fact that ketone (and aldehyde) alpha hydrogens are more acidic than ester alpha hydrogens indicates that a ketone (or aldehyde) carbonyl is more electron-withdrawing than an ester carbonyl. This is so because the OR′ oxygen of the ester is itself *electron-donating* by resonance, which reduces the ester carbonyl's ability to withdraw other electrons.

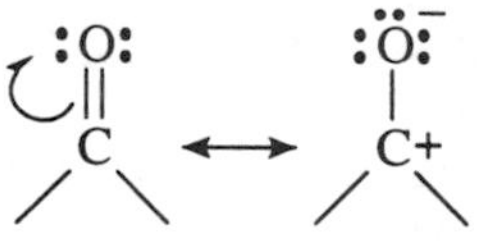

ketone 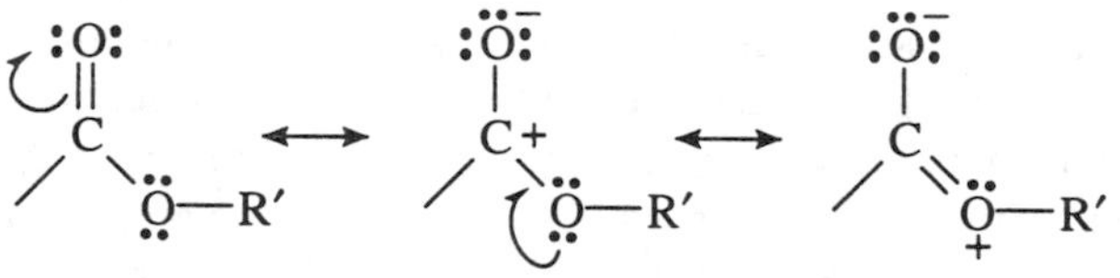   ester

(d) Removal of a methylene H⁺ from cyclopentadiene gives the **cyclopentadienide** ion, which has six electrons in its π system and therefore enjoys aromatic stabilization (Problem 4-14). This ion is far more stable than the allylic anion (Example 4-11) generated by deprotonation of cyclopentene.

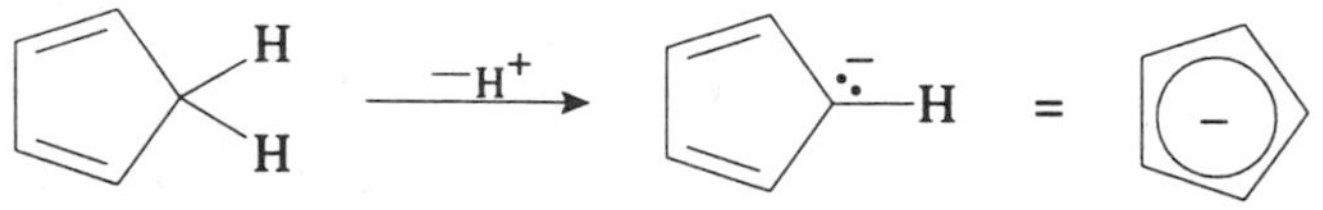

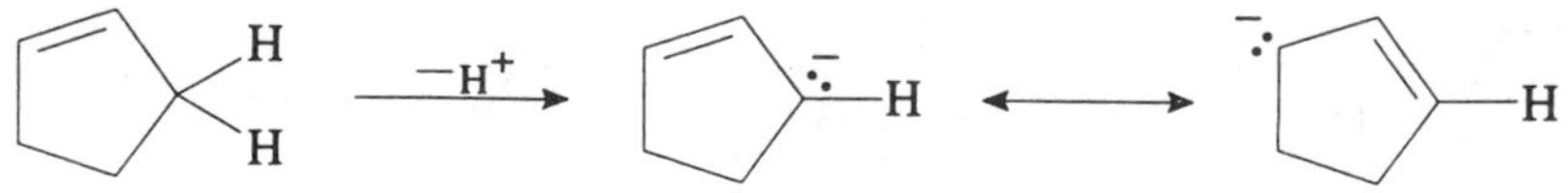

PROBLEM 10-3 Indicate the category (polar, free radical, pericyclic) to which each of these reactions belongs:

(a)

(the Cope rearrangement)

(b)   R—Br + R₃Sn—H $\xrightarrow{\text{initiator}}$ R—H + R₃Sn—Br

(c)
$$CH_3\overset{\overset{\displaystyle O}{\|}}{C}OCH_2CH_3 \ + \ H_2O \ \xrightarrow{\ H^+ \ catalyst\ } \ CH_3CO_2H \ + \ HOCH_2CH_3$$

(d)
$$(CH_3)_3CCl \ + \ OH^- \ \longrightarrow \ (CH_3)_2C\!=\!CH_2 \ + \ H_2O \ + \ Cl^-$$

Solution

(a) This reaction can be viewed as a cyclic reorganization of electrons, so it is likely to be pericyclic:

(b) The requirement of an initiator suggests a free-radical (chain) mechanism. (See Problem 10-6.)

(c) The involvement of polar bonds and an acid catalyst suggests a polar reaction.

(d) The presence of a strong base (OH⁻) among the reactants, as well as ions among the reactants and products, suggests a polar mechanism.

PROBLEM 10-4 Write the equilibrium equation for the autodissociation of acetic acid, indicating which species are serving as Brønsted acids and which are Brønsted bases.

Solution

$$\mathbf{CH_3CO_2H \ + \ CH_3CO_2H \ \rightleftharpoons \ CH_3CO_2H_2{}^+ \ + \ CH_3CO_2{}^-}$$

acid ⟵——————— conjugates ———————⟶ base

base ⟵— conjugates ⟶ acid

PROBLEM 10-5 Indicate the oxidation number (ON) for each atom in each reaction below. Indicate which atoms are involved in oxidation–reduction.

(a)
$$H_2C\!=\!CH_2 \ + \ R\overset{\overset{\displaystyle}{\underset{\displaystyle O}{\|}}}{C}\!-\!OOH \ \longrightarrow \ H_2\underset{\diagdown \ \ \diagup}{\overset{}{C}}\!-\!\underset{O}{CH_2} \ + \ R\overset{\overset{\displaystyle}{\underset{\displaystyle O}{\|}}}{C}\!-\!OH$$
(a peracid)

(b)
$$HC\!\equiv\!CH \ + \ H_2O \ \xrightarrow{\ H^+/Hg^{2+} \ catalyst\ } \ H_3C\overset{\overset{\displaystyle}{\underset{\displaystyle O}{\|}}}{C}\!-\!H$$

Solution Rewrite each structure as if all bonds were ionic:

(a)

$$\begin{array}{ccc} H^+ & H^+ & :\!\ddot{O}\!:^{2-} \\ :\!\overset{\cdot\cdot}{C}\!:^{2-} \ :\!\overset{\cdot\cdot}{C}\!:^{2-} \ + \ R\!\cdot \ \cdot\overset{\cdot\cdot}{C}^{3+} \ :\!\overset{\cdot\cdot}{O}\!\cdot \ \cdot\overset{\cdot\cdot}{O}\!:^{1-} H^+ & & \\ H^+ & H^+ & \end{array}$$

$$\longrightarrow$$

$$\begin{array}{ccc} H^+ & H^+ & :\!\ddot{O}\!:^{2-} \\ \cdot\overset{\cdot\cdot}{C}\!\cdot \ \cdot\overset{\cdot\cdot}{C}\!\cdot^- \ + \ R\!\cdot \ \cdot\overset{\cdot\cdot}{C}^{3+} \ :\!\ddot{O}\!:^{2-} \ H^+ & & \\ H^+ & H^+ & \\ & :\!\ddot{O}\!:^{2-} & \end{array}$$

Thus both ethylene carbons are oxidized by one electron, while both peracid oxygens are reduced by one electron.

(b)
$$H^+ \ :\!\overset{-}{\underset{\cdot}{C}}\!\overline{} \ :\!\overset{-}{\underset{\cdot}{C}}\!\overline{} \ H^+ \ + \ H^+ \ :\!\ddot{O}\!:^{2-} \ H^+ \ + \ H^+ \ :\!\overset{\cdot\cdot}{C}^{3-} \ \cdot\overset{}{C}\!:^+ \ H^+$$
$$\begin{array}{cc} & H^+ \quad :\!\ddot{O}\!:^{2-} \\ & H^+ \end{array}$$

Thus one acetylenic carbon is oxidized by two electrons; the other is reduced by two electrons.

PROBLEM 10-6 Suggest a free-radical chain mechanism for reaction (b) in Problem 10-3. Assume that initiating radical In· is formed in the first initiation step. Show all subsequent initiation and propagation steps, and indicate which species are the chain carriers.

Solution

initiation: In· + R₃′Sn—H ⟶ R₃′Sn· + In—H

propagation: { R₃′Sn· + R—Br ⟶ R₃′Sn—Br + R·
R· + R₃′Sn—H ⟶ R₃′Sn· + R—H

The chain carriers are R· and R₃′Sn·. (Alternatively, the initiation step could be In· + RBr → InBr + R·).

PROBLEM 10-7 Predict the predominant product for the reaction below; describe your reasoning.

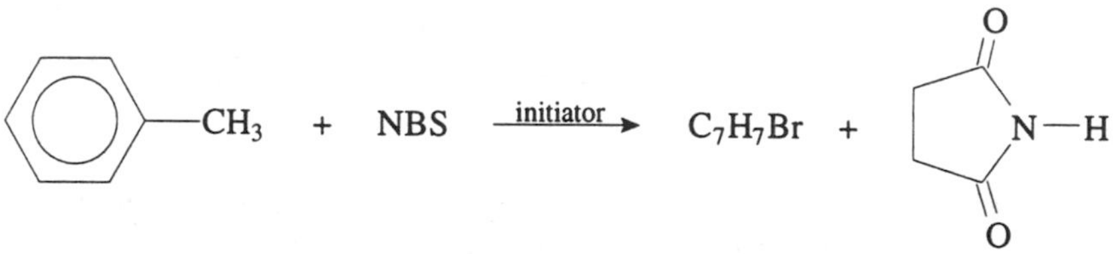

Solution We know from our previous experience with NBS (Section 10-6) and from the presence of an initiator that this is a free-radical reaction. The preferred brominated product will be the one that arises from the most stable free-radical precursor. The four possible products and their precursor radicals are:

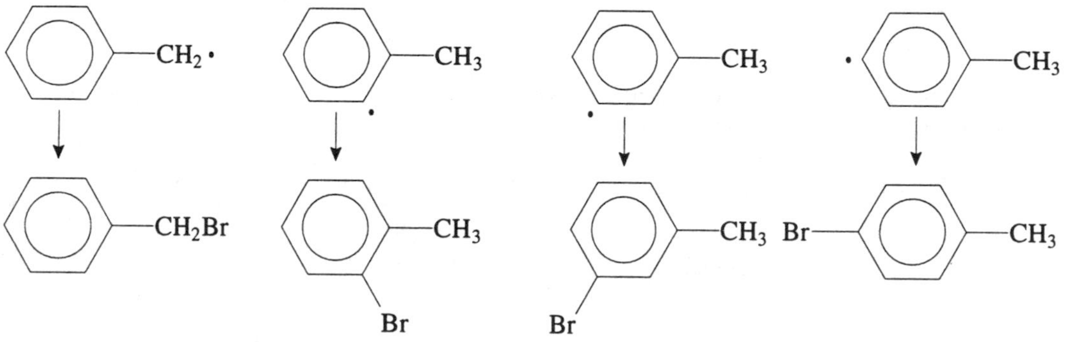

Of these, only the first radical has an odd electron capable of being resonance stabilized:

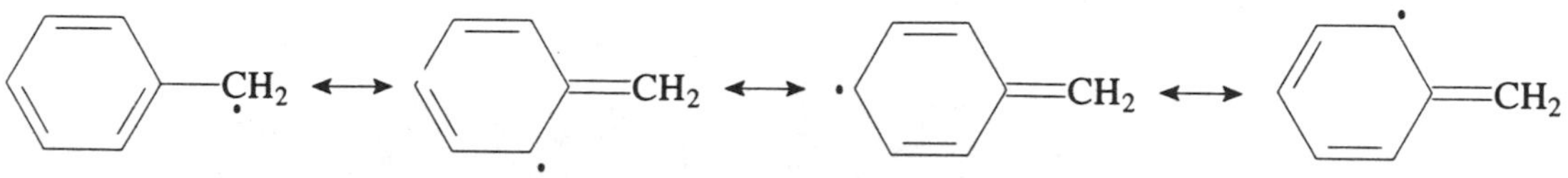

Therefore, the first product shown above is the predominant one, and the reaction is described as involving *benzylic* bromination.

PROBLEM 10-8 Predict the predominant product(s) of each concerted pericyclic reaction below:

(a)

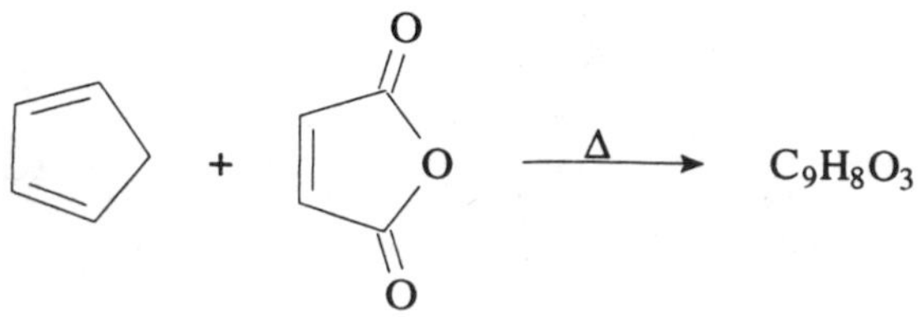

(b)

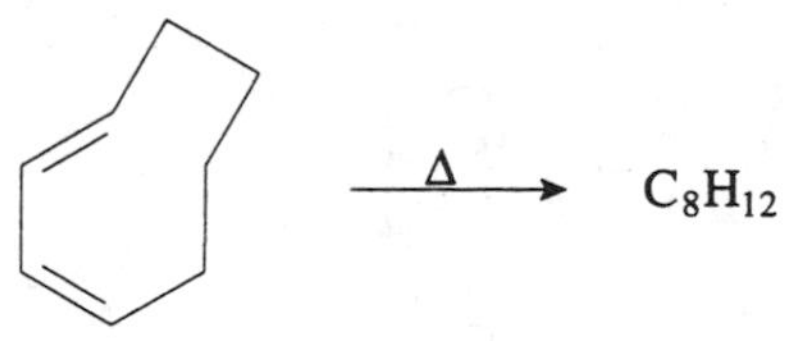

$\xrightarrow{\Delta}$ C$_8$H$_{12}$

(c)

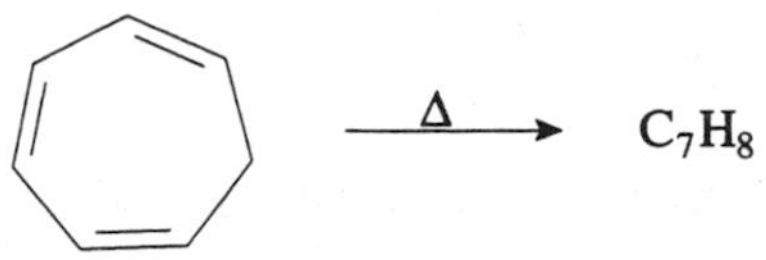

$\xrightarrow{\Delta}$ C$_7$H$_8$

Solution

(a)

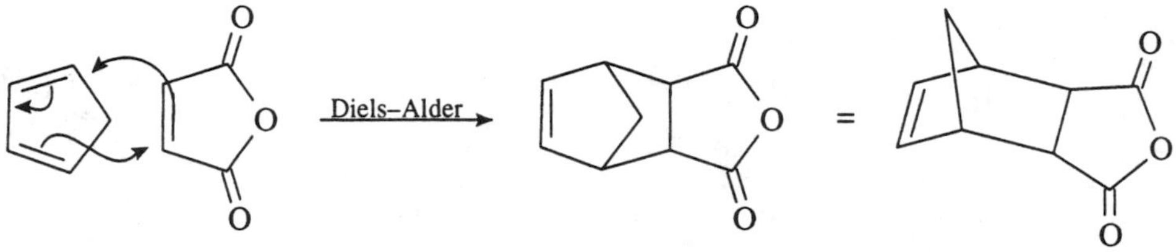

$\xrightarrow{\text{Diels–Alder}}$

$=$

(b)

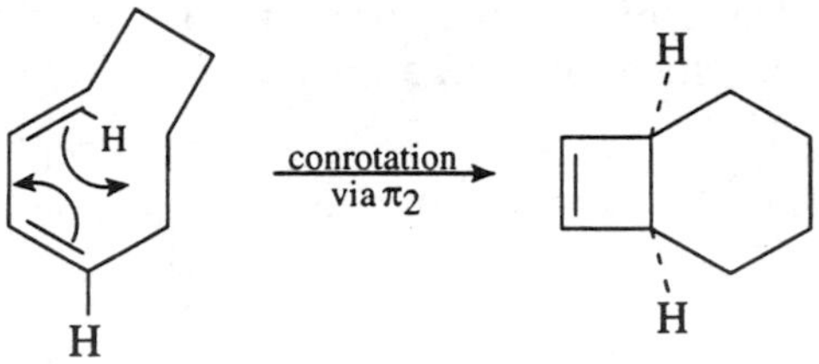

$\xrightarrow[\text{via } \pi_2]{\text{conrotation}}$

(c)

$\xrightarrow[\text{via } \pi_3]{\text{disrotation}}$

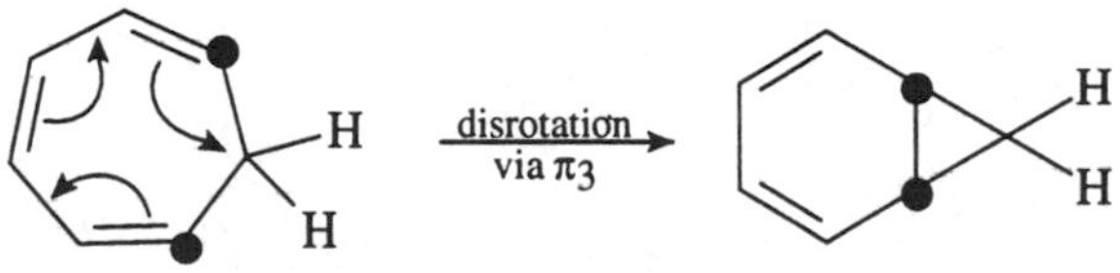

PROBLEM 10-9   **(a)** Specifically, what type of cycloaddition is the reaction below? **(b)** Will this reaction be thermally or photochemically allowed? Explain.

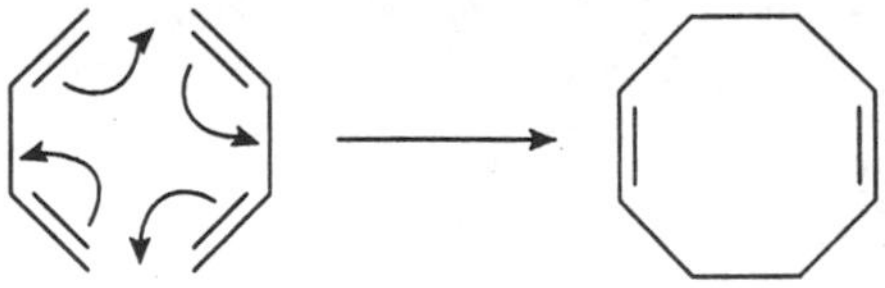

Solution   **(a)** $\pi_4 + \pi_4$. **(b)** For the thermal reaction, we'll compare the HOMO of one fragment (π_2) with the LUMO of the other (π_3):

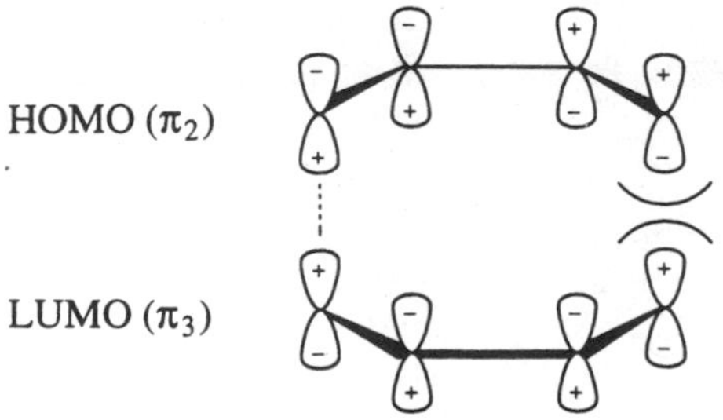

HOMO (π_2)

LUMO (π_3)

Clearly, these do not permit bonding overlap between both sets of termini in this all-suprafacial approach. For the photochemical reaction the HOMO is now π_3, which matches perfectly with the LUMO of the other partner:

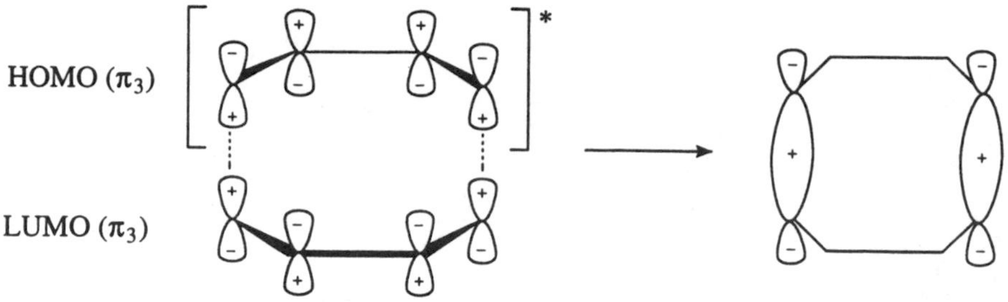

HOMO (π_3)

LUMO (π_3)

So, the all-suprafacial reaction is photochemically allowed and thermally forbidden.

PROBLEM 10-10 Which product(s) will be formed during the ring-opening reaction below? Explain.

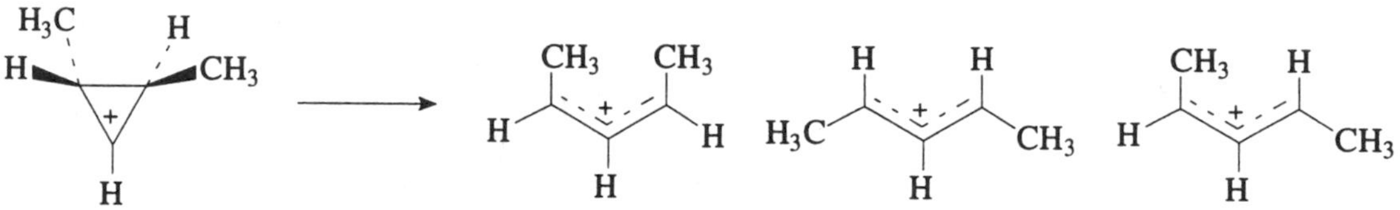

Solution We'll examine the HOMO of the open form, in this case the allyl cation. The HOMO of this cation is π_1 (Section 4-5), which looks like this:

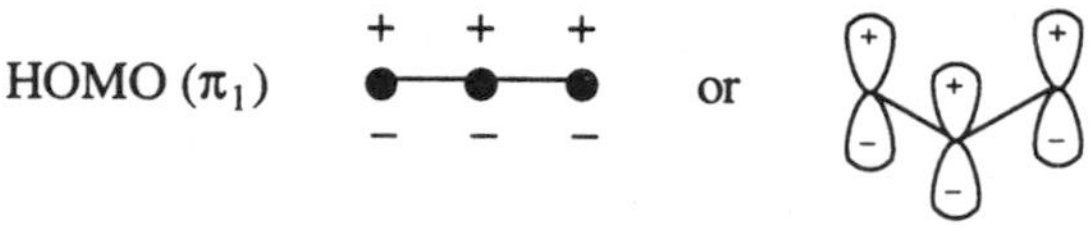

HOMO (π_1)   or

Thus ring opening to this ion can only occur by a disrotatory path, which will give only the last of the three products:

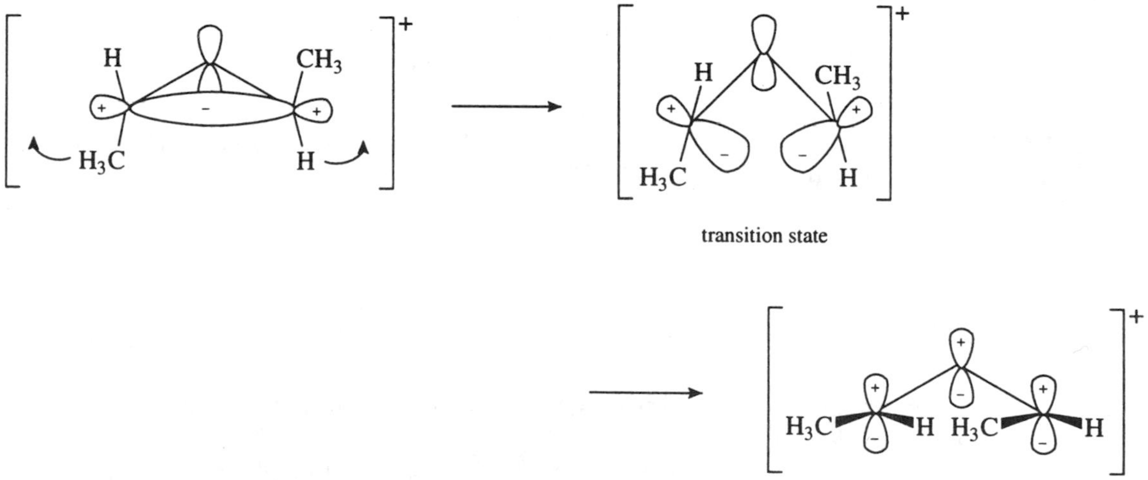

transition state

FINAL EXAM
(Chapters 1–10)

1. A certain unknown compound has the following elemental composition by weight: C, 30.6%; H, 3.9%; Cl, 45.2%. You may assume the remaining mass is due to oxygen.
   (a) Calculate the empirical formula of this compound.
   (b) The molecular *ions* for this compound are found at nominal masses (m/e) of 78 and 80 amu (ratio, 3:1). Give the molecular formula for this compound and explain why there are two molecular ions. What is the I value for this molecular formula, and what does it represent?
   (c) Draw structures for all possible isomers (structural *and* stereo) for this molecular formula, showing accurate geometries and all nonbonding pairs. For each structure give a complete systematic name and indicate the functional group class(es) represented.
   (d) The IR spectrum of this unknown shows no evidence for C=O or O–H bonds, but the compound is found to be dextrorotatory. Can you propose a unique structure for the unknown? Would NMR data be of help? Why?

2. Enols are generally unstable with respect to isomerization to the corresponding ketone (or aldehyde):

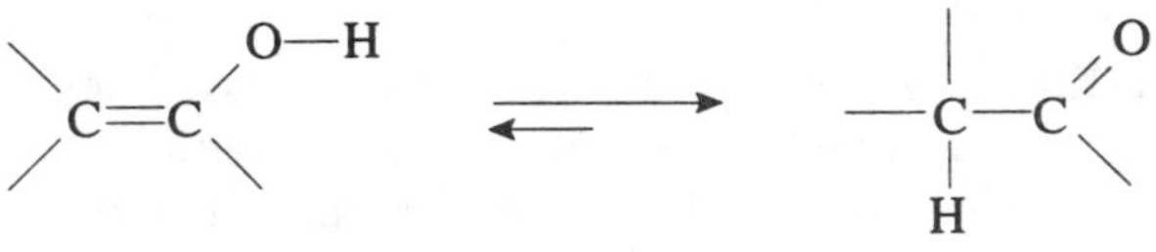

   However, in the reaction below, the *enol* is far more stable than the ketone.

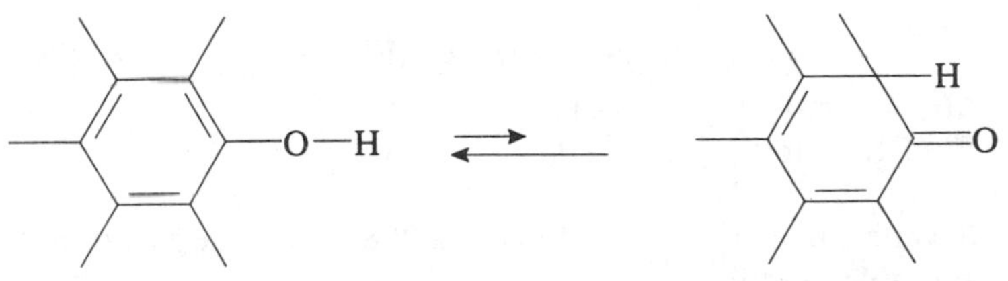

   (a) Why is the enol more stable?
   (b) Draw at least five resonance structures of phenol, using arrows to show electron movement. Is the hydroxy group electron-donating or -withdrawing by resonance? At which ring position(s) is the resonance effect most pronounced? In what direction (+ or –) is the inductive effect of an OH group?
   (c) The pK_a of *m*-nitrophenol is 8.28, while that of *p*-nitrophenol is 7.15. Which compound is more acidic? Why is it more acidic? Explain your answer in terms of resonance.

3. Consider the structure of LSD (lysergic acid, diethyl amide), shown below:

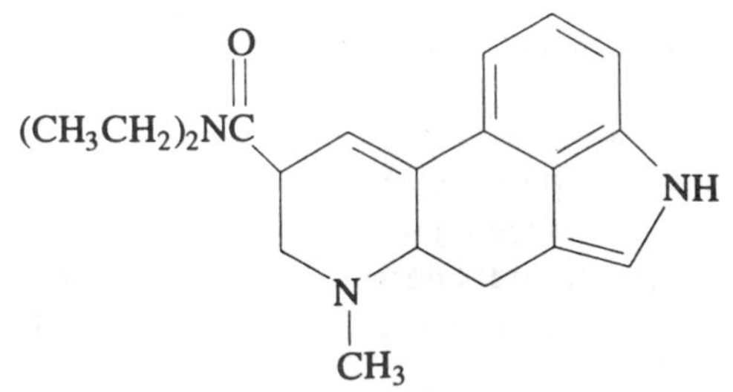

(a) What functional group classes are represented in LSD?

(b) Indicate with an asterisk (*) each chiral center in LSD. How many stereoisomers of LSD are possible?

(c) Pure natural lysergic acid has $[\alpha]_D = +40°$. A certain *dl* mixture of lysergic acid has $[\alpha]_D = +10°$. Calculate the enantiomeric excess (optical purity) of the mixture, as well as the fraction of each enantiomer.

4. (a) Draw out the condensed structural formula for NCCH$_2$CHCHCOOH (the trans isomer) in such a way as to accurately depict the shape around each internal atom. Show all nonbonding pairs and indicate the hybridization at each carbon.

(b) From your structure give the nominal value of each bond angle below:

N–C–C _____ ; O–C–O _____ ; H–C–H _____ ; C(H)–C(H)–C(O$_2$H) _____

(c) What functional group classes are represented in your structure?

(d) How many of each of the following are there in your structure:

σ bonds _____   σ electrons _____   σ* orbitals _____
π bonds _____   π electrons _____   π* orbitals _____

5. *cis*-2,2,5,5-Tetramethyl-3-hexene is much less stable than the trans isomer because of steric crowding between the *tert*-butyl groups.

(a) From the heats of hydrogenation for both compounds (cis, −151 kJ mol^{-1}; trans, −112 kJ mol^{-1}), determine $\Delta H°$ for the isomerization of the cis isomer to the trans isomer.

(b) Assuming that the entropies of both isomers are comparable, calculate the equilibrium constant for the cis-to-trans equilibrium. How much of the cis isomer is present at equilibrium?

(c) Suppose this equilibration could be catalyzed by Brønsted acids. Give a mechanism for the reaction, and draw a complete reaction profile diagram for your mechanism. What intermediate(s) is (are) involved? To which mechanistic class of reaction does this belong?

6. Consider the reaction and the thermodynamic data below:

	1,3-butadiene	+ ethylene	⟶	cyclohexene
$\Delta H_f°$ (kJ mol^{-1})	112	52.3		−5.4
$S°$ (J K^{-1} mol^{-1})	279	220		311

(a) Calculate the values of $\Delta H°$, $\Delta S°$, and $\Delta G°$ (at 298 K) for the reaction. Is this a thermodynamically favorable reaction?

(b) Are the signs of $\Delta H°$ and $\Delta S°$ reasonable for this reaction? Why?

(c) What is the name of this type of reaction, and to which mechanistic class does it belong?

(d) Predict the rate law for this reaction. What is the order in each reactant, and what is the overall order?

(e) Draw a complete reaction profile diagram for this reaction. Do you expect this reaction to have an early or late transition state? Why?

7. (a) Draw a side view (perpendicular to the C–C bond) of the structure of 1,2-dibromoethane in the conformation where the bromines are eclipsed.

(b) Draw the Newman projection of this conformation, and indicate the dihedral angle between the two C–Br bonds.

(c) Specify what symmetry elements (if any) this conformation has. Indicate which (if any) of the atoms are equivalent.

(d) Draw structures (side views and Newman projections) of other conformations of this molecule with dihedral angles between the two C–Br bonds of 60°, 120°, and 180°. Which of these conformations are eclipsed and which are staggered? Which, if any, are chiral?

(e) Draw a graph of the conformational energy of 1,2-dibromoethane as a function of dihedral angle, using your estimates of the relative energy of each of the four conformations. (Refer to Problem 7-10.)

8. Consider the net reaction below (where ether is the solvent):

$$4 \; CH_3\overset{\overset{\displaystyle O}{\|}}{C}CH_3 \; + \; LiAlH_4 \; \xrightarrow{\;\text{ether}\;} \; [Product_1] \; \xrightarrow{\;4 \, H_2O\;} \; 4 \; CH_3\overset{\overset{\displaystyle OH}{|}}{C}HCH_3 \; + \; Al(OH)_3 \; + \; LiOH$$

(a) Does this overall reaction involve oxidation and reduction? If so, what is oxidized and what is reduced? Is $LiAlH_4$ an oxidizing or reducing agent?

(b) How could you determine which of the hydrogens (CH or OH) in the alcohol product came from the water and which came from the $LiAlH_4$? Assuming that the OH hydrogens come from water, suggest a structure for $Product_1$.

(c) Do you think this reaction involves a polar mechanism or a free-radical mechanism? Why?

Answers to Final Exam

1. (a) See Section 7-1. First, we'll calculate the percent oxygen:

$$\% \, O = 100\% - (30.6 + 3.9 + 45.2) = 20.3\%$$

Next, we divide each element's percentage by its atomic weight to generate mole ratios:

C: 30.6/12.0 = 2.55   Cl: 45.2/35.5 = 1.27
H: 3.9/1.0 = 3.9   O: 20.3/16.0 = 1.27

Finally, we divide each ratio by the smallest one (1.27) to get the smallest integer ratio of the atoms: C, 2; H, 3; Cl, 1; O, 1. This corresponds to the empirical formula C_2H_3ClO, which has an empirical formula weight (*EFW*) = 2(12.0) + 3(1.0) + 1(35.5) + 1(16.0) = 78.5 amu.

(b) To find the molecular formula we need to determine the value of *n* in $(C_2H_3ClO)_n$ by comparing the *EFW* with the molecular weight *MW* (*n = MW/EFW*, Section 7.1). Normally, the *MW* is the same as the *m/e* of the molecular ion in the compound's mass spectrum (Section 7-2). However, in the present case there are *two* molecular ions, which reminds us that chlorine consists of *two* isotopes, ^{35}Cl and ^{37}Cl, in the ratio of 3:1 (Example 1-6). So, the two molecular ions are $C_2H_3{}^{35}ClO^+$ (*MW* 78) and $C_2H_3{}^{37}ClO^+$ (*MW* 80) in the ratio of 3:1, to give an average *MW* of 78.5.

Therefore, *n* = 1, and the molecular formula is the same as the empirical formula, C_2H_3ClO). The *I* value for this formula (Section 6-6) is given by

$$I = \frac{2(2) + 2 - 3 - 1}{2} = 1$$

This *I* value indicates the presence of one ring *or* one π bond.

(c)

Structure	Name	Class
	acetyl chloride (ethanoyl chloride)	acid chloride
	chloroacetaldehyde (chloroethanal)	chloro aldehyde
	1-chloro-1-ethenol	chloro alcohol (chloro enol)
	cis-2-chloro-1-ethenol	chloro alcohol (chloro enol)
	trans-2-chloro-1-ethenol	chloro alcohol (chloro enol)
	ethenyl hypochlorite*	hypochlorite*

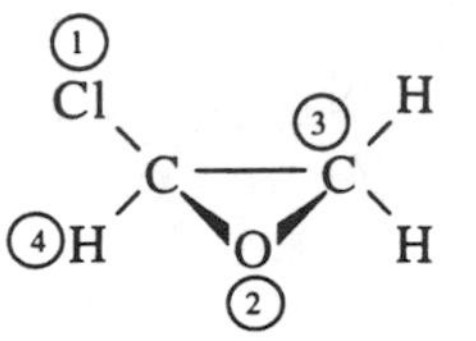

(S)-2-chlorooxirane

chloro ether
(chloro oxirane,
chloro epoxide)

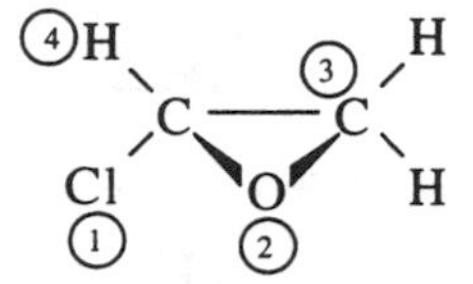

(R)-2-chlorooxirane
[(R)-chloroethylene oxide]

chloro ether
(chloro oxirane)

*This is a new functional group we've not seen before.

(d) The IR spectrum rules out the first five structures, and the fact that it is optically active, rules out the hypochlorite (which is achiral). This leaves only the two enantiomers of the oxirane. However, because we have no additional information, we cannot predict whether the positive optical rotation belongs to the R or the S enantiomer (Section 8-5). Both of these structures will have identical NMR spectra, consisting of two carbon signals and three hydrogen signals, since there are no symmetry-equivalent atoms in either structure.

2. (a) For a typical enol–ketone equilibrium, the total bond strength of the ketone (C=O + C–C + C–H) is greater than that of the enol (C=C + C–O + O–H) by 66 kJ mol^{-1} (Example 9-1). In the case of phenol, however, the C=C bond is part of an aromatic system, the stability of which more than offsets the bond-strength comparison. The resonance stabilization of an aromatic ring must therefore be at least 66 kJ mol^{-1} and is actually closer to 150 kJ mol^{-1}.

(b)

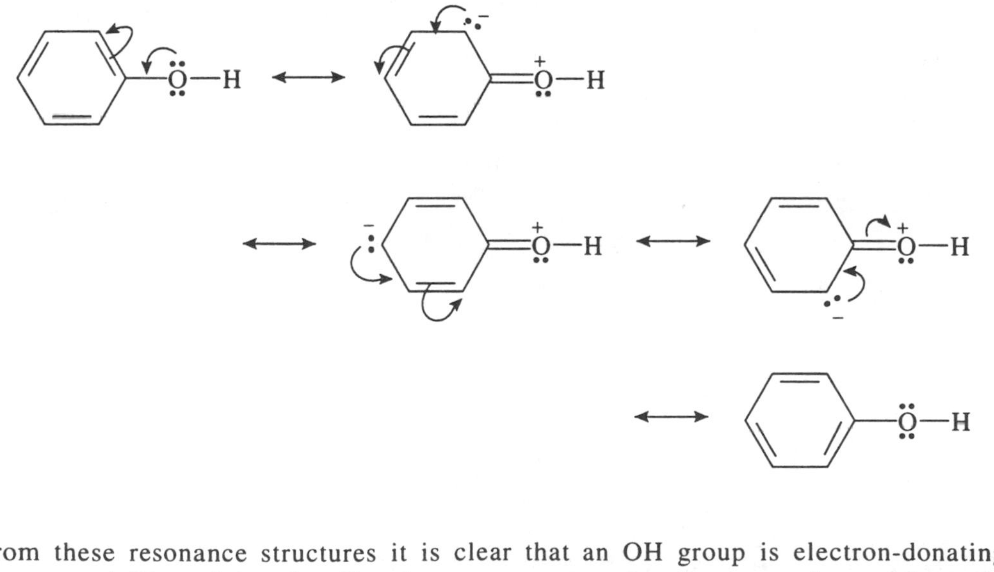

From these resonance structures it is clear that an OH group is electron-donating by resonance (+R), and that the effect is most pronounced at the ring carbons ortho and para to the OH group. By contrast, the *inductive* effect of an OH group is electron-*withdrawing* (–I) by virtue of oxygen's high electronegativity (relative to carbon and hydrogen), attracting nearby electrons toward itself. Because these effects are in opposite directions, it is important to note that where resonance *is* possible, the +R effect dominates over the –I effect.

(c) Remember that the greater the pK_a the *lower* the acidity, so the para isomer is the stronger acid. This is because a *p*-nitro group is able to accept the charge from the phenoxy oxygen onto its own oxygens by "through resonance," while the *m*-nitro group cannot.

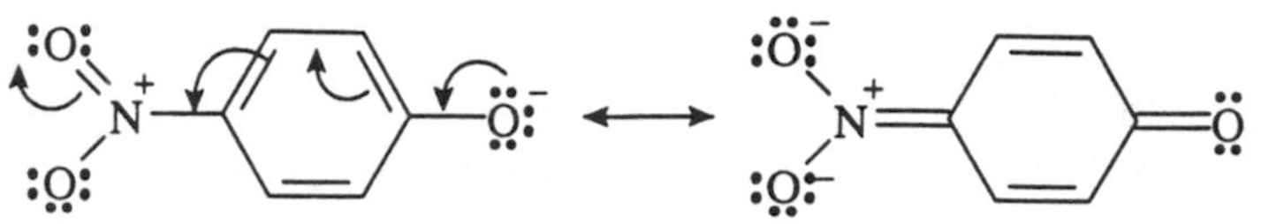

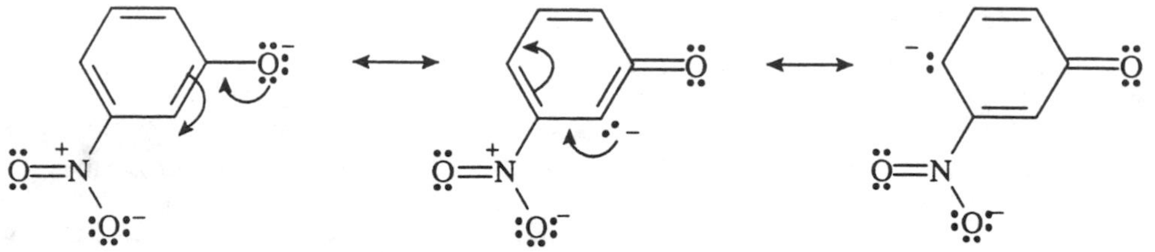

3. (a) 3° amide, 3° amine, 2° amine, aromatic ring, alkene (C=C bonds, 2).

(b)

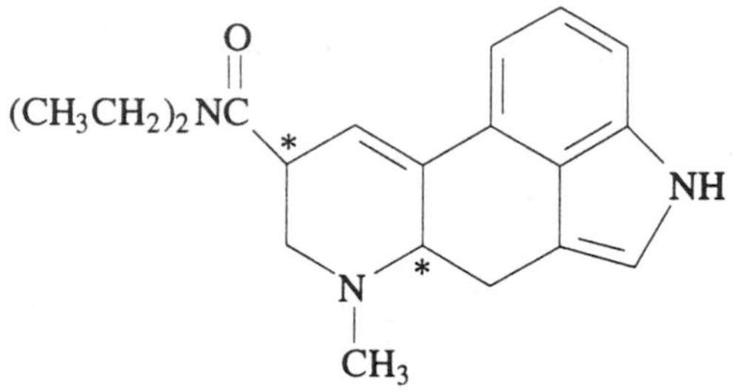

The two chiral centers give rise to $2^2 = 4$ stereoisomers, only one of which has the noted physiological effects.

(c) See Section 8-6.

$$ee = op = \frac{[\alpha]_{net}}{[\alpha]_{max}} = \frac{10°}{40°} = 0.25$$

Since $ee = 2f_d - 1$,

$$2f_d - 1 = 0.25$$
$$f_d = 0.63$$
$$f_l = 1.00 - f_d = 0.37$$

4. (a)

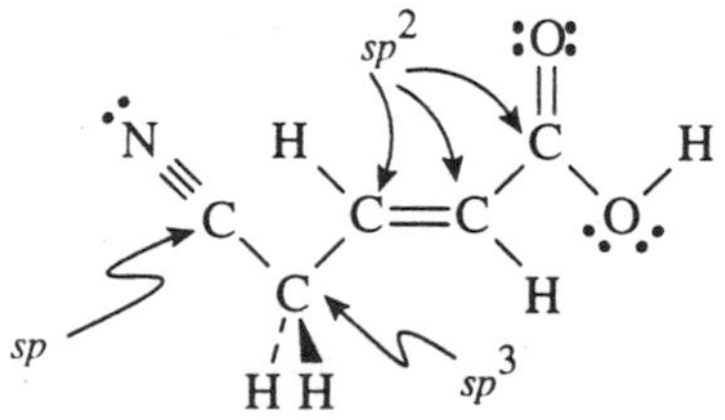

(b) 180°; 120°; 109.5°; 120°

(c) nitrile, carboxylic acid, alkene

(d) 12 σ bonds, 24 σ electrons, 12 σ* orbitals, 4 π bonds, 8 π electrons, 4 π* orbitals

(e) Yes! The C=O bond is in resonance with both the C=C bond and the nonbonding pair on the other oxygen:

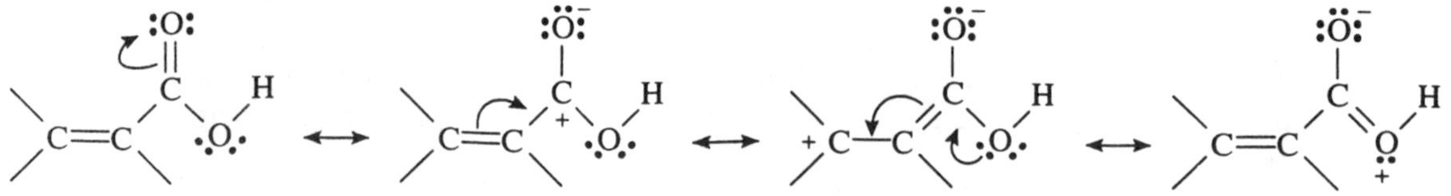

5. **(a)** Because hydrogenation of both isomers lead to the same compound, 2,2,5,5-tetramethylhexane, the difference between the heats of hydrogenation is exactly equal to $\Delta H°$ for the cis-to-trans isomerization (see Figure FE-1).

$$-151 \text{ kJ mol}^{-1} - (-112 \text{ kJ mol}^{-1}) = -39 \text{ kJ mol}^{-1}$$

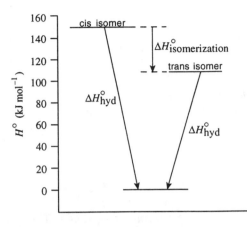

FIGURE FE-1. Enthalpy diagram comparing the heats of hydrogenation of *trans-* and *cis-*2,2,5,5-tetramethyl-3-hexene.

(b) If the entropies of both isomers are comparable, then $\Delta S°$ for the cis-to-trans isomerization is zero and therefore $\Delta G° (= \Delta H° - T\Delta S°) = -39 \text{ kJ mol}^{-1}$. From this we can calculate K, using Eq. (9-7):

$$K = \frac{[\text{trans isomer}]}{[\text{cis isomer}]} = e^{-\Delta G°/RT}$$

$$= e^{(39,000 \text{ J mol}^{-1})/[(8.31 \text{ J K}^{-1} \text{ mol}^{-1})(298 \text{ K})]}$$

$$= e^{15.75}$$

$$= 6.9 \times 10^6$$

Thus there will be only 1 molecule of the cis isomer for every 7 million molecules of the trans isomer at equilibrium.

(c)

See Figure FE-2 for the reaction profile. The intermediate is a carbocation, and the mechanism is a polar one, involving acid catalysis and heterolytic cleavages.

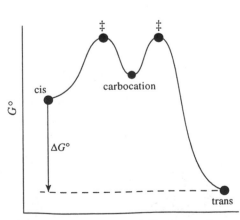

FIGURE FE-2. Reaction profile diagram for the isomerization of *cis-*2,2,5,5-tetramethyl-3-hexene to its trans isomer, catalyzed by acid.

6. (a) See Sections 9-1, -2, and -3.

$$\Delta H^\circ = \sum \Delta H^\circ_{f(products)} - \sum \Delta H^\circ_{f(reactants)}$$
$$= [-5.4 - (112 + 52.3)] \text{ kJ mol}^{-1}$$
$$= -170 \text{ kJ mol}^{-1}$$

$$\Delta S^\circ = \sum S^\circ_{products} - \sum S^\circ_{reactants}$$
$$= [311 - (279 + 220)] \text{ J K}^{-1} \text{ mol}^{-1}$$
$$= -188 \text{ J K}^{-1} \text{ mol}^{-1}$$

$$\Delta G^\circ = \Delta H^\circ - T\Delta S^\circ$$
$$= -170 \text{ kJ mol}^{-1} - [(298 \text{ K})(-0.188 \text{ J K}^{-1} \text{ mol}^{-1})]$$
$$= -114 \text{ kJ mol}^{-1}$$

Because ΔG° is negative, this is a thermodynamically favorable reaction.

(b) The negative sign of ΔH° is expected for a reaction that, in effect, converts two π bonds to two stronger σ bonds. The negative sign of ΔS° is consistent for a process where two molecules combine to form one, decreasing the disorder of the system.

(c) This is an example of a Diels–Alder reaction, which belongs to the pericyclic class.

(d) See Section 9-7. Because this is a concerted (one-step) reaction, where one molecule of each reactant is involved in the (only) transition state, the rate law will be first order in each reactant, second order overall: *Rate* = k[1,3-butadiene][ethylene].

(e) See Figure FE-3. Because this reaction has a negative ΔG°, the transition state will resemble the reactants, i.e., an early transition state.

FIGURE FE-3. Reaction profile diagram for the concerted reaction of 1,3-butadiene and ethylene.

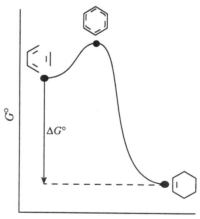

7. (a)   **(b)**   dihedral angle = 0°   **(c)**

There is one plane of symmetry (σ_1) bisecting the C–C bond, and another (σ_2) containing both carbons and both bromines. There is also a C_2 axis at the intersection of the planes.

σ_1 renders the two carbons, the two bromines, the two front hydrogens, and the two back hydrogens equivalent, while σ_2 renders the front hydrogens equivalent to the back ones. Thus, all four hydrogens are equivalent.

(d)

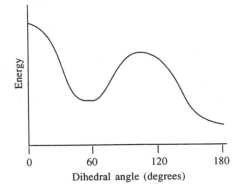

The eclipsed conformations are those with dihedral angles of 0° and 120°, while the staggered ones have dihedral angles of 60° and 180°. The 0° (bromines s-cis) and 180° (bromines s-trans) conformations each have at least one plane of symmetry, so they are achiral. The 60° (bromines gauche) and 120° conformations, on the other hand, have only C_2 axes and no planes, so they *are* both chiral!

(e) See Figure FE-4. The relative energy of the four conformations is 0° > 120° > 60° > 180°.

FIGURE FE-4. The conformational energy of 1,2-dibromoethane as a function of the dihedral angle between the two C–Br bonds.

8. (a) Begin by determining the oxidation number of each atom:

The only atoms that change are on the carbonyl carbon (which is reduced by two electrons) and the hydride hydrogens (those attached to aluminum), which are each oxidized by two electrons. Thus $LiAlH_4$ is a reducing agent (which itself is oxidized), and acetone is the oxidizing agent (which itself is reduced).

(b) The easiest way would be to perform an isotope labeling experiment using either $LiAlD_4$ and H_2O, or $LiAlH_4$ and D_2O. Product$_1$ has the following structure:

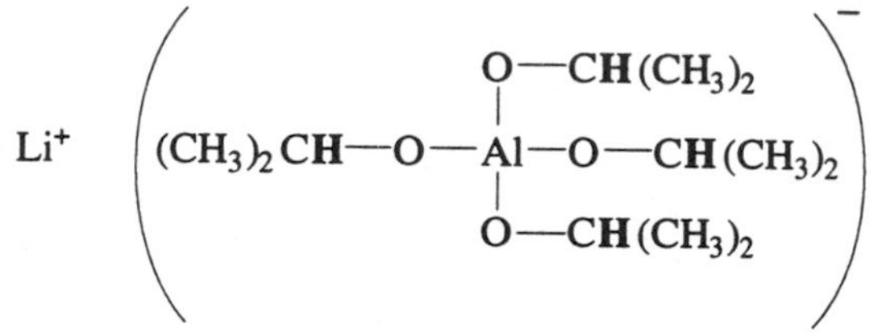

$$Li^+ \left((CH_3)_2CH-O-\underset{\underset{O-CH(CH_3)_2}{|}}{\overset{\overset{O-CH(CH_3)_2}{|}}{Al}}-O-CH(CH_3)_2 \right)^-$$

The boldfaced hydrogens are those delivered by the $LiAlH_4$.

(c) Because highly polar bonds (Al–H, C=O, O–H) are involved, and there is no need for initiation, we can be quite certain that this reaction follows a polar mechanism.

INDEX

Page numbers in italics refer to solved problems.

GLOSSARY

(Consult the Index for references to text discussions.)

A

Ac, Acetyl ($CH_3C=O$)

α (alpha), Designation of position in a molecule. An α carbon is directly attached to the functional group of interest, while an α hydrogen is attached to the α carbon

Ar, Symbol for the generic aryl (aromatic) ring or substituent

B

β (beta), Designation of position in a molecule. A β carbon is one carbon away from the functional group of interest, while a β hydrogen is attached to the β carbon

B:, Symbol for the generic base

C

CFC, Chlorofluorocarbon

cmc, Critical micelle concentration

C_n, An *n*-fold axis of symmetry

D

DABCO, Diazobicyclo[2.2.2]octane

DIBAH, Diisobutylaluminum hydride

DME, 1,2-Dimethoxyethane

DMF, Dimethylformamide

DMSO, Dimethyl sulfoxide

(2,4-)DNP, 2,4-Dinitrophenylhydrazone derivative of a ketone or aldehyde

E

E(+), Symbol for the generic electrophile

E (as in E1, E2, etc.), Elimination

E (as opposed to Z), Stereochemical designator

EAS, Electrophilic aromatic substitution

H

[H], Symbol for generic reducing agent

HMPA, Hexamethyl phosphoramide

HOMO, Highest (energy) occupied molecular orbital

HONO, Nitrous acid (HNO_2)

I

In, Symbol for the generic free-radical initiator

IUPAC, International Union of Pure and Applied Chemistry

K

k, Symbol for rate constant

K, Symbol for equilibrium constant

K_a, Acid dissociation constant

K_b, Base dissociation constant

kie, Kinetic isotope effect

L

LDA, Lithium diisopropylamide

LUMO, Lowest unoccupied molecular orbital

M

M, Symbol for the generic metal

mCPBA, *m*-Chloroperbenzoic acid

MO, Molecular orbital

MW, Molecular weight

N

NAD(H), Nicotinamide adenine dinucleotide (reduced form)

NAS, Nucleophilic aromatic substitution

NBS, *N*-Bromosuccinimide

NE, Neutralization equivalent weight (of an acid or base)

Nu:, Symbol for the generic nucleophile

O

[O], Symbol for generic oxidizing agent

P

PAH, Polycyclic aromatic hydrocarbon

PCC, Pyridinium chlorochromate

Ph, Phenyl group

pK_a, $-\log(K_a)$

pK_b, $-\log(K_b)$

R

R, Symbol for the generic alkyl group

R (as opposed to S), Stereochemical designator of absolute configuration

RX, Symbol for the generic alkyl halide

S

S (as in S_N1, S_N2, etc.), Substitution

S (as opposed to R), Stereochemical designator of absolute configuration

S (as in $\Delta S°$), Entropy

SET, Single electron transfer

$(Sia)_2BH$, Disiamylborane (diisoamylborane)

T

THF, Tetrahydrofuran

THP, Tetrahydropyranyl derivative (of an alcohol)

TMS, Trimethylsilyl (group)

Tf, Trifluoromethanesulfonyl group (CF_3SO_2)

Ts, *p*-Toluenesulfonyl group (p-CH_3-C_6H_4-SO_2)

V

VSEPR, Valence shell electron shell repulsion

X

X, Often used as the symbol for the generic halide (F, Cl, Br, I)

Z

Z, Sometimes used as a symbol for the generic nucleofuge

Z (as opposed to E), Stereochemical designator